MW00709543

CODING THEORY AND CRYPTOLOGY

LECTURE NOTES SERIES
Institute for Mathematical Sciences, National University of Singapore

Series Editors: Louis H. Y. Chen and Yeneng Sun
Institute for Mathematical Sciences
National University of Singapore

Published

Vol. 1   Coding Theory and Cryptology
edited by Harald Niederreiter

Lecture Notes Series, Institute for Mathematical Sciences,
National University of Singapore

CODING THEORY
AND CRYPTOLOGY

Editor

Harald Niederreiter

National University of Singapore

SINGAPORE UNIVERSITY PRESS
NATIONAL UNIVERSITY OF SINGAPORE

World Scientific
New Jersey • London • Singapore • Hong Kong

Published by

Singapore University Press
Yusof Ishak House, National University of Singapore
31 Lower Kent Ridge Road, Singapore 119078

and

World Scientific Publishing Co. Pte. Ltd.
5 Toh Tuck Link, Singapore 596224
USA office: Suite 202, 1060 Main Street, River Edge, NJ 07661
UK office: 57 Shelton Street, Covent Garden, London WC2H 9HE

British Library Cataloguing-in-Publication Data
A catalogue record for this book is available from the British Library.

CODING THEORY AND CRYPTOLOGY

ISBN 981-238-132-5

This book is printed on acid-free paper.

Contents

Foreword   vii

Preface   ix

Extremal Problems of Coding Theory
  A. Barg   1

Analysis and Design Issues for Synchronous Stream Ciphers
  E. Dawson and L. Simpson   49

Quantum Error-Correcting Codes
  K. Feng   91

Public Key Infrastructures
  D. Gollmann   143

Computational Methods in Public Key Cryptology
  A. K. Lenstra   175

Detecting and Revoking Compromised Keys
  T. Matsumoto   239

Algebraic Function Fields Over Finite Fields
  H. Niederreiter   259

Authentication Schemes
  D. Y. Pei   283

Exponential Sums in Coding Theory, Cryptology and Algorithms
  I. E. Shparlinski   323

Distributed Authorization: Principles and Practice
  V. Varadharajan   385

Introduction to Algebraic Geometry Codes
  C. P. Xing   435

Foreword

The Institute for Mathematical Sciences at the National University of Singapore was established on 1 July 2000 with funding from the Ministry of Education and the University. Its mission is to provide an international center of excellence in mathematical research and, in particular, to promote within Singapore and the region active research in the mathematical sciences and their applications. It seeks to serve as a focal point for scientists of diverse backgrounds to interact and collaborate in research through tutorials, workshops, seminars and informal discussions.

The Institute organizes thematic programs of duration ranging from one to six months. The theme or themes of each program will be in accordance with the developing trends of the mathematical sciences and the needs and interests of the local scientific community. Generally, for each program there will be tutorial lectures on background material followed by workshops at the research level.

As the tutorial lectures form a core component of a program, the lecture notes are usually made available to the participants for their immediate benefit during the period of the tutorial. The main objective of the Institute's Lecture Notes Series is to bring these lectures to a wider audience. Occasionally, the Series may also include the proceedings of workshops and expository lectures organized by the Institute. The World Scientific Publishing Company and the Singapore University Press have kindly agreed to publish jointly the Lecture Notes Series. This volume on "Coding Theory and Cryptology" is the first of this Series. We hope that through regular publication of lecture notes the Institute will achieve, in part, its objective of promoting research in the mathematical sciences and their applications.

October 2002

Louis H. Y. Chen
Yeneng Sun
Series Editors

Preface

The inaugural research program of the Institute for Mathematical Sciences at the National University of Singapore took place from July to December 2001 and was devoted to coding theory and cryptology. The program was split into three parts of about equal length: (i) mathematical foundations of coding theory and cryptology; (ii) coding and cryptography; (iii) applied cryptology.

As part of the program, tutorials for graduate students and junior researchers were given by leading experts. These tutorials covered fundamental aspects of coding theory and cryptology and were meant to prepare for original research in these areas. The present volume collects the expanded lecture notes of these tutorials. In the following, we give a brief indication of the range of topics that is represented in this volume. The 11 articles can roughly be classified into four groups, corresponding to mathematical foundations, coding theory, cryptology, and applied cryptology.

Coding theory and cryptology require several sophisticated mathematical tools which are covered in the articles by Lenstra, Niederreiter, and Shparlinski. The lecture notes of Lenstra present a detailed review of those parts of computational number theory that are relevant for the implementation and the analysis of public-key cryptosystems, such as fast arithmetic, prime generation, factoring, discrete logarithm algorithms, and algorithms for elliptic curves. It is important to note in this context that all public-key cryptosystems of current commercial interest rely on problems from computational number theory that are believed to be hard. The article by Niederreiter provides a quick introduction to the theory of algebraic function fields over finite fields. This theory is crucial for the construction of algebraic-geometry codes and has also found recent applications in cryptography. Exponential sums form another powerful number-theoretic tool in coding theory and they have recently come to the fore in cryptology as well. The lecture notes of Shparlinski offer an engaging introduction to the theory and the applications of exponential sums.

The articles by Barg, Feng, and Xing cover topics of great current interest in coding theory. Barg presents a selection of results on extremal problems of coding theory that are centered around the concept of a code as a packing of the underlying metric space. The results include combinatorial bounds on codes and their invariants, properties of linear codes, and applications of the polynomial method. The lecture notes of Feng describe the mathematical techniques in the rapidly developing subject of quantum error-correcting codes. The article contains also a useful review of classical error-correcting codes. The paper by Xing offers a brief introduction to algebraic-geometry codes and a description of some recent constructions of algebraic-geometry codes.

The articles by Dawson and Simpson, Matsumoto, and Pei deal with important mathematical aspects of cryptology. Dawson and Simpson provide a detailed account of current issues in the design and analysis of stream ciphers. The topics include Boolean functions, correlation attacks, the design of keystream generators, and implementation issues. The paper by Matsumoto is devoted to key management problems in group communication and describes recently developed key distribution schemes and protocols for such systems. The lecture notes of Pei present a detailed exposition of the author's recent work on optimal authentication schemes, both without and with arbitration. The essence of this work is to meet information-theoretic bounds by combinatorial constructions.

The lecture notes of Gollmann and Varadharajan treat topics in applied cryptology. Gollmann contributes to the intense debate on public-key infrastructures with an incisive examination of the security problems that public-key infrastructures claim to address. The article contains also a useful summary of current standards for public-key infrastructures. Varadharajan presents a detailed account of current principles for authorization policies and services in distributed systems. The article also outlines the constructs of a language that can be used to specify a range of commonly used access policies.

I want to take this opportunity to thank Professor Louis H.Y. Chen, the Director of the Institute for Mathematical Sciences, for his guidance and leadership of the IMS and for the invaluable advice he has so freely shared with the organizers of the research program. The expertise and the dedication of all the IMS staff were crucial for the success of the program. I am very grateful to my colleagues San Ling and Chaoping Xing of the Department of Mathematics at the National University of Singapore for the tremendous help they have given in running the program and to the overseas advisers Eiji Okamoto (Toho University, Japan), Igor E. Shparlinski (Macquarie University, Australia), and Neil J.A. Sloane (AT&T Shannon Lab, USA) for their support and suggestions. The financial well-being of the research program was guaranteed by a generous grant from DSTA in Singapore, which is herewith acknowledged with gratitude. Finally, I would like to thank World Scientific Publishing, and especially Kim Tan and Ye Qiang, for the professionalism with which they have produced this volume.

Singapore, August 2002   Harald Niederreiter

EXTREMAL PROBLEMS OF CODING THEORY

Alexander Barg

Bell Labs, Lucent Technologies, 600 Mountain Ave., Rm. 2C-375
Murray Hill, NJ 07974, USA
and
IPPI RAN, Moscow, Russia
E-mail: abarg@research.bell-labs.com

This article is concerned with properties of codes as packings of metric spaces. We present a selection of results on extremal problems of geometric coding theory.

Contents

1.	Introduction	3
2.	Metric Spaces	4
	2.1. Asymptotics	6
	2.1.1. Hamming space	6
	2.1.2. Binary Johnson space $\mathscr{J}^{n,w}$	6
	2.1.3. The sphere $\mathscr{S}^n$	7
	2.1.4. Grassmann space $G_{n,k}(L)$	7
3.	Codes	7
	3.1. Distance distribution	8
	3.2. Asymptotic parameters	9
4.	Average Properties of Codes	9
5.	Averaging over Translations	13
6.	Volume Bounds	15
	6.1. Basic volume bounds	15
	0.2. Elias-type bounds as sphere-packing bounds	19
7.	Linear Codes	19
	7.1. Shortening of unrestricted codes	20
	7.2. Supports, ranks, subcodes	21
	7.3. Rank distribution of a linear code	22
	7.4. Higher weights and weight enumerators	23

8.   Decoding   24
9.   Error Probability   25
     9.1.   A bound on $P_{rmde}(\mathcal{C})$ 27
10. Upper Bound on the Error Probability   29
     10.1. Geometric view of Theorem 10.1   31
     10.2. Explication of the random coding bounds   32
     10.3. Spherical codes   33
11. Polynomial Method   34
12. Krawtchouk Polynomials   36
     12.1. Jacobi polynomials   39
13. Christoffel-Darboux Kernel and Bounds on Codes   40
14. Bounding the Distance Distribution   42
     14.1. Upper bounds   42
     14.2. Lower bounds   43
15. Linear Codes with Many Light Vectors   44
16. Covering Radius of Linear Codes   45
References   46

Notation

$A_w(\mathcal{C}), A_w$	element of the distance distribution of a code $\mathcal{C}$
$A_{\mathcal{C}}(x,y), A(x,y)$	distance enumerator of a code $\mathcal{C}$
$a_w(\mathcal{C}), a_w$	distance profile
$\sharp(A), |A|$	number of elements of a finite set A
$\mathcal{B}_w(X,x), \mathcal{B}_w(x)$	ball of radius w in X with center at x
B_w	$\mathrm{vol}(\mathcal{B}_w)$
$\mathcal{C}$	code
$\mathcal{C}'$	dual code of a linear code $\mathcal{C}$
$D(a\|b)$	information distance between two binomial distributions, Section 2.1
$d(\mathcal{C})$	distance of the code $\mathcal{C}$
$\mathrm{d}(x,y)$	distance between x and y
$\delta_{\mathrm{GV}}(R)$	relative Gilbert-Varshamov distance, Definition 4.3
$\mathscr{H}_q^n$	Hamming space, the n-fold direct product of a q-set
$h_q(x)$	entropy function, Section 2.1
$\mathscr{J}_q^{n,w}$	Johnson space, sphere of radius w about zero in $\mathscr{H}_q^n$
$\mathscr{J}^{n,w}$	binary Johnson space

$K_k(x)$	Krawtchouk polynomial of degree k
$M(X; d)$	maximum size of a d-code in X
$P_{de}(\mathcal{C})\ [P_{ue}(\mathcal{C})]$	probability of decoding [undetected] error of the code $\mathcal{C}$
$p_{i,j}^k(\mathcal{H}_q^n)$	intersection number of $\mathcal{H}_q^n$, see Section 2
$\mathscr{S}^n$	unit sphere in $\mathbb{R}^n$
$\mathcal{S}_w(X, x), \mathcal{S}_w(x)$	sphere of radius w in X with center at x
S_w	$\mathrm{vol}(\mathcal{S}_w)$
$\|x\|$	Hamming norm of x
$\|x\|$	ℓ_2-norm of x
$\chi(x)$	indicator function of a subset in a set X
$f(n) \cong g(n)$	exponential equivalence of functions, Section 2.1

1. Introduction

This article is devoted to results of coding theory that are centered around the concept of a code as a packing of the corresponding metric space. We derive a few results in several rather diverse directions such as combinatorial bounds on codes and their invariants, properties of linear codes, error exponents, and applications of the polynomial method. A common goal of the problems considered is to establish bounds on natural combinatorial parameters of a code. The primary aim of this article is to explain the basic ideas that drive this part of coding theory. In particular, we do not strive to explain the best known result for each problem that we discuss. Our motivation is that, as in each living mathematical discipline, the current best results are often of an *ad hoc* nature and do not add to our understanding of the central ideas. Pointers to the literature that develops the subjects of this article are supposed to compensate for this.

The title of this article refers to estimates of code parameters for codes of large length. This approach helps us to highlight fundamental properties of codes. There is also a vast literature devoted to beating current records for parameters of short codes in various metric spaces; this will not be discussed below (see [41]).

Let X be a metric space. $\mathcal{B}_w(c) = \mathcal{B}_w(X, c)$ denotes the ball and $\mathcal{S}_w(c) = \mathcal{S}_w(X, c)$ the sphere of radius w with center at the point $c \in X$. The volume of a subset $Y \subseteq X$ will be denoted by $\mathrm{vol}\,Y$ (we only deal with the counting measure for finite spaces and the Lebesgue measure in $\mathbb{R}^n$).

2. Metric Spaces

In this section we list the main examples of metric spaces occurring in coding theory.

A. *Hamming space* $\mathcal{H}_q^n$. Let Q be a finite set of size q.

$$\mathcal{H}_q^n = \{(x_1, \ldots, x_n), x_i \in Q\}.$$

Another definition: $\mathcal{H}_q^n = (K_q)^n$, where K_q is a complete graph on q vertices; in this context $\mathcal{H}_q^n$ is also called the Hamming graph.

We denote the elements of Q by $0, 1, \ldots, q-1$. If q is a power of a prime, then we assume that Q is the finite field $\mathbb{F}_q$, and then $\mathcal{H}_q^n$ is an n-dimensional linear space over Q. We call elements of $\mathcal{H}_q^n$ words, or points, or vectors.

The *Hamming norm* or the Hamming weight of a point $x = (x_1, \ldots, x_n) \in \mathcal{H}_q^n$ is defined as

$$|x| = \#\{i : x_i \neq 0\}.$$

The distance induced by this norm is called the *Hamming distance*.

A ball $\mathcal{B}_w(\mathcal{H}_q^n, c)$ of radius w with center at any point $c \in \mathcal{H}_q^n$ has the volume

$$B_w = \text{vol}(\mathcal{B}_w) = \sum_{i=0}^{w} \binom{n}{i}(q-1)^i.$$

Intersection numbers $p_{i,j}^k$ of the space by definition are

$$p_{i,j}^k(\mathcal{H}_q^n) = \#\{z \in \mathcal{H}_q^n : d(z,x) = i, d(z,y) = j; d(x,y) = k\}.$$

Thus, $p_{i,j}^k$ is the number of triangles with fixed vertices x and y, distance k apart, and a floating vertex z that obeys the distance conditions. Explicitly,

$$p_{i,j}^k(\mathcal{H}_q^n) = \sum_{\alpha=0}^{\lfloor \frac{i+j-k}{2} \rfloor} \binom{k}{j-\alpha}\binom{j-\alpha}{k+\alpha-i}\binom{n-k}{\alpha}(q-1)^\alpha (q-2)^{i+j-k-2\alpha}.$$

In particular,

$$p_{ij}^k(\mathcal{H}_2^n) = \binom{k}{(1/2)(j-i+k)}\binom{n-k}{(1/2)(j+i-k)}\chi(j-i+k \in 2\mathbb{Z}).$$

For any two vectors x, y define their *support* as

$$\text{supp}(x,y) = \{i : x_i \neq y_i\}.$$

If $y = 0$, we write supp(x) instead of supp($x, 0$). If $A \subseteq \mathcal{H}_q^n$, then

$$\text{supp}(A) = \bigcup_{a, a' \in A} \text{supp}(a, a').$$

We note that the Hamming metric is not the only interesting distance on $\mathcal{H}_q^n$. Generally the set Q can support various algebraic structures such as groups and rings. Even for $q = 4$ this already leads to nonequivalent metrics on $\mathcal{H}_q^n$: the Hamming distance and the Lee distance. This diversity increases for larger q; eventually the taxonomy of norms and norm-like functions itself becomes a subject of study.

B. *Johnson space.*

$$\mathcal{J}_q^{n,w} = \{ x \in \mathcal{H}_q^n : |x| = w \}.$$

The metric in $\mathcal{J}_q^w$ is the Hamming distance.

If $q = 2$, then we write $\mathcal{J}^{n,w}$ and omit the lower index. In this case it is sometimes more convenient to use the Johnson metric $d_J = (1/2) \, d_H$, where d_H is the Hamming distance.

A ball $\mathcal{B}_r(\mathcal{J}^{n,w})$ has the volume

$$B_r = \sum_{i=0}^{\lfloor r/2 \rfloor} \binom{w}{i} \binom{n-w}{i}.$$

C. *Unit sphere in $\mathbb{R}^n$.*

$$\mathcal{S}^n = \{ x \in \mathbb{R}^n : \|x\| = 1 \}.$$

The distance in $\mathcal{S}^n$ is defined by the angle between the vectors:

$$\theta(x, y) = \arccos(x, y).$$

It is often convenient to use the inner product $t = (x, y)$ instead of θ.

D. *Projective spaces and beyond.* Coding theory is mostly concerned with $X = P^{n-1}\mathbb{R}$, $P^{n-1}\mathbb{C}$, and $P^{n-1}\mathbb{H}$. The distance between $x, y \in X$ is measured by the angle $\theta = \arccos |(x, y)|$ or by the absolute value of the inner product $t = |(x, y)|$.

One generalization of these projective spaces has recently gained attention in coding theory. Let $X = G_{n,k}(L)$ be a Grassmann space, i.e., the manifold of k-planes ($k < n/2$) through the origin in the n-space over L (here $L = \mathbb{R}$ or $\mathbb{C}$). To define the distance between two planes we need to introduce principal angles. Let p and q be two planes in X. The absolute value of the inner product $|(x, y)|, x \in p, y \in q$, as a function of $2k$ variables has k stationary points $\rho_1, \ldots, \rho_k$. Define the principal angles $\theta_1, \ldots, \theta_k$ between p and q by their cosines: $\theta_i = \arccos \rho_i$. There are several justifiable ways of defining the distance between p and q. So far the following definition received most attention in coding theory:

$$d(p, q) = \sqrt{\sin^2 \theta_1 + \cdots + \sin^2 \theta_k}.$$

2.1. *Asymptotics*

We next turn to asymptotic formulas for sphere volume in metric spaces of interest to coding theorists. They will be used to derive volume bounds on codes in Section 6.

Let $f(n)$ and $g(n)$ be two functions of $n \in \mathbb{N}$. We write $f \cong g$ if $\lim_{n \to \infty} \frac{1}{n} \log \frac{f(n)}{g(n)} = 1$. The base of the logarithms and exponents is 2 throughout.

2.1.1. *Hamming space*

Let $X = \mathcal{H}_q^n$, where q is fixed, $n \to \infty$ and let $w = \omega n$. We have, for $\omega, p \in (0, (q-1)/q)$,

$$\sum_{i=0}^{w} \binom{n}{i} (q-1)^i \left(\frac{p}{q-1}\right)^i (1-p)^{n-i} \cong \exp[-nD(\omega\|p)],$$

where the information divergence between two binomial distributions, $D(\omega\|p)$, equals

$$D(\omega\|p) = \omega \log \frac{\omega}{p} + (1-\omega) \log \frac{1-\omega}{1-p}.$$

In particular, with $p = (q-1)/q$ we obtain an asymptotic formula for the volume of the ball:

$$B_{\omega n} \cong \exp[nh_q(\omega)],$$

where $h_q(y)$ is the entropy function defined by

$$h_q(y) = -y \log \frac{y}{q-1} - (1-y) \log(1-y)$$

for $y \in (0, 1)$ and extended by continuity to $y = 0, y = 1$.

2.1.2. *Binary Johnson space $\mathcal{J}^{n,w}$*

Let $w = \omega n$. The volume of the ball is given by

$$B_{\rho n} \cong \exp\left[n\left(\omega h_2\left(\frac{\rho}{2\omega}\right) + (1-\omega)h_2\left(\frac{\rho}{2(1-\omega)}\right)\right)\right].$$

2.1.3. *The sphere* $\mathscr{S}^n$

A ball in $X = \mathscr{S}^n$,

$$\mathcal{B}_\theta(X; x) = \{y \in X : \angle(x, y) \le \theta\},$$

is the spherical cap cut on the surface of X by the circular cone $\mathrm{Con}\,(x, \theta)$ with apex at the origin and axis along x. Let $\Omega(\theta) = \mathrm{vol}(\mathcal{B}_\theta(X; x))$. We have

$$n^{-1} \log \Omega(\theta) = \frac{1}{2} \log \frac{2e\pi \sin^2 \theta}{n} (1 + o(1)) \quad (0 \le \theta \le \pi/2).$$

2.1.4. *Grassmann space* $G_{n,k}(L)$

Let $\mathcal{B}_\delta(G_{n,k})$ be a ball in the Grassmann manifold of radius δ with respect to the distance $\mathrm{d}(p, q)$. The volume of the ball of radius δ is given by

$$B_\delta = \left(\frac{\delta}{\sqrt{k}}\right)^{\beta nk + o(n)} \quad (\beta = 1 \text{ if } L = \mathbb{R};\ \beta = 2 \text{ if } L = \mathbb{C}).$$

All the results of Section 2.1 except the last one are standard. The volume of the ball in $G_{n,k}$ is computed in [10] (see [17] for a discussion of sphere packings in $G_{n,k}$).

3. Codes

Let X be a finite or compact infinite metric space. A code $\mathcal{C}$ is a finite subset of X. The *distance* of the code is defined as $\mathrm{d}(\mathcal{C}) = \min_{x,y \in \mathcal{C}; x \neq y} \mathrm{d}(x, y)$.

Let $M = |\mathcal{C}|$ be the size of (i.e., the number of points in) the code. The rate of the code, measured in bits, is

$$R(\mathcal{C}) = n^{-1} \log M,$$

where n is the dimension of X, clear from the context. The *relative distance* of $\mathcal{C}$ is defined as

$$\delta(\mathcal{C}) = \frac{d(\mathcal{C})}{n}.$$

The argument $\mathcal{C}$ will often be omitted.

A code $\mathcal{C} \subseteq \mathscr{H}_q^n$ of size M and distance d is denoted by $\mathcal{C}(n, M, d)$. If the distance of a code $\mathcal{C}$ is d, we sometimes call it a d-code.

The distance between a point $x \in X$ and a subset $Y \subseteq X$ is defined as

$$\mathrm{d}(x, Y) = \min_{y \in Y} \mathrm{d}(x, y).$$

3.1. *Distance distribution*

Let X be a finite space of diameter D.

Definition 3.1: The *distance distribution* of a code $\mathcal{C} \subseteq X$ is the vector $A = (A_0, A_1, \ldots, A_D)$, where

$$A_i = |\mathcal{C}|^{-1} \sharp \{(x, y) \in \mathcal{C} \times \mathcal{C} : d(x, y) = i\}.$$

Thus $A_0 = 1$.

Let $X = \mathcal{H}_q^n$. If $\mathcal{C}$ is a linear code in $\mathcal{H}_q^n$ (where q is a prime power), then its distance distribution is equal to the weight distribution $(A_0, A_d, \ldots, A_n)$, where

$$A_i = \sharp \{x \in \mathcal{C} : |x| = i\}.$$

Let $w = \omega n$ and let $\alpha_\omega(\mathcal{C}) = n^{-1} \log A_{\omega n}$. The $(n+1)$-vector $[\alpha_\omega(\mathcal{C}), \omega = n^{-1}(0, 1, \ldots, n)]$ is called the *distance (weight) profile* of $\mathcal{C}$. The main use of the distance profile is in asymptotic problems, where it is represented by a real (usually, continuous) function.

The *distance enumerator* of a code $\mathcal{C} \subseteq X$ is the polynomial

$$A_\mathcal{C}(x, y) = \sum_{i=0}^{n} A_i x^{n-i} y^i.$$

For codes in infinite spaces we use a slightly more convenient definition of the distance distribution. For instance, let $\mathcal{C} \subseteq \mathcal{S}^n$. The distance distribution of $\mathcal{C}$ is given by

$$b(s, t) = |\mathcal{C}|^{-1} \sharp \{(x, y) \in \mathcal{C} \times \mathcal{C} : s \le (x, y) < t\}.$$

We have $|\mathcal{C}| = \int_{-1}^{1} db(x)$, where $db(x)$ is a discrete measure defined from $b(s, t)$ in a standard way.

The main applications of the distance distribution are in combinatorial properties of codes, bounds on error probability of decoding (Sections 9, 10), and other extremal problems of coding theory (*e.g.*, Section 16).

Definition 3.2: The *covering radius* of a code $\mathcal{C} \in X$ is defined as

$$r(\mathcal{C}) = \max_{x \in X} d(x, \mathcal{C}).$$

3.2. *Asymptotic parameters*

Let X be one of the metric spaces introduced above. Let

$$M(X;d) = \max_{\mathcal{C} \in X, d(\mathcal{C})=d} |\mathcal{C}|;$$

$$R(\delta) = R(X;\delta) = \lim_{n \to \infty} [n^{-1} \log M(X;d)].$$

In the cases of interest to coding theory, this limit is not known to exist. Therefore, it is usually replaced by $\limsup$ and $\liminf$ as appropriate, and the corresponding quantities are denoted by $\overline{R}(\delta), \underline{R}(\delta)$. We write X in the notation only if the underlying space is not clear by the context.

A *notational convention*: $R(\mathcal{H}_q^n; \delta)$, for instance, means the highest achievable rate of a sequence of codes of relative distance δ in the Hamming space; here $\mathcal{H}_q^n$ is used as a notation symbol in which n has no particular value. This agreement is used throughout the text.

Analogously,

$$\delta(R) = \delta(X;R) = \lim_{n \to \infty} \max_{\mathcal{C} \in X; R(\mathcal{C}) \geq R} d(\mathcal{C})/n.$$

4. Average Properties of Codes

This section is concerned with estimates of codes' parameters obtained by various averaging arguments. In many cases, the existence bounds thus obtained are the best known for large code length. We will establish the average rate, distance, and the distance distribution of unrestricted codes and linear codes in the Hamming space.

Theorem 4.1: *Let* $X = \mathcal{H}_q^n$. *Let* M *be such that*

$$M(M-1) < \frac{q^n}{B_{d-1}}.$$

Then there exists an (n, M, d) *code.*

Proof: Let $\mathcal{C} = \{x_1, \ldots, x_M\}$ be an ordered collection of points. Call $\mathcal{C}$ bad if $d(\mathcal{C}) \leq d-1$ and call a point $x_i \in \mathcal{C}$ bad if it has neighbors in $\mathcal{C}$ at distance $\leq d-1$. If the points $x_2, \ldots, x_M$ are fixed, then x_1 is bad in at most $(M-1)B_{d-1}$ codes. The points $x_2, \ldots, x_M$ can be chosen in $q^{n(M-1)}$ ways, so there are no more than $(M-1)B_{d-1}q^{n(M-1)}$ codes in which point x_1 is bad. This is true for any point $x_i, 1 \leq i \leq M$; thus, there are no more than $M(M-1)B_{d-1}q^{n(M-1)}$ bad codes. If this number is less than the total number of codes q^{nM}, i.e., if

$$M(M-1)B_{d-1} < q^n,$$

then there exists a good code. $\qquad\qquad\qquad\qquad\qquad\qquad\qquad\square$

This result can be improved by the so-called Gilbert procedure (Section 6). However, for large n, Theorem 4.1 accurately describes the parameters of typical codes in $\mathscr{H}_q^n$. More formally, we have the following result.

Theorem 4.2: Let $X = \mathscr{H}_2^n$ and $n \to \infty$. For all codes in X of rate R except for a fraction of codes that decreases exponentially with n, the relative distance approaches the bound

$$2R = 1 - h_2(\delta).$$

Proof: Consider the Shannon ensemble $\mathcal{A}$ of 2^{nM} binary codes, $M = 2^{Rn}$, where every code has probability 2^{-nM}. Or, what is the same, consider a random code formed of M independently chosen vectors, where all the coordinates of every vector are i.i.d. Bernoulli r.v.'s with $P(0) = P(1) = 1/2$.

Let us assume that δ is chosen to satisfy $2R = 1 - h_2(\delta) + \varepsilon$, where $\varepsilon > 0$. We will prove that with probability approaching 1 a random (n, M) code $\mathcal{C} \in \mathcal{A}$ contains a pair of vectors at distance δn or less. Let $x_1, x_2, \ldots, x_M$ be an ordered (multi)set of independent random vectors such that $\Pr[x_i = y] = 2^{-n}$ for any $y \in \{0, 1\}^n$. Let $\nu_{i,j}, 1 < j < i < M$, be the indicator random variable of the event $d(x_i, x_j) = \delta n$. The $\nu_{i,j}$ are pairwise-independent random variables, each with mean

$$\mathsf{E}\,\nu_{i,j} = \Pr[\nu_{i,j} = 1]$$

and variance

$$\mathrm{Var}[\nu_{i,j}] = \mathsf{E}\,\nu_{i,j}^2 - (\mathsf{E}\,\nu_{i,j})^2 = \mathsf{E}\,\nu_{i,j} - (\mathsf{E}\,\nu_{i,j})^2 < \mathsf{E}\,\nu_{i,j}.$$

Consider the number $N_{\mathcal{C}}(d) = \sum_{j<i} \nu_{i,j}$ of unordered pairs of codewords (x_i, x_j) with $i \neq j$ in $\mathcal{C}$ at distance $d = \delta n$ apart. We have

$$\mathsf{E}\,N_{\mathcal{C}}(d) = \binom{M}{2} \mathsf{E}\,\nu_{i,j} \cong 2^{n(2R-1+h_2(\delta))}$$

$$\mathrm{Var}[N_{\mathcal{C}}(d)] = \binom{M}{2} \mathrm{Var}[\nu_{i,j}] < \mathsf{E}\,N_{\mathcal{C}}(d).$$

For any $\alpha > 0$ by the Chebyshev inequality we have

$$\Pr[|N_{\mathcal{C}}(d) - \mathsf{E}\,N_{\mathcal{C}}(d)| \geq \mathsf{E}\,N_{\mathcal{C}}(d)^{(1+\alpha)/2}] \leq (\mathsf{E}\,N_{\mathcal{C}}(d))^{\alpha}$$

$$\cong 2^{\alpha n(1 - 2R - h_2(\delta))} = 2^{-\alpha n \varepsilon} \to 0.$$

Thus, in particular, with probability tending to 1 we have $N_{\mathcal{C}}(d) > 0$, or, in other words, a random code contains a pair of vectors at distance $d = \delta n$. Since $\varepsilon > 0$ can be taken arbitrarily small, this proves an upper bound $\delta \leq h_2^{-1}(1 - 2R)$ on the relative distance of almost all codes in $\mathcal{A}$.

On the other hand, for any δ such that $2R = 1 - h_2(\delta) - \varepsilon$ the average number of codeword pairs with relative distance $d = \delta n$ decreases exponentially with n. Then

$$\Pr[N_{\mathcal{C}}(d) > 1] \leq \mathsf{E}\,N_{\mathcal{C}}(d) \to 0;$$

hence with probability tending to 1 a random code $\mathcal{C}$ has distance $\geq \delta n$. □

This theorem implies that for $R > 1/2$ the relative distance of almost all long codes converges to zero. Thus, unrestricted codes on the average are much worse than linear codes (cf. Theorem 4.4 below).

Definition 4.3: The *relative Gilbert-Varshamov distance* $\delta_{\mathrm{GV}}(R)$ is defined by the equation

$$R = \log q - h_q(\delta) \qquad (0 \leq \delta \leq 1 - 1/q).$$

Theorem 4.4: *Let* $X = \mathcal{H}_q^n$ *and* $n \to \infty$. *For all linear codes in* X *of rate* R *except for a fraction of codes that decreases exponentially with* n, *the relative distance approaches* $\delta_{\mathrm{GV}}(R)$.

Proof: (outline) Consider the ensemble $\mathcal{L}$ of random $[n, k = Rn]$ linear binary codes defined by $(n - k) \times n$ parity-check matrices whose elements are chosen independently with $P(0) = P(1) = 1/2$. If N_w is the random variable equal to the number of vectors of weight $w > 0$ in a code $\mathcal{C} \in \mathcal{L}$, then $\mathsf{E}\,N_w = \binom{n}{w}(q-1)^w/q^{n-k}$ and $\mathrm{Var}\,N_w \leq \mathsf{E}\,N_w$. Thus, $\mathsf{E}\,N_w$ grows exponentially in n for $\omega := \frac{w}{n} > \delta_{\mathrm{GV}}(R)$. Thus, the relative distance of a random linear code approaches $\delta_{\mathrm{GV}}(R)$ as n grows, and the fraction of codes whose relative distance deviates from δ_{GV} by ε tends to 0 exponentially in n for any $\varepsilon > 0$. □

Theorem 4.5: *There exists a linear* $[n, k]$ *code* $\mathcal{C}$ *with* $A_0 = 1$,

$$A_w(\mathcal{C}) \leq \begin{cases} n^2 q^{k-n} S_w, & w \text{ such that } \log S_w \geq (n - k) \log q - 2 \log n, \\ 0, & w : \log S_w < (n - k) \log q - 2 \log n. \end{cases}$$

Proof: Consider linear codes defined in the same way as in the proof of Theorem 4.4. A vector of weight $w > 0$ lies in the kernel of $q^{(n-1)(n-k)}$ matrices. All the $S_w = \binom{n}{w}(q-1)^w$ vectors of weight w are annihilated by at most $S_w q^{(n-1)(n-k)}$ matrices. Thus, on the average the number of vectors of weight w in the code does not exceed $S_w q^{-(n-k)}$ and the fraction of matrices for which this number is $\geq n^2 S_w q^{-(n-k)}$ (call them bad) is at most n^{-2}. Even if the sets of bad matrices for different $w = 1, 2, \ldots, n$ are disjoint, this leaves us with a fraction of $1 - n^{-1}$ of good matrices; any good matrix defines a code $\mathcal{C}$ of dimension $\dim \mathcal{C} \geq k$ over $\mathbb{F}_q$ with

$$A_w(\mathcal{C}) \leq n^2 q^{k-n} S_w, \quad 1 \leq w \leq n.$$

Writing the right-hand side as $\exp[(k-n)\log q + \log S_w + 2\log n]$, we see that once w is such that the exponent becomes negative, we obtain $A_w < 1$. Since $\mathcal{C}$ is linear, this implies that $A_w = 0$ for these values of w. □

Corollary 4.6: *For any $R < \log q - h_q(\delta)$ there exists a sequence of linear codes of growing length n with weight profile α_0, where $\alpha_{0,0} = 0$,*

$$\alpha_{0,w} \leq R - \log q + h_q(w) \quad (\delta_{GV}(R) < w < 1 - \delta_{GV}(R)),$$

$$\alpha_{0,w} = -\infty \qquad\qquad (0 < w < \delta_{GV}(R)).$$

Linear codes that satisfy Theorem 4.5 or Corollary 4.6 will be called *random*.

Theorem 4.7: *(Average value of the distance) Let $\mathcal{C}$ be an (n, M) code. Then, provided in each case that the denominator is positive,*

$$M \leq \frac{d}{d - \frac{q-1}{q}n} \qquad\qquad \mathcal{C} \subseteq \mathcal{H}_q^n,\ d = d(\mathcal{C}),$$

$$M \leq \frac{nd}{nd - 2wn + \frac{q}{q-1}w^2} \qquad \mathcal{C} \subseteq \mathcal{J}_q^{n,w},\ d = d(\mathcal{C}),$$

$$M \leq \frac{1-t}{-t} \qquad\qquad \mathcal{C} \subseteq \mathcal{S}^n,\ t = t(\mathcal{C}).$$

Here $t(\mathcal{C})$ is the maximal inner product of the code $\mathcal{C} \subseteq \mathcal{S}^n$.

This result is called the *Plotkin* bound for $\mathcal{H}_q^n$, the *Johnson* bound for $\mathcal{J}_q^{w,n}$, and the *Rankin* bound for $\mathcal{S}^n$. It is proved by computing the average distance between pairs of points in $\mathcal{C}$ [18, 38].

Thus, for large values of the code distance $d(\mathcal{C})$ the value of M cannot grow exponentially with n, and so for any family of codes $\mathcal{C}$ the rate $R \to 0$.

For instance, for $X = \mathcal{H}_q^n$ the code size M is at most $O(n)$ if $\delta = d(\mathcal{C})/n > \frac{q-1}{q}$. Below in asymptotic problems we always assume the reverse inequality.

Already for general unrestricted codes the technique presented in this section produces weak results. In more complicated problems of coding theory one resorts to more refined averaging methods, such as averaging over the choice of subsets rather than individual vectors, etc. [11].

5. Averaging over Translations

This section presents another averaging technique which is useful for deriving upper bounds on code parameters and linking the Hamming and Johnson spaces.

Lemma 5.1: *Let x, y be two vectors in $\mathcal{H}_q^n$ with $\mathrm{d}(x, y) = u$. The number of vectors $z \in \mathcal{H}_q^n$ such that $x - z \in \mathcal{J}_q^{n,w}$ and $y - z \in \mathcal{J}_q^{n,w}$ equals $p_{ww}^u(\mathcal{H}_q^n)$.*

Proof: Obvious. □

Lemma 5.2: [34] *Let $\mathcal{C} \subseteq X \subseteq \mathcal{H}_q^n$ be a code and $Y, Z \subseteq \mathcal{H}_q^n$ be arbitrary subsets. Then*

$$\sum_{c \in \mathcal{C}} |(Y - c) \cap Z| = \sum_{z \in Z} |(\mathcal{C} + z) \cap Y|.$$

Proof:

$$\sum_{c \in \mathcal{C}} |(Y - c) \cap Z| = \sum_{c \in \mathcal{C}} \sum_{y \in Y} \sum_{z \in Z} \chi\{y - c = z\} = \sum_{z \in Z} \sum_{y \in Y} \sum_{c \in \mathcal{C}} \chi\{c + z = y\}$$

$$= \sum_{z \in Z} |(\mathcal{C} + z) \cap Y|.$$

vskip-5mm □

Corollary 5.3: *Let $\mathcal{C} \subseteq \mathcal{J}_q^{n,v}$ be a d-code. Then*

$$|\mathcal{C}| p_{u,w}^v < S_u M(\mathcal{J}_q^{n,w}; d).$$

Proof: In Lemma 5.2 take $X = S_v(0)$, $Y = S_w(0)$, $Z = S_u(0)$. Let $y \in Y, c \in \mathcal{C}$, then $y - c \in Z$ if and only if $d(y, c) = u$. The number of $y \in Y$ with this property for a fixed c equals $p_{u,w}^v$. On the right-hand side we observe that the set $(\mathcal{C} + z) \cap Y$ is a d-code in Y. □

Corollary 5.4: *Let* $\mathcal{C}, Y$ *be subsets of* $\mathcal{H}_q^n$. *Then*

$$|Y||\mathcal{C}| = \sum_{z \in \mathcal{H}_q^n} |(\mathcal{C} + z) \cap Y|.$$

Proof: Follows by putting $X = Z = \mathcal{H}_q^n$ in Lemma 5.2. $\square$

Lemma 5.5: [36] *Let* $\mathcal{C}$ *be code in* $\mathcal{H}_q^n$. *Then*

$$|\mathcal{C}|A_i(\mathcal{C})p_{w,w}^i = \sum_{x \in \mathcal{H}_q^n} |\mathcal{C}(x,w)|A_i(\mathcal{C}(x,w)),$$

where $p_{w,w}^i$ *is the intersection number and* $\mathcal{C}(x,w) = (\mathcal{C} + x) \cap \mathcal{J}_q^{n,w}$.

Proof: Count in two ways the total number of pairs of codewords in $\mathcal{C}(x,w)$ distance i apart for all $x \in \mathcal{H}_q^n$. By definition, this is the right-hand side of the claimed identity. On the other hand, every pair of codewords in $\mathcal{C}$ at a distance i falls in $\mathcal{J}_q^{n,w}$ in $p_{w,w}^i$ shifts of $\mathcal{C}$ (Lemma 5.1). $\square$

Any of last two results implies the well-known *Bassalygo-Elias inequality:*

$$M(\mathcal{H}_q^n; d)\binom{n}{w}(q-1)^w \leq M(\mathcal{J}_q^{n,w}; d)q^n \tag{5.1}$$

(take $i = 0$ in Lemma 5.5 or take $Y = \mathcal{J}_q^{n,w}$ in Corollary 5.4).

Theorem 5.6: (*Elias-type bounds*)

$$R(\mathcal{H}_q^n; \delta) \leq \log q - h_q\big(\lambda(1 - \sqrt{1 - \delta/\lambda})\big) \quad (\lambda = 1 - q^{-1}),$$

$$R(\mathcal{J}^{n,\omega n}; \delta) \leq h_2(\omega) - h_2\left(\frac{1}{2} - \frac{1}{2}\sqrt{1 - 2\delta}\right),$$

$$R(\mathcal{S}^n; \theta) \leq -\log(\sqrt{2}\sin(\theta/2)).$$

Proof: Consider the Hamming case. From Theorem 4.7 for the Johnson space we see that when $w = \omega n$ approaches the (smaller) root of the denominator,

$$\omega_0 = \lambda - \sqrt{\lambda(\lambda - \delta)},$$

the quantity $n^{-1}\log M(\mathcal{J}_q^{n,w}; d, w) \to 0$. Substituting ω_0 into inequality (5.1) and computing logarithms completes the proof. The other two cases follow by some modifications of this argument. $\square$

Solving the inequality of the theorem for Hamming space with respect to δ, we obtain the bound

$$\delta(R) \le \delta_E(R),$$

where

$$\delta_E(R) := 2\delta_{GV}(R)(1 - \delta_{GV}(R)/2\lambda)$$

is sometimes called the *Elias radius*. The bound itself is called the Bassalygo-Elias bound (the Hamming case) and the Coxeter-Rankin bound (the spherical case). For the Johnson space the result can be proved analogously to the Hamming case; see [33], where a nonasymptotic version of this bound is also derived. In Section 6.2 we give a proof based on a volume argument.

6. Volume Bounds

This section is devoted to the standard bounds proved via a volume, or packing argument. We begin with the standard Gilbert-Varshamov and Hamming bounds, which basically say that if the spheres of radius d are disjoint, then their centers form a d-code, and that the size of a d-code is bounded above by the number of spheres of radius $d/2$ that can be packed into X. The second part of Section 6.1 deals with an improvement of the Hamming bound for the Johnson space. The ideas developed there will also be central in the derivation of error exponents in Section 9.

6.1. *Basic volume bounds*

Let X be one of the metric spaces discussed above with distance d and volume form vol. Let $B_d = \text{vol}(\mathcal{B}_d)$ be the volume of the ball of radius d in X.

Theorem 6.1: (*Gilbert bound*) *If M is any number such that $MB_d <$ vol(X), then X contains a code of size $M+1$ and distance d. If the distance d takes only natural values, then d can be replaced with $d-1$.*

For $\mathcal{H}_q^n$ this bound was given in Theorem 4.4 and Corollary 4.6.

Theorem 6.2: (*Shannon bound* [45]) *For any*

$$R < -\log\sin\theta$$

there exists a number n_0 such that for every $n \ge n_0$ the sphere $\mathcal{S}^n$ contains a code of size 2^{Rn} and distance θ.

Proof: Follows on substituting the volume of the spherical cap from Section 2.1.3 into the Gilbert bound. □

Theorem 6.3: *(Hamming bound) Let $\mathcal{C}$ be a code of size M in X. Then $M \le \mathrm{vol}(X)/B_{d/2}$.*

Concrete expressions of this bound for various metric spaces can be obtained using the formulas of Section 2.1. The Hamming bound is usually good for small values of d and weak otherwise. It can be improved for $\mathcal{J}^{n,w}$ by making use of the embedding $\mathcal{J}^{n,w} \subseteq \mathcal{H}_2^n$. The idea is to notice [12] that for some v, a ball of radius $d/2$ with center on $\mathcal{S}_w(\mathcal{H}_2^n, 0) = \mathcal{J}^{n,w}$ intersects the sphere $\mathcal{S}_v(\mathcal{H}_2^n, 0)$ by a larger subset than it intersects the sphere $\mathcal{S}_w$ itself.

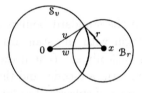

$\mathcal{B}_r(x)$ intersects $\mathcal{S}_v(0)$ by a large subset

Lemma 6.4: *Let $w = \omega n \le n/2, r = \rho n \le v$. Then*

$$\sum_{x \in \mathcal{S}_w(0)} |\mathcal{B}_r(x) \cap \mathcal{S}_v(0)| = \binom{n}{w} \sum_{i=w-v}^{r} p_{i,v}^w \le r \binom{n}{r} \binom{n}{w - j_0},$$

where $j_0/n \to \gamma_0 = \rho\frac{1-2\omega}{1-2\rho}$. This inequality is asymptotically tight, i.e., for $w - v \sim j_0$ it turns into an equality in the $\cong$ sense.

Corollary 6.5: *Let*

$$N(w,r) = \sum_{x \in \mathcal{S}_w(0)} |\mathcal{B}_r(x) \cap \mathcal{S}_r(0)|.$$

Then $N(w,r) \le r\binom{n}{r}^2(1 + o(1))$ with equality $N(w,r) \cong \binom{n}{r}^2$ only for $w/n \sim 2\rho(1-\rho)$.

Proof: (of Lemma 6.4 and Corollary 6.5) Let $V = \sum\limits_{\substack{x \in \mathcal{H}_2^n \\ |x|=w}} |\mathcal{B}_r(x) \cap \mathcal{S}_v(0)|$.

Let $j = w - v$. The equality

$$V = \binom{n}{w} \sum_{i=j}^{r} p_{i,v}^w$$

follows by definition. To prove the inequality, observe that

$$p_{r,v}^w \le \sum_{i=j}^{r} p_{i,v}^w = \sum_{i=j}^{r} \binom{w}{\frac{1}{2}(i+j)} \binom{n-w}{\frac{1}{2}(i-j)}.$$

As is easily verified, the summation term in the last sum grows on i. Substituting $i = r$, we obtain

$$\sum_{i=j}^{r} p_{i,v}^w \le r \binom{w}{\frac{1}{2}(r+j)} \binom{n-w}{\frac{1}{2}(r-j)} = r p_{r,v}^w.$$

Therefore,

$$\binom{n}{w} p_{r,v} \le V \le r \binom{n}{w} p_{r,v}^w.$$

Further,

$$\binom{n}{w} p_{r,v}^w = \binom{n}{r} p_{v,w}^r = \binom{n}{r} \binom{r}{\frac{1}{2}(r+j)} \binom{n-r}{w - \frac{1}{2}(r+j)}.$$

In the last expression, for a fixed vector y of weight r, we are counting vectors x of weight w with $d(x,y) = n(\omega - \gamma)$, where $\gamma = j/n$. Their number is maximized if x is a typical vector obtained after n independent drawings from the binomial probability distribution given by $P(1) = \omega - \gamma$, $P(0) = 1 - \omega + \gamma$. We obtain the following condition on the maximizing value of γ:

$$\rho - \frac{1}{2}(\rho + \gamma) = \rho(\omega - \gamma).$$

In other words, the maximizing value of j is attained for $j/n \to \gamma_0$. Note that, at least for large n, $j_0 = \gamma_0 n$ satisfies the condition on $j = w - v$ implied by the restriction on v in the Lemma, so the choice $j = j_0$ is consistent. Thus,

$$\max_{j:j\le r}\binom{r}{\frac{1}{2}(r+j)}\binom{n-r}{w-\frac{1}{2}(r+j)}$$

$$\cong \exp\left[n\left(\rho h_2\left(\frac{\rho+\gamma_0}{2\rho}\right)+(1-\rho)h_2\left(\frac{2\omega-\rho-\gamma_0}{1-\rho}\right)\right)\right]$$

$$\cong \exp[h_2(\omega-\gamma_0)]\cong\binom{n}{w-j_0},$$

where the last line follows by substituting the value of γ_0 and simplifying.
The corollary follows on substituting $v=r$ into the lemma.   □

Theorem 6.6: *Let* $n\to\infty, w/n\to\omega, 0<\omega<1/2, \omega>\delta>0$. *Then*

$$M(\mathcal{J}^{n,w};\delta n)\le a(n)\frac{\mathrm{vol}(\mathcal{S}_w)}{\mathrm{vol}(\mathcal{B}_{\delta n/2})},\tag{6.1}$$

where $n^{-1}\log a(n)\to 0$ *as* $n\to\infty$. *Therefore,*

$$R(\mathcal{J}^{n,\omega n};\delta)\le h_2(\omega)-h_2(\delta/2).$$

Proof: Let $\mathcal{C}\subseteq\mathcal{J}^{n,\omega n}$ be an (n,M,d) code and $r=\lfloor(d-1)/2\rfloor$. We have

$$|\mathcal{B}_r(c)\cap\mathcal{S}_v(0)|=\sum_{i=w-v}^{r}p_{i,v}^w.$$

Also $\mathcal{B}_r(c)\cap\mathcal{B}_r(c')=\emptyset$ for two different codewords c and c'. Therefore for any $v, w-r\le v\le w+r$,

$$M\le\frac{\mathrm{vol}(\mathcal{S}_v)}{\sum_{i=w-v}^{r}p_{i,v}^w}.$$

Take $v=\frac{\omega-\delta/2}{1-\delta}n$. Then using Lemma 6.4, we obtain

$$M\le\frac{\binom{n}{w}S_v}{\binom{n}{w}\sum_{i=w-v}^{r}p_{i,v}^w}\cong\frac{\binom{n}{w}S_v}{\binom{n}{r}\binom{n}{v}}=\frac{\binom{n}{w}}{\binom{n}{r}}.$$

                                                                                □

The denominator in the estimate (6.1) is an exponentially greater quantity than the volume computed in 2.1.2; hence the estimate itself is asymptotically better than the Hamming bound. In particular, by Theorem 6.6,

$R(\mathscr{J}^{n,\omega n};\delta) = 0$ for $\delta \geq 2\omega$ while the Hamming bound implies this conclusion only for $\delta \geq 4\omega(1 - \omega)$. The actual answer is

$$R(\mathscr{J}^{n,\omega n};\delta) \begin{cases} > 0 & 0 \leq \delta < 2\omega(1 - \omega) \\ = 0 & \delta \geq 2\omega(1 - \omega) \end{cases}$$

by combining the Gilbert-Varshamov and Elias bounds.

6.2. Elias-type bounds as sphere-packing bounds

The Hamming bound is not the best bound obtainable by the volume argument for large code length. Namely, suppose that the balls of radius $r > d/2$ around the codewords intersect, but the intersection volume grows as a polynomial function $p(n)$ of the code length. Then

$$|\mathcal{C}|B_r \leq \text{vol}(X)p(n).$$

Letting $n \to \infty$, we obtain a bound better than the Hamming bound. For instance, let $\mathcal{C}$ be an $(n, M = q^{Rn}, \delta n)$ code in $\mathscr{H}_q^n$ and let $\mathcal{B}_{\omega n}$ be a ball of radius ωn. By a slight modification of Theorem 4.7 we conclude that for $\omega < \omega_{\text{crit}} := \lambda - \sqrt{\lambda(\lambda - \delta)}$ the number of codewords inside the ball grows at most polynomially in n. Therefore, for any codeword c a point $x \in \mathscr{H}_q^n$ with $d(x, c) < n\omega_{\text{crit}}$ can be distance $n\omega_{\text{crit}}$ or less away from at most $p(n)$ codewords, where $p(n)$ is some polynomial. We then have

$$\frac{1}{n}\log(MB_{\omega n}) = R - \log q + h_q(\omega_{\text{crit}}) + o(1) < (\log p(n))/n,$$

which again proves Theorem 5.6 for $X = \mathscr{H}_q^n$. Other parts of this theorem can be proved by a similar argument.

7. Linear Codes

This section deals with combinatorial and linear-algebraic properties of codes. The technique used here is based on an interplay of the rank distributions and weight distributions of linear codes. Readers familiar with matroids will immediately notice a connection with representable matroids and their invariants.

Let q be a prime power. A linear $[n, k, d]$ code $\mathcal{C}$ is a subspace of $\mathbb{F}_q^n$ of dimension k and distance d. A matrix $\mathbf{G}$ whose rows form a basis of $\mathcal{C}$ is called the *generator matrix* of the code. The linear $[n, n - k, d']$ code $\mathcal{C}' = \{x : \forall_{c \in \mathcal{C}}(c, x) = 0\}$ is called the dual code of $\mathcal{C}$. The generator matrix $\mathbf{H}$ of $\mathcal{C}'$ is called the *parity-check matrix* of $\mathcal{C}$. For any matrix $\mathbf{G}$ with n

columns and a subset $E \subseteq \{1, 2, \ldots, n\}$ we denote by $\mathbf{G}(E)$ the subset of columns of $\mathbf{G}$ indexed by the elements of E.

The goal of this section is to derive in a simple way some combinatorial identities related to the famous MacWilliams theorem.

Definition 7.1: *Puncturings and shortenings.* Let $\mathcal{C}$ be an $[n, k, d]$ code. Puncturing it results in an $[n-1, k, d-1]$ code. More generally, let $E \subseteq \{1, 2, \ldots, n\}, |E| = n - t$. The projection $\mathcal{C}_E = \mathrm{proj}_E \, \mathcal{C}$ is a linear subcode of $\mathcal{C}$ of length $n - t$ and dimension equal to $\mathrm{rk}\, \mathbf{G}(E)$.

A 1-shortening of $\mathcal{C}$ on coordinate i is formed of $|\mathcal{C}|/q$ codewords $c \in \mathcal{C}$ such that $c_i = 0$; this is a linear $[n, k-1, d]$ subcode. Successively applying this operation, we obtain a t-shortening of $\mathcal{C}$ on the coordinates in $\{1, 2, \ldots, n\} \setminus E$, where E is some t-subset. This is a linear subcode $\mathcal{C}^E \subseteq \mathcal{C}$ such that $\mathrm{supp}\, \mathcal{C}^E \subseteq E$.

7.1. Shortening of unrestricted codes

Before proceeding further, we give an application of shortenings to properties of unrestricted (i.e., linear or not) codes. In general, for an (n, M, d) code $C \subseteq \mathcal{H}_q^n$, shortening is defined as follows: out of the M codewords at least M/q coincide in a given coordinate i. Consider the code formed of these codewords with the ith coordinate deleted. This gives an $(n-1, \geq M/q, d)$ code. Iterating, we get

Lemma 7.2: *Let $\mathcal{C}$ be an (n, M, d) code. For any $t \leq n - d$, we have*

$$M \leq q^t M(\mathcal{H}_q^{n-t}, d).$$

Proof: Let $E \subseteq \{1, 2, \ldots, n\}, |E| = n - t \geq d$. Shortening of $\mathcal{C}$ on the coordinates outside E gives a code $\mathcal{C}^E(n - t, \geq q^{-t}M, d)$. $\qquad\square$

Theorem 7.3: $\overline{R}(\delta)$ *is continuous.*

Proof: From the previous lemma we obtain

$$\overline{R}(\delta) \leq \tau + (1 - \tau)\overline{R}\left(\frac{\delta}{1 - \tau}\right).$$

Assume that $\tau < 1 - 2\delta$. Let $\eta = \delta/(1 - \tau)$, then $0 < \delta < \eta < 1/2$. We have

$$0 \leq \overline{R}(\delta) - (1 - \tau)\overline{R}(\eta) \leq \tau.$$

Letting $\tau \to 0$ proves the claim. $\qquad\square$

The same claim is also valid for $\overline{R}(\delta)$ for codes on S^{n-1}.

7.2. *Supports, ranks, subcodes*

Theorem 7.4: (i) *If $t \leq d(\mathcal{C}) - 1$, then $\mathcal{C}_E$ is an $[n - t, k, \geq d(\mathcal{C}) - t]$ code.*
(ii) $\dim \mathcal{C}^E = |E| - \mathrm{rk}\,\mathbf{H}(E)$.
(iii) $\mathcal{C}_E \cong \mathcal{C}/\mathcal{C}^{\bar{E}}$.
(iv) $(\mathcal{C}_E)' = (\mathcal{C}')^E$; $(\mathcal{C}^E)' = (\mathcal{C}')_E$.

Proof: (i) will follow by Lemma 7.5. (ii)-(iii) are obvious. Let us prove the first part of (iv). Let $a \in (\mathcal{C}')^E$, then $\mathbf{G}(E)a^T = 0$, so $a \in (\mathcal{C}_E)'$. Further, by (ii)

$$\dim(\mathcal{C}')^E = |E| - \mathrm{rk}(\mathbf{G}(E))$$

$$= |E| - \dim \mathcal{C}_E = \dim(\mathcal{C}_E)'.$$

The second part of (iv) is analogous. $\qquad\qquad\square$

Lemma 7.5: $|E| - \mathrm{rk}(\mathbf{H}(E)) = k - \mathrm{rk}(\mathbf{G}(\bar{E}))$.

Proof: Let $\mathcal{C}_E = \mathrm{proj}_E\,\mathcal{C}$ be the projection of $\mathcal{C}$ on the coordinates in E. Clearly, $\dim \mathcal{C}_E = \mathrm{rk}(\mathbf{G}(E))$. On the other hand, $\mathcal{C}_E \cong \mathcal{C}/\mathcal{C}^{\bar{E}}$ by Theorem 7.4(iii); hence by (ii)

$$\dim \mathcal{C}_E = k - \dim \mathcal{C}^{\bar{E}} = k - |E| + \mathrm{rk}(\mathbf{H}(\bar{E})).$$

$\qquad\qquad\square$

Lemma 7.6: (The MacWilliams identities)

$$\sum_{i=0}^{n-u} A_i' \binom{n-i}{u} = |\mathcal{C}'|q^{-u} \sum_{i=0}^{u} A_i \binom{n-i}{n-u}. \tag{7.1}$$

Proof: We have the following chain of equalities:

$$\sum_{i=0}^{n-u} A_i' \binom{n-i}{u} = \sum_{|E|=n-u} |(\mathcal{C}')^E|$$

$$= \sum_{|E|=n-u} q^{n-u-\mathrm{rk}(\mathbf{G}(E))}$$

$$= q^{n-k-u} \sum_{|E|=n-u} q^{u-\mathrm{rk}(\mathbf{H}(\bar{E}))}$$

$$= q^{n-k-u} \sum_{i=0}^{u} A_i \binom{n-i}{n-u}. \tag{7.2}$$

Here the first equality follows by counting in two ways the size of the set

$$\{(E,c) : \sharp E = n - u \text{ and } c \in (\mathcal{C}')^E, |c| \leq n - u\},$$

the second one is straightforward, the third one (the central step in the proof) is implied by Lemma 7.5, and the final step follows by the same argument as the first one. □

Theorem 7.7:

$$|\mathcal{C}|A'_j = \sum_{i=0}^{n} A_i K_j(i), \qquad (7.3)$$

where

$$K_j(i) = \sum_{\ell=0}^{i} (-1)^\ell \binom{i}{\ell} \binom{n-i}{j-\ell} (q-1)^{j-\ell}$$

is the Krawtchouk number.

Proof: Multiply both sides of (7.1) by $(-1)^u \binom{u}{\ell}$, sum over u from 0 to n, and use the fact that $\sum_{u \in \mathbb{Z}} (-1)^u \binom{n-i}{u}\binom{u}{s} = (-1)^{n-i}\delta_{n-i,s}$. This gives

$$A_j = |\mathcal{C}|^{-1} \sum_{i=0}^{n} A'_i \sum_{u=i}^{n} (-1)^{u-n+j} q^{n-u} \binom{u}{n-j}\binom{n-i}{n-u}.$$

The sum on u in the last expression is just another form for $K_j(i)$. □

7.3. Rank distribution of a linear code

Rewrite (7.2) by collecting on the right-hand side subsets of one and the same rank. Namely, let

$$U_u^v = |\{E \subseteq \{1, 2, \ldots, n\} \mid |E| = u, \mathrm{rk}(\mathbf{G}(E)) = v\}|.$$

Then by (7.2) and Theorem 7.4(ii) we have

$$\sum_{i=0}^{w} \binom{n-i}{n-w} A_i = \sum_{v=0}^{n-k} q^{w-v} (U')_w^v, \qquad (7.4)$$

where the numbers U' are the rank coefficients of $\mathcal{C}'$. Further, Lemma 7.5 implies that

$$(U')_u^{u-k+v} = U_{n-u}^v.$$

The last two equations relate the weight enumerator of $\mathcal{C}$ and its *rank distribution* $(U_u^v, 0 \leq u \leq n, 0 \leq v \leq k)$.

Example 7.8: (MDS codes) An (n, M, d) code is called *maximum distance separable* (MDS) if $M = q^{n-d+1}$. Let $\mathcal{C}$ be an $[n, k, n - k + 1]$ q-ary linear MDS code with a parity-check matrix $\mathbf{H}$. Then $\rho^* E = \text{rk}(\mathbf{H}(E)) = \min\{|E|, n - k\}$ for all $E \subseteq S$, $0 \leq |E| \leq n - k$. Therefore

$$(U')_u^v = \begin{cases} \binom{n}{u} & \text{if } (0 \leq v = u \leq n - k) \text{ or } (u \geq n - k + 1, \ v = n - k); \\ 0 & \text{otherwise.} \end{cases}$$

This enables us to compute the weight spectrum of $\mathcal{C}$. Substituting the values of $(U')_u^v$ into (7.4), we obtain $A_0 = 1$, $A_i = 0$ for $1 \leq i \leq n - k$, and

$$A_{n-k+\ell} = \binom{n}{k-\ell} \sum_{j=0}^{\ell-1} (-1)^j \binom{n-k+\ell}{j} (q^{\ell-j} - 1) \quad (1 \leq \ell \leq k). \quad (7.5)$$

Clearly, the dual code $\mathcal{C}'$ is also MDS of dimension $n - k$.

Definition 7.9: The *rank polynomial* of a linear code $\mathcal{C}$ is

$$U(x, y) = \sum_{u=0}^{n} \sum_{v=0}^{k} U_u^v x^u y^v.$$

Relations of the rank polynomial of $\mathcal{C}$, its dual code $\mathcal{C}'$, and the weight polynomial of $\mathcal{C}$ are given by the following results.

Theorem 7.10:

$$U'(x, y) = x^n y^{\dim \mathcal{C}'} U\left(\frac{1}{xy}, y\right).$$

Theorem 7.11:

$$A(x, y) = y^n |\mathcal{C}| \ U\left(\frac{x-y}{y}, \frac{1}{q}\right) = (x - y)^n U'\left(\frac{qy}{x-y}, \frac{1}{q}\right).$$

7.4. *Higher weights and weight enumerators*

Let $\mathcal{C}$ be an $[n, k, d]$ q-ary linear code. Define the r-th support weight distribution of $\mathcal{C}$, $0 \leq r \leq k$, as the vector $(A_i^r, 0 \leq i \leq n)$, where

$$A_i^r = \sharp\{\mathcal{D} : \mathcal{D} \text{ is a linear subcode of } \mathcal{C}, \dim \mathcal{D} = r, |\text{supp}(\mathcal{D})| = i\}.$$

Theorem 7.12: (*Generalized MacWilliams identities* [46])

$$\sum_{i=0}^{w} \binom{n-i}{n-w} A_i^r = \sum_{v=0}^{n-k} \begin{bmatrix} w - v \\ r \end{bmatrix} (U')_w^v \quad (0 \leq w \leq n, \ 0 \leq r \leq k).$$

This theorem is proved similarly to (7.4).

Theorem 7.13: [30] *Let*

$$D_{\mathcal{C}}^r(x,y) = \sum_{i=0}^{n} \left(\sum_{m=0}^{r} [r]_m A_i^m \right) x^{n-i} y^i,$$

then

$$D_{\mathcal{C}}^r(x,y) = q^{-r(n-k)} (y + (q^r - 1)x)^n D_{\mathcal{C}'}^r \left(\frac{y-x}{y+(q^r-1)x} \right), \quad r \geq 0.$$

Here $[r]_m = \prod_{u=0}^{m-1}(q^m - q^i)$.

A simple way to prove this theorem is to realize that $D_{\mathcal{C}}^r(x,y)$ is the Hamming weight enumerator of the code $\mathcal{C}^{(r)} = \mathbb{F}_{q^r} \otimes_{\mathbb{F}_q} \mathcal{C}$.

Concluding remarks

The ideas of this section can be developed in several directions. First, it is possible to consider different versions of rank polynomials and of support weight distributions such as, for instance,

$$A_i^{(r)} = \sum_{E \subseteq \{1,\ldots,n\}, |E|=i} \#\{\{c_1, c_2, \ldots, c_r\} \subseteq \mathcal{C}, \operatorname{supp}(c_1, c_2, \ldots, c_r) = E\}$$

$$(i = 0, 1, \ldots, n).$$

The corresponding generating functions, as a rule, satisfy MacWilliams-type identities. This line of thought leads to matroid invariants that we mentioned in the beginning of this section. See [7, 15] for more on this subject.

Another avenue is to study alphabets with some algebraic structure such as abelian groups, finite rings, and modules over them. This enables one to define various norm-like functions on the alphabet and study weight enumerators of codes with respect to these functions [27]. When duality is appropriately defined, these enumerators typically also satisfy MacWilliams-type identities.

8. Decoding

Definition 8.1: Let $\mathcal{C} \subseteq X$ be a code in a metric space X. A (partial) mapping $\psi_t : X \to \mathcal{C}$ is called *decoding* if for any y such that $d(y, \mathcal{C}) \leq t$,

$$\psi(y) = \arg\min_{c \in \mathcal{C}} d(c, y).$$

For $y \notin \cup_{c \in \mathcal{C}} \mathcal{B}_t(c)$ the value of $\psi_t(y)$ is undefined.

We will only consider the two extremes: *complete decoding* and *error detection*. Under complete decoding, $t = r(\mathcal{C})$ (see Definition 3.2), *i.e.*, $X \subseteq \cup_{c \in \mathcal{C}} B_t(c)$. Under error detection, $t = 0$. Error detection will be briefly considered in the beginning of Section 9; otherwise we will focus on complete decoding, denoted by ψ hereafter.

To justify the term "decoding", assume that $\mathcal{C} \subseteq \mathcal{H}_q^n$ is used for transmission over a q-ary symmetric channel (qSC) given by a random mapping $Q \to Q$ such that

$$P(b|a) = (1 - p)\delta_{a,b} + \frac{p}{q - 1}(1 - \delta_{a,b}),$$

where p is called the crossover probability of the channel, $p < (q - 1)/q$. Suppose that the transmitted codeword c is received as $y \in \mathcal{H}_q^n$. The event $\psi(y) \neq c$ is called a decoding error. Let $P_{\mathrm{de}}(c)$ be its probability. As it turns out, the complete decoder ψ is a good choice for minimizing $P_{\mathrm{de}}(c)$.

Definition 8.2: Let $\mathcal{C}$ be a code. The *Voronoi domain* of a codeword c with respect to $\mathcal{C}$ is the set

$$D(c, \mathcal{C}) = \{x \in X : \forall_{c' \in \mathcal{C}} \ \mathrm{d}(x, c) \leq \mathrm{d}(x, c')\}.$$

Lemma 8.3: *Let $\mathcal{C}$ be a linear code and x a vector in $\mathcal{H}_q^n$. The complete decoding of x can be defined as follows:*

$$\psi(x) = x - \ell(\mathcal{C} - x),$$

where $\ell(\mathcal{C} - x)$ is a vector of lowest weight in the coset $\mathcal{C} - x$.

9. Error Probability

In this and the next section we are concerned with upper bounds on the error probability of complete decoding of the best possible codes used on a binary symmetric channel (BSC). The analysis performed for this channel offers a simple model for results on error exponents for arbitrary memoryless channels. For the BSC it is possible to derive the error exponent bound starting with a transparent geometric description of error events. The ideas developed below are central to (single-user) information theory. Although they are several decades old, they continue to attract attention of researchers to this date. In particular, the main problem in this area, that of the exact error exponent, is still unsolved.

Let $X = \mathcal{H}_q^n$. Let $\mathcal{C}$ be used for transmission over a qSC with crossover probability p. If c is the transmitted codeword, then the channel defines a probability distribution on X given by

$$P(y|c) = \left(\frac{p}{q-1}\right)^{d(y,c)} (1-p)^{n-d(y,c)}.$$

The error probability of decoding for a given vector $c \in \mathcal{C}$ is defined as

$$P_{\text{de}}(c) = \Pr\{X \backslash D(c, \mathcal{C})\} = \sum_{y \in X \backslash D(c,\mathcal{C})} P(y|c).$$

The average error probability for the code $\mathcal{C}$ is

$$P_{\text{de}}(\mathcal{C}) = |\mathcal{C}|^{-1} \sum_{c \in \mathcal{C}} P_{\text{de}}(c).$$

The probability of undetected error $P_{\text{ue}}(\mathcal{C})$ for the code $\mathcal{C}$ is defined analogously.

Theorem 9.1:

$$P_{\text{ue}}(\mathcal{C}) = A\left(1-p, \frac{p}{q-1}\right) - (1-p)^n.$$

Proof: Let $\pi(i) = (\frac{p}{q-1})^i (1-p)^{n-i}$ and let $A_i(c) = \sharp\{c' \in \mathcal{C} : d(c, c') = i\}$. We calculate

$$P_{\text{ue}}(\mathcal{C}) = |\mathcal{C}|^{-1} \sum_{c \in \mathcal{C}} \sum_{c' \in \mathcal{C} \backslash \{c\}} \pi(d(c, c')) = |\mathcal{C}|^{-1} \sum_{c \in \mathcal{C}} \sum_{i=1}^{n} A_i(c)\pi(i)$$

$$= \sum_{i=1}^{n} A_i(\mathcal{C})\pi(i)$$

$$= A\left(1-p, \frac{p}{q-1}\right) - (1-p)^n.$$
$\square$

Definition 9.2: The *error exponent* for the Hamming space (known also as the *reliability function* of the q-ary symmetric channel) is defined as follows:

$$E(R, p, n) = -n^{-1} \log(\min_{\mathcal{C}:R(\mathcal{C}) \geq R} P_{\text{de}}(\mathcal{C})),$$

$$E(R, p) = \lim_{n \to \infty} E(R, p, n).$$

Conventions made after the formula for $R(\delta)$ in Section 3.2 apply to this definition as well.

Analogously to this definition one defines the exponent $E_{ue}(R, p)$ of the probability of undetected error. It is easy to derive a lower bound on this exponent.

Theorem 9.3:

$$E_{ue}(R, p) \geq D(\delta_{GV}(R)\|p) + \log q - R \qquad 0 \leq R \leq \delta_{GV}(p),$$

$$E_{ue}(R, p) \geq \log q - R \qquad\qquad \delta_{GV}(p) \leq R \leq 1.$$

Proof: Follows by combining Theorem 9.1 and Corollary 4.6. $\qquad\square$

9.1. *A bound on* $P_{rmde}(\mathcal{C})$

We proceed with bounds on $P_{de}(\mathcal{C})$.

$$P_{de}(\mathcal{C}) \leq |\mathcal{C}|^{-1} \sum_{c \in \mathcal{C}} \sum_{c' \in \mathcal{C} \setminus \{c\}} P(c \to c'),$$

where

$$P(c \to c') := \sum_{y \in X : \, d(y,c') \leq d(y,c)} P(y|c)$$

is the probability, under the binomial distribution, of the half-space cut out by the median hyperplane between c and c'. Note that $P(y|c)$ depends only on the Hamming weight of the error vector $x = y - c$.

Lemma 9.4: *Let* $P_{de}(\mathcal{C}, x \in U)$ *be the joint probability of decoding error and the event* $x \in U \subseteq X$. *Then for any* $r = 0, 1, \ldots, n$,

$$P_{de}(\mathcal{C}, p) \leq P(\mathcal{C}, x \in \mathcal{B}_r(0)) + P(x \notin \mathcal{B}_r(0)).$$

Let us specialize this result using the distance distribution of the code.

Lemma 9.4: *Let* $\mathcal{C}$ *be a d-code with distance distribution* $(A_0, A_d, \ldots, A_n)$. *Then for any* $r = 0, 1, \ldots, n$,

$$P_{de}(\mathcal{C}, p) \leq P_1 + P_2, \tag{9.1}$$

where

$$P_1 = \sum_{w=d}^{2r} A_w \sum_{e=\lceil w/2 \rceil}^{r} |\mathcal{B}_e(c) \cap \mathcal{S}_e(0)| p^e (1-p)^{n-e} \tag{9.2}$$

$$= \sum_{w=d}^{2r} A_w \sum_{i=\lceil w/2 \rceil}^{r} \sum_{\ell=0}^{r-i} \binom{w}{i} \binom{n-w}{\ell} p^{i+\ell} (1-p)^{n-i-\ell}, \tag{9.3}$$

where c is a code vector with $|c| = w$, and

$$P_2 = \sum_{e=r+1}^{n} \binom{n}{e} p^e (1-p)^{n-e}. \tag{9.4}$$

Proof: Let c be the transmitted codeword, let x be the error vector in the channel, $|x| = e$, and let $y = c + x$ be the received vector. A decoding error occurs if $v = d(y, c') \le d(y, c) = e$.

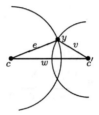

Let $d(c, c') = w$ and suppose that $A_w(c)$ is the number of codewords in $\mathcal{C}$ at distance w from c. Since all the error vectors of one and the same weight are equiprobable, we have

$$P_{\mathrm{de}}(c, x \in \mathcal{B}_r(0)) \le A_w(c) \sum_{e=\lceil w/2 \rceil}^{r} |\mathcal{B}_e(c') \cap \mathcal{S}_e(0)| p^e (1-p)^{n-e}.$$

Computing the average value of $P_{\mathrm{de}}(c, x \in \mathcal{B}_r(0))$ over the code, we obtain (9.2). To obtain (9.3), observe that, as in Lemma 6.4,

$$P_1 = \sum_{w=d}^{2r} A_w \sum_{e=\lceil w/2 \rceil}^{r} p^e (1-p)^{n-e} \sum_{v=0}^{r} p_{e,v}^w. \tag{9.5}$$

Let $i = |\mathrm{supp}(x) \cap \mathrm{supp}(c, c')|$ and $\ell = e - i$. Now (9.3) follows by substituting the definition of $p_{e,v}^w$ and renaming the summation indexes.

The expression for P_2 is immediate. $\qquad \Box$

The choice of $U = \mathcal{B}_r(0)$ in the last two lemmas is justified by the fact that the noise is spherically symmetric. It makes sense to choose the radius r so that the bound on P_{de} is minimized. An interesting fact is that for random codes of rate R this minimum is attained for $r/n \to \delta_{\mathrm{GV}}(R)$.

Lemma 9.6: *The minimum of the bound (9.1) is attained for* $r = r_0$, *where*

$$r_0 = \min\left\{ r : \sum_{w=d}^{2r} A_w \sum_{e=\lceil w/2 \rceil}^{r} |\mathcal{B}_e(c) \cap \mathcal{S}_e(0)| \geq S_r \right\},$$

where $c \in \mathcal{C}, |c| = w$. *Further, if* $(\mathcal{C}_i, i = 1, 2, \ldots)$ *is a sequence of random linear codes of growing length, then* $\lim_{n \to \infty} \frac{r_0}{n} = \delta_{GV}(R)$.

Proof: The first part of the claim is obvious since P_1 is a growing and P_2 a falling function of r. Let us prove the second part. Let A_w be an element of the weight distribution of a random linear code of length n. We then have (see Theorem 4.5)

$$A_w \leq n^2 \binom{n}{w} 2^{Rn-n}.$$

Using Corollary 6.5, we now obtain

$$n^2 2^{Rn-n} \sum_{w=d}^{2r} \binom{n}{w} \sum_{e=\lceil w/2 \rceil}^{r} |\mathcal{B}_e(c) \cap \mathcal{S}_e(0)| \cong 2^{Rn-n} \binom{n}{r}^2.$$

To find r_0 for large n, we equate this to $S_r = \binom{n}{r}$ to obtain

$$2^{Rn-n} \binom{n}{r_0} \cong 1,$$

i.e., $r_0/n \to \delta_{GV}(R)$, as claimed. $\qquad\square$

10. Upper Bound on the Error Probability

The error probability of decoding of good codes is known to fall exponentially when the code length increases, for all code rates less than the capacity $\mathscr{C}$ of the channel. The following theorem gives a bound on this exponent.

Theorem 10.1: $E(R, p) \geq E_0(R, p)$, *where for* $0 \leq R \leq R_x$

$$E_0(R, p) = E_x(R, p) = -\delta_{GV}(R) \log 2\sqrt{p(1-p)}, \qquad (10.1)$$

for $R_x \leq R \leq R_{crit}$

$$E_0(R, p) = E_r(R, p) = D(\rho_0 \| p) + R_{crit} - R, \qquad (10.2)$$

and for $R_{crit} \leq R \leq \mathscr{C} = 1 - h_2(p)$

$$E_0(R, p) = E_{sp}(R, p) = D(\delta_{GV}(R) \| p); \qquad (10.3)$$

here

$$R_x = 1 - h_2(\omega_0), \qquad (10.4)$$

$$R_{\text{crit}} = 1 - h_2(\rho_0), \qquad (10.5)$$

$$\rho_0 = \frac{\sqrt{p}}{\sqrt{p} + \sqrt{1-p}}, \qquad \omega_0 := 2\rho_0(1-\rho_0) = \frac{2\sqrt{p(1-p)}}{1 + 2\sqrt{p(1-p)}}.$$

For $\mathscr{C} \le R \le 1$, $E(R,p) = 0$.

Proof: Suppose we transmit with random $[n, Rn, d]$ linear codes, $d/n \to \delta_{\text{GV}}(R)$. We use (9.1), (9.2), and (9.4) to bound the error probability

$$P_1 \le n^2 2^{Rn-n} \sum_{w=d}^{2r} \binom{n}{w} \sum_{e=\lceil w/2 \rceil}^{r} |\mathcal{B}_e(c) \cap \mathcal{S}_e(0)| p^e (1-p)^{n-e}. \qquad (10.6)$$

Letting $e = \rho n$, we obtain from Corollary 6.5 the estimate

$$P_1 \cong \max_{\delta_{\text{GV}}(R)/2 \le \rho \le \delta_{\text{GV}}(R)} \exp[-n(D(\rho\|p) + (1-R) - h_2(\rho))].$$

The unconstrained maximum on ρ on the right-hand side (the minimum of the exponent) is attained for $\rho = \rho_0$, and, again by Corollary 6.5, the unconstrained maximum on $\omega = w/n$ in (10.6) is attained for $\omega = \omega_0$. The three cases (10.1)-(10.3) are realized depending on how ω_0 and ρ_0 are located with respect to the optimization limits.

The case (10.2) corresponds to ω_0, ρ_0 within the limits: $\delta_{\text{GV}}(R)/2 \le \rho_0 \le \delta_{\text{GV}}(R), \omega_0 \ge \delta_{\text{GV}}(R)$. Then the exponent of P_1 is

$$D(\rho_0\|p) + h_2(\delta_{\text{GV}}(R)) - h_2(\rho_0), \qquad (10.7)$$

i.e., the random coding exponent of (10.2). We need to compare the exponent of P_1 with the exponent $D(\delta_{\text{GV}}(R)\|p)$ of P_2. Under the assumption $\rho_0 < \delta_{\text{GV}}(R)$ their difference is

$$[D(\rho_0\|p) - h_2(\rho_0)] - [D(\delta_{\text{GV}}(R)\|p) - h_2(\delta_{\text{GV}}(R))] < 0$$

since $D(x\|p) - h_2(x)$ is an increasing function of x for $x > \rho_0$. This proves that the dominating exponent for $R \le R_{\text{crit}}$ is given by (10.7).

Suppose now that $\omega_0 \ge \delta_{\text{GV}}(R)$ and $\rho_0 \ge \delta_{\text{GV}}(R)$, i.e., $R \ge R_{\text{crit}}$. In this case the exponent of P_1 is dominated by the term with $\rho = \delta_{\text{GV}}(R)$. Then we conclude that the exponents of P_1 and P_2 are both equal to the sphere-packing exponent of (10.3).

If $\omega_0 \leq \delta_{\mathrm{GV}}$, i.e., $R \leq R_x$, the maximum on ω is attained for $\omega = \delta_{\mathrm{GV}}$, and we get for the right-hand side of (10.6) the following expression:

$$\sum_{e \geq d/2} \binom{n\delta_{\mathrm{GV}}}{n\delta_{\mathrm{GV}}/2}\binom{n(1-\delta_{\mathrm{GV}})}{e-(1/2)n\delta_{\mathrm{GV}}}p^e(1-p)^{n-e}.$$

This is maximized when $e - (1/2)n\delta_{\mathrm{GV}} = n(1-\delta_{\mathrm{GV}})p$, i.e., for

$$\rho = \frac{e}{n} = (1-\delta_{\mathrm{GV}})p + \frac{\delta_{\mathrm{GV}}}{2}.$$

Substituting, we obtain the "expurgation exponent" $E_x(R,p)$ of (10.1). To finish off this case we need to show that the exponent $D(\delta_{\mathrm{GV}}\|p)$ of the term $P[\mathrm{w}(y) \geq d]$ is greater for $\omega_0 \leq \delta_{\mathrm{GV}} \leq 1/2$ than $E_x(R,p)$. This is confirmed by a straightforward calculation.

The proof of the equality $E(R,0) = 0$ for $R \geq \mathscr{C}$ (the "converse coding theorem") will be omitted. $\qquad\square$

In the next two subsections we study a geometric interpretation of this theorem and provide background and intuition behind it.

10.1. *Geometric view of Theorem 10.1*

A close examination of the proof reveals the intuition behind decoding error events for random codes. The capacity region of the BSC is given on the (R,p)-plane by $(0 \leq R, 0 \leq p \geq 1/2, R + h_2(p) \leq 1)$. According to the three cases in the theorem, this region can be partitioned naturally into the regions of low noise A, moderate noise B, and high noise C, where

$$(10.1) \qquad A = \{(R,p) : R \leq 1 - h_2(\omega_0)\},$$
$$(10.2) \qquad B = \{(R,p) : 1 - h_2(\omega_0) \leq R \leq 1 - h_2(\rho_0)\},$$
$$(10.2) \qquad C = \{(R,p) : 1 - h_2(\rho_0) \leq R \leq 1 - h_2(p)\}.$$

As n increases, within each region the error events are dominated by a particular (relative) weight ω_{typ} of incorrectly decoded codewords. Moreover, the relative weight ρ_{typ} of error vectors that form the main contribution to the error rate also converges to a particular value. We have, for the regions A, B, and C, respectively,

$$\omega_0 < \delta_{\mathrm{GV}}, \qquad\qquad \rho_{\mathrm{typ}} = (1-\delta_{\mathrm{GV}})p + \tfrac{1}{2}\delta_{\mathrm{GV}}, \quad \omega_{\mathrm{typ}} = \delta_{\mathrm{GV}},$$
$$\omega_0 \geq \delta_{\mathrm{GV}}, \rho_0 < \delta_{\mathrm{GV}}, \quad \rho_{\mathrm{typ}} = \rho_0, \qquad\qquad\qquad \omega_{\mathrm{typ}} = 2\rho_0(1-\rho_0),$$
$$\omega_0 \geq \delta_{\mathrm{GV}}, \rho_0 \geq \delta_{\mathrm{GV}}, \quad \rho_{\mathrm{typ}} = \delta_{\mathrm{GV}}, \qquad\qquad\quad \omega_{\mathrm{typ}} = 2\delta_{\mathrm{GV}}(1-\delta_{\mathrm{GV}}).$$

When the code is used in the low-noise region, the typical relative weight of incorrectly decoded codewords is $\delta_{\mathrm{GV}}(R)n$, i.e., it does not depend on the channel. In the moderate-noise region, the typical weight of incorrect codewords is ρ_0 and in the high-noise region it is $\delta_{GV}(R)$. We observe therefore that for $R > R_x$ the error probability does not depend on the minimum distance of the code.

The geometry of decoding for $R < R_{crit}$ and for $R > R_{crit}$ is of very different nature. Consider an error event that corresponds to the moderate-noise region. Its probability is dominated by errors y of relative weight ρ_0. From the proof of the theorem and Corollary 6.5 it can be seen that the number of points of the sphere $S_{\rho_0 n}$ that are decoded incorrectly behaves exponentially as

$$\frac{\binom{n}{\rho_0 n}}{\binom{n}{\delta_{GV} n}}\binom{n}{\rho_0 n};$$

hence their fraction has the same exponent as $\binom{n}{\rho_0 n}/\binom{n}{\delta_{GV} n}$. We see that for $\rho_0 < \delta_{GV}$ an exponentially small fraction of error vectors y of weight $\rho_0 n$ leads to an incorrect codeword c'. We have $|c'|/n \to w_0, d(y, c')/n \to \rho_0$.

Moreover, with some additional arguments it is possible to show that an error vector y on the sphere of radius $\rho_0 n$ around the transmitted codeword typically falls in at most one ball $B_{\rho_0 n}(c')$ around an incorrect codeword c'. Hence, the union bound of (9.2), (10.6) *for random linear codes* is exponentially tight. Error vectors y that are incorrectly decoded to a codeword c' occupy one and the same fraction

$$\cong \frac{\binom{n}{\rho_0 n}}{\binom{n}{\delta_{GV} n}}\frac{\binom{n}{\delta_{GV} n}}{\binom{n}{w}} = \frac{\binom{n}{\rho_0 n}}{\binom{n}{w}}$$

of the sphere $S_{\rho_0 n}$ for almost every incorrect codeword c', $|c'| = w_0 n$.

On the other hand, once R exceeds R_{crit} or $\rho_0 > \delta_{GV}$, almost every point y on the sphere $S_{\rho_0 n}(c)$ leads to a decoding error; the only relief comes from the fact that such points are received from the channel in an exponentially small fraction of transmissions. For $R > R_{crit}$ almost every incorrectly decoded error vector y will have exponentially many codewords that are same distance or nearer to it as the transmitted word.

10.2. Explication of the random coding bounds

Random coding bounds were a central topic in information theory in the first few decades of its development. They can be derived by a variety of approaches. Following the methods developed in this chapter, we single out two related, though not identical ways to random coding bounds, both with long lineage. The first one is writing an estimate for the error probability of complete decoding of a code and then averaging this expression over an ensemble of random codes. The strongest results on this way were obtained by Gallager whose treatise [26] also gives the most comprehensive account of this method. A recent informal discussion of it by one of the main contributors is found in [13].

The second approach suggests first to establish properties of a good code in the ensemble such as distance distribution and then to estimate the error probability for this particular code. This idea was also suggested by Gallager [25]. This is what we did when we first proved Theorem 4.5 and then used the code whose existence is proved in it, in Theorem 10.1.

There are two reasons for which the second approach may prevail. First, under it, Theorem 10.1 generalizes without difficulty to *arbitrary discrete memoryless channels*. Of course, in this case the Hamming weight and distances do not describe the effect of noise adequately; their role is played by the composition ("type") codewords and information distance between types, respectively [20]. Remarkably, it is possible to extend some of the geometric intuition of Section 10.1 to this very general case [19].

Apart from this generalization, in engineering applications it is more important to be able to bound the error probability for a specific code than for an ensemble of codes. Constructing good bounds on this probability received much attention in the last decade (see [42, 43]). This problem has also revived interest in deriving random coding bounds via estimating the error probability for a specific code.

There is also a minor technical problem with computing the average error probability for some code ensembles: for low rates typical codes are often poor (cf. Theorem 4.2). To deal with this, one usually employs "expurgation" of the code, i.e., removing from it codewords for which the error probability $P_{de}(C)$ is large. This issue is discussed in more detail in [26, 8].

Other methods in this area include hypothesis testing [14] and, lately, applications of statistical mechanics.

10.3. *Spherical codes*

Similar results can be derived for codes on the sphere $\mathscr{S}^n$ in $\mathbb{R}^n$ used over a Gaussian channel with additive white noise. In particular, an analog of Theorem 10.1 was proved by Shannon in [45]. As it is seen now [6], the idea of Shannon's proof is qualitatively similar to the proof of Theorem 10.1, though the analytic part is somewhat more involved.

11. Polynomial Method

This section is concerned with one application of harmonic analysis to extremal problems of coding theory. The ideas are primarily due to MacWilliams [37] and Delsarte [21]. We will provide details for codes in the Hamming space; however, it should be kept in mind that a similar theory can be developed for a large class of finite and infinite metric spaces [22, 28].

Theorem 11.1: [21] *Let* $x \in \mathcal{H}_q^n, |x| = i$, *be a vector. Then the Krawtchouk number* $K_k(i)$ *equals*

$$K_k(i) = \sum_{y \in \mathcal{H}_q^n, |y| = k} \varphi_x(y),$$

where $\varphi_x(y) = \exp(2\pi i(x, y)/q)$ *is a character of the additive group* $(\mathbb{Z}_q)^n$.

Definition 11.2: Let $\mathbf{K} = \|K_k(i)\|$ be the $(n+1) \times (n+1)$ Krawtchouk matrix with rows numbered by i and columns by k, and let $A = (A_0, A_1, \dots, A_n)$ be the distance distribution of an (n, M) code $\mathcal{C}$ in $\mathcal{H}_q^n$. The vector

$$(A_0', A_1', \dots, A_n') = M^{-1}A\mathbf{K}$$

is called the *dual distance distribution* of $\mathcal{C}$.

    The number $d' = \min(i \geq 1 : A_i > 0)$ is called the *dual distance* of the code $\mathcal{C}$. If $\mathcal{C}$ is linear, then d' is the minimum distance of the dual code $\mathcal{C}'$.

    The code $\mathcal{C}$ is called a *design* of strength t if $d'(\mathcal{C}) = t + 1$.

Theorem 11.3: [21] *The components of the dual distance distribution of any code are nonnegative, and* $A_0' = 1$, $\sum A_i' = q^n/M$.

    The main application of this result to extremal problems of coding theory is given in the following theorem.

Theorem 11.4: *Let* $\mathcal{C}$ *be code with distance* d *and dual distance* d'. *To bound*

$$F(g) = \sum_{i=d}^{n} g(i) A_i(\mathcal{C})$$

below, choose

$$f(x) = \sum_{k=0}^{n} f_k K_k(x), \ f_k \geq 0, d' \leq k \leq n,$$

so that $f(i) \le g(i), d \le i \le n.$ *Then*

$$F(g) \ge |\mathcal{C}| f_0 - f(0).$$

To bound $F(g)$ *above, choose*

$$h(x) = \sum_{k=0}^{n} h_k K_k(x), \ h_k \le 0, d' \le k \le n,$$

so that $h(i) \ge g(i), d \le i \le n.$ *Then*

$$F(g) \le |\mathcal{C}| h_0 - h(0).$$

Proof: For instance, let us prove the second part. We have

$$\sum_{i=0}^{n} h(i) A_i = \sum_{i=0}^{n} A_i \sum_{j=0}^{n} h_j K_j(i) = \sum_{j=0}^{n} h_j |\mathcal{C}| A'_j \le h_0 |\mathcal{C}|.$$

Here the second step is by definition of A'_j and the final step is justified by the Delsarte inequalities. Hence

$$F(g) = \sum_{i=d}^{n} g(i) A_i \le \sum_{i=d}^{n} h(i) A_i \le h_0 |\mathcal{C}| - h(0).$$

The first part of the theorem is established analogously. $\qquad \square$

Corollary 11.5: *Let*

$$f(x) = \sum_{k=0}^{n} f_k K_k(x), \ f_0 > 0, \ f_k \ge 0, k = 1, 2, \ldots, n,$$

be a polynomial such that $f(i) \le 0, \ i = d, d+1, \ldots, n.$ *Then for any* (n, M, d) *code we have*

$$M \le \frac{f(0)}{f_0}.$$

Proof: Take $g \equiv 0$ in the first statement of Theorem 11.4. $\qquad \square$

Theorem 11.4 is essentially due to Delsarte [21] who proved it in the form given in the last corollary. It was realized not long ago [3] that the same method can be used for bounding code parameters other than its size. This approach implies numerous results in extremal problems in coding theory, both "finite" and asymptotic.

Corollary 11.6: *Let* $g(i) = p^i (1-p)^{n-i}$ *and let* $f(x)$ *be a polynomial that satisfies the conditions of Theorem 11.4. Then for any code* $\mathcal{C}$,

$$P_{ue}(\mathcal{C}) \ge |\mathcal{C}| f_0 - f(0).$$

Corollary 11.7: *Let $r \in \{1, 2, \ldots, n\}$. For any code $\mathcal{C}$ there exists a number $i, 1 \leq i \leq r$, such that*

$$A_i(\mathcal{C}) \geq \frac{|\mathcal{C}|f_0 - f(0)}{f(i)}.$$

Results similar to those established here for $\mathcal{H}_q^n$ can be proved for the Johnson space $\mathcal{J}^{n,w}$. In the Johnson space the role of Krawtchouk polynomials is played by one class of Hahn polynomials $H_k(x)$.

Spherical codes. Let $\mathcal{C} \subseteq X = \mathcal{S}^n$ and $b(s, t)$ be its distance (inner product) distribution, see Section 3.1. The analog of Delsarte inequalities has the following form.

Theorem 11.8: [23, 28]

$$\int_{-1}^{1} P_k^{\alpha,\alpha}(x)db(x) \geq 0, \quad k = 0, 1, \ldots,$$

where $\alpha = (n-3)/2$ and $P_k^{\alpha,\alpha}(x)$ is a Gegenbauer polynomial of degree k.

Corollary 11.9: *Let $\mathcal{C}$ be a code with distance distribution $b(s, t)$. To bound*

$$F(g) = \int_{-1}^{t} g(x)db(x),$$

choose

$$f(x) = \sum_{k=0}^{l} f_k P_k^{\alpha,\alpha}(x), \quad f_k \geq 0, \ 1 \leq k \leq l, \ l = 1, 2, \ldots,$$

so that $f(x) \leq g(x), -1 \leq x \leq t$. Then

$$F(g) \geq |\mathcal{C}|f_0 - f(1).$$

12. Krawtchouk Polynomials

Definition 12.1: Krawtchouk polynomials are real polynomials orthogonal on $\{0, 1, \ldots, n\}$ with weight $\mu(i) = q^{-n}\binom{n}{i}(q-1)^i$:

$$\langle K_i, K_j \rangle = \binom{n}{i}(q-1)^i \delta_{i,j},$$

where

$$\langle K_i, K_j \rangle = \sum_{s=0}^{n} K_i(s)K_j(s)\mu(s).$$

Explicitly,

$$K_k(x) = \sum_{\sigma=0}^{k} (-1)^\sigma \binom{x}{\sigma} \binom{n-x}{k-\sigma} (q-1)^{k-j}.$$

Properties

$$\|K_i\|^2 = \binom{n}{i}(q-1)^i; \; K_i(0) = \binom{n}{i}(q-1)^i.$$

$$K_i(s) \le \sqrt{\frac{\binom{n}{i}(q-1)^i}{\mu(s)}}.$$

Any polynomial $f(x)$ of degree $\le n$ can be expanded into the Krawtchouk basis: $f = \sum f_k K_k$, where

$$f_k = \frac{\langle f, K_k \rangle}{\langle K_k, K_k \rangle}. \tag{12.1}$$

In particular, let $f(x) = K_i(x)K_j(x)$, then

$$f(a) = \sum_{k=0}^{n} p_{i,j}^k K_k(a), \; a = 0, 1, \ldots, n,$$

where $p_{i,j}^k$ is the intersection number of $\mathscr{H}_2^n$.

Proof: We treat the case $q = 2$, the general case being analogous.

$$K_i(s)K_j(s) = \sum_{|\mathbf{y}|=i} (-1)^{(\mathbf{x},\mathbf{y})} \sum_{|\mathbf{y'}|=j} (-1)^{(\mathbf{x},\mathbf{y'})} = \sum_{\substack{\mathbf{y},\mathbf{y'} \\ |\mathbf{y}|=i, |\mathbf{y'}|=j}} (-1)^{(\mathbf{x},\mathbf{y}-\mathbf{y'})}. \qquad \square$$

From this we can derive another expression for $p_{i,j}^k$:

$$\left\langle \sum_{k=0}^{n} p_{i,j}^k K_k, K_l \right\rangle = p_{i,j}^l \binom{n}{l} = \langle K_i K_j, K_l \rangle = \sum_{\sigma=0}^{n} K_i(\sigma)K_j(\sigma)K_l(\sigma)\mu(\sigma).$$

The polynomial $K_k(x)$ has k simple zeros $x_{1,k} < x_{2,k} < \cdots < x_{k,k}$ located between 0 and n. Zeros of $K_k(x)$ and $K_{k+1}(x)$ possess the "interlacing" property:

$$0 < x_{1,k+1} < x_{1,k} < x_{2,k+1} < \cdots < x_{k,k} < x_{k+1,k+1} < n.$$

Most of these and many other properties of Krawtchouk polynomials were proved by Delsarte [22]. Given the importance of Krawtchouk polynomials for coding theory, they have received much attention. Comprehensive surveys of their properties are provided in [35, 32].

In the Johnson space $\mathscr{J}^{n,w}$ the role of Krawtchouk polynomials is played by Eberlein polynomials $E_k(x)$ and some Hahn polynomials $H_k(x)$ (they are the p- and q-polynomials of the Johnson association scheme, respectively). The polynomials $H_k(x)$ are orthogonal on the set $\{0, 1, \ldots, w\}$ with weight $\mu(i) = \binom{w}{i}\binom{n-w}{i}/\binom{n}{w}$. Their properties were established in [22], see also [39].

Asymptotics. Asymptotic behavior of extremal zeros and of the polynomials themselves plays a major role in coding-theoretic applications of the polynomial method.

Let $K_{\tau n}(\xi n)$ be a binary Krawtchouk polynomial, $\tau < 1/2$. We are concerned with the asymptotics of the first zero $x_{1,\tau n}$ and of the exponent $k(\tau, \xi) = \lim_{n\to\infty} n^{-1}\log_2 K_t(x)$. Let $\phi(u) = (1/2) - \sqrt{u(1-u)}$.

The zeros of $K_t(x)$ are located inside the segment $[n\phi(\tau), n(1 - \phi(\tau))]$ and for the minimum zero we have

$$n\phi(\tau) \le x_{1,\tau n} \le n\phi(\tau) + t^{1/6}\sqrt{n - t}. \qquad (12.2)$$

The following is known about $k(\tau, \xi)$.

$$k(\tau, \xi) \le k_1(\tau, \xi) = \frac{1}{2}(h_2(\tau) - h_2(\xi) + 1) \quad (0 \le \tau \le 1),$$

$$k(\tau, \xi) \sim k_2(\tau, \xi) = h_2(\tau) + I(\tau, \xi) \quad (0 \le \xi < \phi(\tau)),$$

$$k(\tau, \xi) < k_3(\tau, \xi) = \frac{1}{2}\left[\xi \log_2 \frac{\phi(\tau)}{1 - \phi(\tau)} + \log_2(1 - \phi(\tau)) + h_2(\tau) + 1\right]$$

$$(0 \le \xi \le \phi(\tau)).$$

Here

$$I(\tau, \xi) = \int_0^\xi \log \frac{s + \sqrt{s^2 - 4y(1-y)}}{2 - 2y} dy,$$

where $s = 1 - 2\tau$. While these asymptotic relations cover coding theory needs, much more accurate results on the asymptotic behavior of $K_t(x)$ in the region $x < (1/2)(n-1) - \sqrt{t(n-t)}$ (i.e., outside the oscillatory region) are given in [31].

We note that $k_1(\tau, \xi) \ge k_3(\tau, \xi) \ge k_2(\tau, \xi)$ for $0 \le \xi \le \phi(\tau)$, with equality if and only if $\xi = \phi(\tau)$. Moreover, for this ξ also

$$(k_1(\tau, \xi))'_\xi = (k_2(\tau, \xi))'_\xi = (k_3(\tau, \xi))'_\xi.$$

It is also possible to prove a lower bound on Krawtchouk polynomials.

Theorem 12.3: *Let $x \in [x_{1,t}, n - x_{1,t}]$ be an integer. Then*

$$(K_t(x) + K_{t+1}(x))^2 \geq O(1) \binom{n}{t} / \mu(x).$$

With the exception of the last theorem, similar results are known for the Hahn [36] and Gegenbauer [2] polynomials.

12.1. *Jacobi polynomials*

The role of Gegenbauer polynomials for projective spaces over $\mathbb{R}, \mathbb{C}$, and $\mathbb{H}$ is played by various Jacobi polynomials. Jacobi polynomials $P_k^{\alpha,\beta}(x)$ are orthogonal on the segment $[-1, 1]$ with respect to the measure $d\mu(x) = (1 - x)^\alpha (1 + x)^\beta dx$. We have

$$P_k^{\alpha,\beta}(x) = L_k \prod_{j=1}^{k} (x - t_{j,k}),$$

where

$$L_k = 2^{-k} \sum_{\nu=0}^{k} \binom{k + \alpha}{k - \nu} \binom{k + \beta}{\nu}.$$

The zeros $t_{j,k}$ of P_k are located between -1 and 1; we assume numbering from right to left: $t_{k,k} < t_{k-1,k} < \cdots < t_{1,k}$. We have

$$P_k^{\alpha,\beta}(x) = (-1)^k P_k^{\beta,\alpha}(-x).$$

Zeros of $P_k^{\alpha,\alpha}(x)$ are symmetric with respect to 0, and $t_{k,k} = -t_{1,k}$.

For $k \to \infty, \alpha = ak, \beta = bk$, we have [28]

$$t_{1,k} \to q(a, b) := 4\sqrt{(a + b + 1)(a + 1)(b + 1)} - \frac{a^2 + b^2}{(a + b + 2)^2},$$

$$t_{k,k} \to -q(b, a);$$

in particular, for $\alpha = \beta$ this implies

$$t_{1,k} \to q(a, a) = \frac{\sqrt{1 + 2a}}{1 + a}.$$

The asymptotic exponent of $P_k^{\alpha,\beta}$ has the same qualitative behavior as that of the Krawtchouk and Hahn polynomials, see [2, 31].

13. Christoffel-Darboux Kernel and Bounds on Codes

This section deals with a famous application of the polynomial method to extremal problems. Following [39], consider the polynomial $f(x) = (K_t(a))^2 W_t(x)$, where

$$W_t(x) = \frac{(K_{t+1}(x) + K_t(x))^2}{a - x}.$$

Here a is the smallest root of $K_{t+1}(x) + K_t(x)$ and t is a parameter.

Theorem 13.1: *The polynomial $f(x)$ has the following properties:*
(i) *in the expansion $f(x) = \sum_{k=0}^{2t+1} f_k K_k(x)$ the coefficients f_k are nonnegative and*

$$f_0 = \frac{2}{t+1}(K_t(a))^2 \binom{n}{t};$$

(ii)

$$f(0) = \frac{(K_{t+1}(a))^2}{a} \binom{n}{t}^2 \frac{(n+1)^2}{(t+1)^2};$$

(iii) $f_k \cong \log_2(K_t(a))^2 p_{t,t}^k$.

Proof: We rewrite $f(x)$ as

$$f(x) = \frac{(K_t(a)K_{t+1}(x) - K_t(x)K_{t+1}(a))^2}{a - x}$$

and use the Christoffel-Darboux formula

$$\frac{K_t(y)K_{t+1}(x) - K_t(x)K_{t+1}(y)}{y - x} = \frac{2\binom{n}{t}}{t+1} \sum_{j=0}^{t} \frac{K_j(x)K_j(y)}{\binom{n}{j}}.$$

Then by (12.1)

$$\binom{n}{k} f_k = \left\langle K_t(a)(K_{t+1}(x) + K_t(x)) \frac{2\binom{n}{t}}{t+1} \sum_{j=0}^{t} \frac{K_j(x)K_j(a)}{\binom{n}{j}}, K_k \right\rangle$$

$$= \left\langle \frac{2K_t(a)\binom{n}{t}}{t+1} \sum_{j=0}^{t} \frac{K_j(a)}{\binom{n}{j}} \sum_{i=0}^{n} (p_{t,j}^i + p_{t+1,j}^i) K_i, K_k \right\rangle$$

$$= \frac{2K_t(a)\binom{n}{t}}{t+1} \binom{n}{k} \sum_{j=0}^{t} \frac{K_j(a)}{\binom{n}{j}} (p_{t,j}^k + p_{t+1,j}^k) \geq 0.$$

From the last line we also find f_0; $f(0)$ is found from the definition of $f(x)$. This proves parts (i)-(ii).

From the last line of the above calculation we also find

$$f_k \geq \frac{2K_t(a)}{t+1} p_{t,t}^k.$$

To prove part (iii), we need a matching asymptotic upper bound on f_k. This bound indeed holds true (see [1]), though its proof is somewhat more complicated. □

Theorem 13.2: [39] *Let $\mathcal{C}$ be an (n, M, d) binary code. Then*

$$M \leq \frac{(n+1)^2}{2(t+1)a} \binom{n}{t},$$

where t satisfies $x_{1,t+1} < d < x_{1,t}$.

Proof: The bound follows on verifying that the polynomial $f(x)$ from the previous section satisfies the conditions of Corollary 11.5. □

This result can be improved for all finite n, d by using in Corollary 11.5 another polynomial related to but different from $f_t(x)$ [35].

Theorem 13.3:

$$R(\mathcal{H}^n; \delta) \leq h_2(\phi(\delta)), \tag{13.1}$$

$$R(\mathcal{J}^{n,\omega n}; \delta) \leq h_2\left(\frac{1}{2}\left(1 - \sqrt{1 - (\sqrt{4\omega(1-\omega)} - \delta(2-\delta)} - \delta)^2}\right)\right), \tag{13.2}$$

$$R(\mathcal{S}^n; \theta) \leq \frac{1 + \sin\theta}{2\sin\theta} h_2\left(\frac{1 - \sin\theta}{1 + \sin\theta}\right). \tag{13.3}$$

The bound on $R(\mathcal{H}^n; \delta)$ (the so-called first MRRW bound [39]) follows easily from the previous theorem and (12.2). Bounds (13.2) [39] and (13.3) [28] are proved by repeating the above argument for Hahn and Gegenbauer polynomials, respectively.

Bound (13.1) can be improved for small δ using (13.2) in the Bassalygo-Elias inequality (5.1). The resulting estimate is called the "second MRRW bound"; it is the best upper bound for the Hamming space known to-date. Bound (13.2) is the best known for large δ; however, for small δ it is not as good as the result of Theorem 5.6. Another improvement of (13.2) is given in [44].

Although it is not clear if the second MRRW bound can be improved within the frame of the polynomial method, it is shown in [44] that relying on this method it is not possible to prove tightness of the Gilbert-Varshamov bound. After more than two decades of holding the record, it is believed by many that the second MRRW bound is asymptotically the best obtainable by the polynomial method. Experimental evidence confirming this conjecture is provided in [9].

Asymptotics of the coefficients of the polynomial $f(x)$ in Theorem 13.1 calls for geometric interpretation. It is tempting to conjecture the existence of some packing argument which would lead to the same asymptotic result as Theorem 13.2 and link it to the results of Section 16 on the covering radius, to bounds on constant weight codes and to the union bound on the error probability of decoding. At the time of writing this, any argument of this kind is missing.

14. Bounding the Distance Distribution

14.1. *Upper bounds*

Let $\mathcal{C} \subseteq \mathcal{H}_2^n$ be a code with distance d, dual distance d' and distance distribution $(1, A_d, \ldots, A_n)$. If either d or d' are unknown, below we assume them to be zero.

The nature of the polynomial approach to bounding the distance distribution leads to asymptotic upper bounds that depend either on d or on d'. Lower bounds depend on the code rate R.

Upper bounds in terms of d' also involve R and bound the distance distribution of designs. Since a design in a finite space approximates a uniformly distributed set of points, one expects its distance distribution to approach that of the random code. Therefore in this direction one obtains bounds on the deviation of the distance profile of codes from the function $\alpha_0(R)$ (Section 3.1).

A straightforward way to bound the component A_w above is by saying

$$A_w \leq M(\mathcal{J}^{n,w}; d). \tag{14.1}$$

Better results in many cases are obtained by Theorem 11.4.

Theorem 14.1: [3] *Let $\mathcal{C}$ be a code with distance d and dual distance d':*

$$F(d, d') = \sum_{i=d}^{n} g_w(i) A_i(\mathcal{C}).$$

Let $g_w(i) = (K_w(i))^2$,

$$c = \begin{cases} \dfrac{t+1}{2} \dfrac{\binom{n-d'}{w-d'/2}}{\binom{n-d'}{t-d'/2}}, & \text{if } d'/2 \leq w \leq t, \\ 0, & \text{if } 0 \leq w \leq d'/2. \end{cases}$$

Then for sufficiently large n and $0 \le w < t \le \frac{n}{2}$,

$$F_w(d, d') \le |\mathcal{C}| \left[\binom{n}{w} - c\binom{n}{t} \right] - \binom{n}{w}^2 + \frac{c}{a}\left[\binom{n}{t+1} + \binom{n}{t} \right]^2.$$

The proof is accomplished by choosing in Theorem 11.4

$$f(x) = (K_w(i))^2 - \frac{c}{a-i}(K_{t+1}(i) + K_t(i))^2 \quad (t = \frac{n}{2} - \sqrt{d(n-d)}).$$

Theorem 14.2: [3, 1] *Let $\mathcal{C}$ be a code of distance δn. Its distance profile is bounded above as follows: $A_{\xi n} \le \exp(n(\alpha_\xi + o(1)))$, where*

$$\alpha_\xi \le \begin{cases} h_2(\xi) + h_2(\phi(\delta)) - 1 & \delta \le \xi \le 1 - \delta, \\ -2I(\phi(\delta), \xi) & 1 - \delta \le \xi \le 1. \end{cases}$$

This bound for large δ and ξ is better than (14.1).

Let us summarize the bounds as follows: *There exist sequences of codes of rate R with distance profile $\alpha_{0,w} = h_2(\omega) - h_2(\delta_{GV}(R))$; for no sequence of codes of relative distance δ the weight profile can exceed the bound of Theorem 14.2.*

Theorem 14.3: *Let $\mathcal{C}$ be a code with rate R and relative dual distance δ'. Let $\xi_1 = (1/2)(1 - \sqrt{\delta'(2 - \delta')})$ and ξ_2 be the root of the equation*

$$R = (1 - \delta')h_2\left(\frac{\phi(\xi) - \delta'/2}{1 - \delta'} \right) + 1 + \xi - h_2(\phi(\xi)),$$

or 0, whichever is greater. For any $\xi \in [\min(\xi_1, \xi_2), 1/2]$ and sufficiently large n the distance profile of the code $\mathcal{C}$ approaches $\alpha_{0,\xi}$.

Remark. Similar results can be obtained for $\mathscr{J}^{n,w}$ and $\mathscr{H}_q^n$; for $\mathscr{S}^n$ the dual distance of codes is not well defined.

14.2. *Lower bounds*

We give an example of results in this direction, proved by Corollary 11.7.

Theorem 14.4: [36] *Let $\mathcal{C} \subseteq \mathscr{H}_2^n$ be a code of rate R and let $0 \le \beta \le h_2^{-1}(R)$. For sufficiently large n there exists a value of $\xi \in [0, \phi(\beta)]$ such that the distance profile of $\mathcal{C}$ satisfies*

$$\alpha_\xi \ge R - h_2(\beta) - I(\beta, \xi).$$

The interpretation of this theorem is as follows: For a code of rate R and any $s \geq \phi(h_2^{-1}(R))$ there exists a value ξ of the relative distance such that the average number of neighbors of a codeword is $\exp[n(R-h_2(\beta)-I(\beta,\xi))]$ or greater. Note that $\phi(h_2^{-1}(R))$ is the value of the distance from Theorem 13.3. Further results in this direction are found in [36, 2].

15. Linear Codes with Many Light Vectors

The results of the previous section do not resolve the following question: do there exist sequences of linear codes whose number of codewords of minimum weight grows exponentially in n? It was conjectured in [29] that the answer is negative. This conjecture was disproved in [4], where it is shown that such code families do exist.

Theorem 15.1: *Let*

$$E_q(\delta) := h_2(\delta) - \frac{\log q}{\sqrt{q}-1} - \log \frac{q}{q-1}.$$

Let $q = 2^{2s}, s = 3, 4, \ldots$ be fixed and let $\delta_1 < \delta_2$ be the zeros of $E_q(\delta)$. Then for any $0 < \delta_1 < \delta < \delta_2 < 1/2$ there exists a sequence of binary linear codes $\{\mathcal{C}_i, i = 1, 2, \ldots\}$ of length $n = qN, N \to \infty$, and distance $d_i = n\delta/2$ such that

$$\log A_{d_i} \geq NE_q(\delta) - o(N).$$

The idea of the proof is as follows. Let X be a (smooth projective absolutely irreducible) curve of genus g over $\mathbb{F}_q, q = 2^{2s}$. Let $N = N(X) := \sharp X(\mathbb{F}_q)$ be the number of $\mathbb{F}_q$-rational points of X and suppose that X is such that $N \geq g(\sqrt{q}-1)$ (e.g., X is a suitable modular curve). The set of $\mathbb{F}_q$-rational effective divisors of degree $a \geq 0$ on X is denoted by $Div_a^+(X)$. Recall that $Div_a^+(X)$ is a finite set. For $D \in Div_a^+(X)$ let $\mathcal{C} = \mathcal{C}(D)$ be an $[N, K, d(\mathcal{C})]$ geometric Goppa code constructed in a usual way [49]. Then $K \geq a - g + 1$ and $d(\mathcal{C}) \geq N - a$.

Note that once X and a are fixed, the estimates of the code parameters given above do not depend on the choice of the divisor D. It is conceivable that for some divisors D the code $\mathcal{C}(D)$ will have better parameters. That this is the case was shown by Vlăduţ in 1987 [49]. This result was obtained by computing the average parameters of codes over $Div_a^+(X)$. The same idea applies to the weight spectrum of the Goppa codes: one can compute the average weight spectrum of the code $\mathcal{C}(D), D \in Div_a^+(X)$, and then prove that there exists a code whose weight spectrum is at least as good as the average value. This code is then concatenated with a binary $[n = q-1, 2s, q/2]$ simplex code. This results in a binary code whose number of minimum-weight codewords is given in Theorem 15.1.

16. Covering Radius of Linear Codes

The polynomial method can be used to derive bounds on the covering radius of linear codes.

Theorem 16.1: [16, p. 230] *Let $\mathcal{C}$ be a linear code with dual distance d'. Let r be an integer and $f(x) = \sum_{i=0}^{n} f_i K_i(x)$ be a polynomial such that $f(i) \leq 0, i = r+1, \ldots, n$, and*

$$f_0 > \sum_{j=d'}^{n} |f_j| A'_j.$$

Then $r(\mathcal{C}) \leq r$.

Proof: Let $\mathcal{C}$ be a code of size M and $\mathcal{D}(x) = \mathcal{C} + x$ be the translation of $\mathcal{C}$ by a vector x and let $A(x) = (A_i(x), i = 0, 1, \ldots, n)$ be the *weight distribution* of $\mathcal{D}(x)$. Let $A'(x) = (A'_i(x), i = 0, 1, \ldots, n)$ be the MacWilliams transform of the vector $A(x)$:

$$A'(x) = (M)^{-1} A(x) \mathbf{K},$$

where $\mathbf{K}$ is the Krawtchouk matrix. Note that the components of $A'(x)$ can be negative.

Also let $\mathcal{C}'$ be the dual code of $\mathcal{C}$ and $(A'_i, 0 \leq i \leq n)$ be its weight distribution. It is known that for any x

$$|A'_i(x)| \leq A'_i \quad (i = 0, \ldots, n).$$

Therefore, compute

$$M^{-1} \sum_{i=0}^{n} f(i) A_i(x) = M^{-1} \sum_{i=0}^{n} \sum_{j=0}^{n} f_j K_j(i) A_i(x)$$

$$= \sum_{j=0}^{n} f_j M^{-1} \sum_{i=0}^{n} A_i(x) K_j(i)$$

$$= \sum_{j=0}^{n} f_j A'_j(x) = f_0 + \sum_{j=d'}^{n} f_j A'_j(x)$$

$$\geq f_0 - \sum_{j=d'}^{n} |f_j| A'_j.$$

If the last expression is positive, then so is the sum $\sum_{i=0}^{n} f(i)A_i(x)$. Further, if $f(i) < 0$ for $i = r + 1, \ldots, n$, then there exists an $i \leq r$ such that (for every x) $A_i(x) > 0$, $i.e.$, $r(\mathcal{C}) \leq r$. $\qquad\qquad\square$

Applications of this theorem rely on bounds on the dual weight distribution of a linear code with dual distance d', $i.e.$, of the weight distribution (A_i) of a linear code with distance d'. Since $A_i < M(J^{n,i}; d')$, any upper bound on the size of a constant weight code can be used to obtain a bound on $r(\mathcal{C})$. Good bounds are obtained if we again take the polynomial $W_t(x)$ (Section 13). Another possibility is to use the results of Section 13 and use asymptotics of its coefficients from Theorem 13.1(iii). These methods yield the best asymptotic upper bounds on r in terms of δ' currently known (see [1] and references therein).

Further Reading

In many cases this article provides only a starting point of a large topic in coding and information theory. A lot more information is found in the books and articles referenced in the main text. In addition, the following textbooks or monographs offer a general introduction and more expanded context pertinent to the subjects of this chapter: combinatorial coding theory [5, 38, 50], error exponents [20, 51], algebraic-geometric codes [47, 49], tables of short codes and general reference source [41], covering radius [16], spherical codes and bounds [18, 24], orthogonal polynomials [40, 48].

References

1. A. Ashikhmin and A. Barg, *Bounds on the covering radius of linear codes*, Des. Codes Cryptogr., to appear.
2. A. Ashikhmin, A. Barg, and S. Litsyn, *A new upper bound on the reliability function of the Gaussian channel*, IEEE Trans. Inform. Theory **46** (2000), no. 6, 1945–1961.
3. _____, *Estimates of the distance distribution of codes and designs*, IEEE Trans. Inform. Theory **47** (2001), no. 3, 1050–1061.
4. A. Ashikhmin, A. Barg, and S. Vlăduţ, *Linear codes with exponentially many light vectors*, Journal of Combin. Theory, Ser. A **96** (2001), no. 2, 396–399.
5. E. F. Assmus, Jr. and J. D. Key, *Designs and Their Codes*, Cambridge Tracts in Mathematics, vol. 103, Cambr. Univ. Press, Cambridge, 1992.
6. A. Barg, *On error bounds for spherical and binary codes*, preprint, 2001.
7. _____, *On some polynomials related to the weight enumerators of linear codes*, SIAM J. Discrete Math. **15** (2002), no. 2, 155–164.
8. A. Barg and G. D. Forney, Jr., *Random codes: Minimum distances and error exponents*, IEEE Trans. Inform. Theory **48** (2002), no. 9.

9. A. Barg and D. B. Jaffe, *Numerical results on the asymptotic rate of binary codes*, Codes and Association Schemes (A. Barg and S. Litsyn, eds.), DIMACS series, vol. 56, AMS, Providence, R.I., 2001, pp. 25–32.

10. A. Barg and D. Nogin, *Bounds on packings of spheres in the Grassmann manifolds*, IEEE Trans. Inform. Theory **48** (2002), no. 9.

11. L. A. Bassalygo, S. I. Gelfand, and M. S. Pinsker, *Simple methods for deriving lower bounds in the theory of codes*, Problemy Peredachi Informatsii **27**, no. 4, 3–8 (in Russian) and 277–281 (English transaltion).

12. E. R. Berger, *Some additional upper bounds for fixed-weight codes of specified minimum distance*, IEEE Trans. Inform. Theory **13** (1967), no. 2, 307–308.

13. E. R. Berlekamp, *The performance of block codes*, Notices of the AMS (2002), no. 1, 17–22.

14. R. E. Blahut, *Hypothesis testing and information theory*, IEEE Trans. Inform. Theory (1974), no. 4, 405–417.

15. T. Britz, *MacWilliams identities and matroid polynomials*, Electron. J. Combin. **9** (2002), #R19.

16. G. Cohen, I. Honkala, S. Litsyn, and A. Lobstein, *Covering Codes*, Elsevier Science, Amsterdam, 1997.

17. J. H. Conway, R. H. Hardin, and N. J. A. Sloane, *Packing lines, planes, etc.: Packings in Grassmannian spaces*, Experimental Mathematics **5** (1996), no. 2, 139–159.

18. J. H. Conway and N. J. A. Sloane, *Sphere Packings, Lattices and Groups*, Springer-Verlag, New York-Berlin, 1988.

19. I. Csiszár, *The method of types*, IEEE Trans. Inform. Theory **44** (1998), no. 6, 2505–2523, Information theory: 1948–1998.

20. I. Csiszár and J. Körner, *Information Theory: Coding Theorems for Discrete Memoryless Channels*, Akadémiai Kiadó, Budapest, 1981.

21. P. Delsarte, *Bounds for unrestricted codes, by linear programming*, Philips Res. Rep. **27** (1972), 272–289.

22. _____, *An algebraic approach to the association schemes of coding theory*, Philips Research Repts Suppl. **10** (1973), 1–97.

23. P. Delsarte, J. M. Goethals, and J. J. Seidel, *Spherical codes and designs*, Geometriae Dedicata **6** (1977), 363–388.

24. T. Ericson and V. Zinoviev, *Codes on Euclidean Spheres*, Elsevier Science, Amsterdam e. a., 2001.

25. R. G. Gallager, *Low-Density Parity-Check Codes*, MIT Press, Cambridge, MA, 1963.

26. _____, *Information Theory and Reliable Communication*, John Wiley & Sons, New York e a., 1968.

27. M. Greferath and S. E. Schmidt, *Finite-ring combinatorics and MacWilliams' equivalence theorem*, J. Combin. Theory Ser. A **92** (2000), no. 1, 17–28.

28. G. Kabatyansky and V. I. Levenshtein, *Bounds for packings on the sphere and in the space*, Problemy Peredachi Informatsii **14** (1978), no. 1, 3–25.

29. G. Kalai and N. Linial, *On the distance distribution of codes*, IEEE Trans. Inform. Theory **41** (1995), no. 5, 1467–1472.

30. T. Kløve, *Support weight distribution of linear codes*, Discrete Mathematics **106/107** (1992), 311–316.

31. I. Krasikov, *Nonnegative quadratic forms and bounds on orthogonal polynomials*, J. Approx. Theory **111** (2001), no. 1, 31–49.

32. I. Krasikov and S. Litsyn, *Survey of binary Krawtchouk polynomials*, Codes and Association Schemes (Piscataway, NJ, 1999) (A. Barg and S. Litsyn, eds.), Amer. Math. Soc., Providence, RI, 2001, pp. 199–211.

33. V. I. Levenshtein, *Upper-bound estimates for fixed-weight codes*, Problemy Peredachi Informatsii **7** (1971), no. 4, 3–12.

34. _____, *On the minimal redundancy of binary error-correcting codes*, Information and Control **28** (1975), no. 4, 268–291, Translated from the Russian (Problemy Peredachi Informatsii **10** (1974), no. 2, 26–42).

35. _____, *Krawtchouk polynomials and universal bounds for codes and designs in Hamming spaces*, IEEE Trans. Inform. Theory **41** (1995), no. 5, 1303–1321.

36. S. Litsyn, *New upper bounds on error exponents*, IEEE Trans. Inform. Theory **45** (1999), no. 2, 385–398.

37. F. J. MacWilliams, *A theorem in the distribution of weights in a systematic code*, Bell Syst. Techn. Journ. **42** (1963), 79–94.

38. F. J. MacWilliams and N. J. A. Sloane, *The Theory of Error-Correcting Codes*, 3rd ed., North-Holland, Amsterdam, 1991.

39. R. J. McEliece, E. R. Rodemich, H. Rumsey, and L. R. Welch, *New upper bound on the rate of a code via the Delsarte-MacWilliams inequalities*, IEEE Trans. Inform. Theory **23** (1977), no. 2, 157–166.

40. A. F. Nikiforov, S. K. Suslov, and V. B. Uvarov, *Classical Orthogonal Polynomials of a Discrete Variable*, Springer-Verlag, Berlin, 1991.

41. V. Pless and W. C. Huffman (eds.), *Handbook of Coding Theory*, vol. 1,2, Elsevier Science, Amsterdam, 1998, 2169 pp.

42. G. Sh. Poltyrev, *Bounds on the decoding error probability of binary linear codes via their spectra*, IEEE Trans. Inform. Theory **40** (1994), no. 4, 1284–1292.

43. _____, *On coding without restrictions for the AWGN channel*, IEEE Trans. Inform. Theory **40** (1994), no. 4, 409–417.

44. A. Samorodnitsky, *On the optimum of Delsarte's linear program*, Journal of Combin. Theory, Ser. A **96** (2001), no. 2, 261–287.

45. C. E. Shannon, *Probability of error for optimal codes in a Gaussian channel*, Bell Syst. Techn. Journ. **38** (1959), no. 3, 611–656.

46. J. Simonis, *MacWilliams identities and coordinate partitions*, Linear Alg. Appl. **216** (1995), 81–91.

47. S. A. Stepanov, *Codes on Algebraic Curves*, Kluwer Academic, New York, 1999.

48. G. Szegö, *Orthogonal Polynomials*, Colloquium Publications, vol. 23, AMS, Providence, RI, 1975.

49. M. Tsfasman and S. Vlăduţ, *Algebraic-Geometric Codes*, Kluwer, Dordrecht, 1991.

50. J. H. van Lint, *Introduction to Coding Theory*, 3rd ed., Springer-Verlag, Berlin e. a., 1999, (1st ed. 1981).

51. A. J. Viterbi and J. K. Omura, *Principles of Digital Communication and Coding*, McGraw-Hill, 1979.

ANALYSIS AND DESIGN ISSUES
FOR SYNCHRONOUS STREAM CIPHERS

Ed Dawson and Leonie Simpson

Information Security Research Centre, Queensland University of Technology
GPO Box 2434, Brisbane, Queensland 4001, Australia
E-mail: {e.dawson, lr.simpson}@qut.edu.au

An overview of various types of synchronous stream ciphers is presented. Basic cryptographic properties of synchronous stream ciphers are described, as are some of the most common types of attack. Implementation issues for stream ciphers, including hardware and software efficiency, and resynchronisation proposals are considered. LILI-II, a new stream cipher that is both efficient and secure against known attacks, is described.

1. Introduction

Stream ciphers are the standard form of encryption over communications channels such as mobile telephone and the Internet. Stream ciphers operate by breaking a message into successive characters and encrypting each character under a time-varying function of the key. Thus, for stream ciphers, the same plaintext character will not necessarily be encrypted to the same ciphertext character. For synchronous stream ciphers, the keystream is generated independently of the plaintext message and the ciphertext, using a keystream generator (commonly a pseudorandom number generator). Synchronous stream ciphers require the keystream generators used for encrypting and decrypting to be synchronised, in order to recover the plaintext from the ciphertext. Loss of synchronisation is detected at decryption, as the plaintext message will be garbled from that point on. To retrieve the entire message, resynchronisation and retransmission may be necessary.

The plaintext message characters used for stream ciphers operating on electronic data are usually binary digits (bits) or bytes, as are both the keystream and ciphertext characters. The most common form of synchronous stream cipher is the binary additive stream cipher, where both the

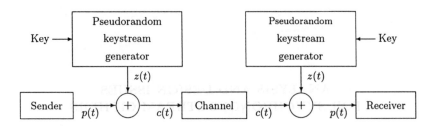

Fig. 1.   Binary additive stream cipher.

plaintext and the keystream are sequences of bits, and the ciphertext is the bitwise modulo two addition of the two streams. A major advantage of the binary additive stream cipher is that encryption and decryption can be performed by identical devices. Figure 1 illustrates the encryption and decryption operations using a binary additive synchronous stream cipher. The key is usually the initial state of the pseudorandom keystream generator. Each secret key used as input to the pseudorandom keystream generator corresponds to a longer pseudorandom output sequence, the keystream. If the same key is used for the keystream generators both on the sending and receiving ends of the transmission, the same keystream will be produced by both. Let $p(t)$, $z(t)$ and $c(t)$ denote the plaintext, keystream and ciphertext bits, respectively, at time t, for some $t \geq 0$. Under encryption, the tth bit in the ciphertext stream is formed as $c(t) = p(t) \oplus z(t)$, where $\oplus$ denotes addition modulo 2. For decryption, the plaintext bitstream is recovered by adding, bitwise modulo 2, the keystream to the ciphertext bitstream. The tth bit in the plaintext stream is given by $p(t) = c(t) \oplus z(t)$.

Stream ciphers offer a number of advantages to the user, including high speed encryption, immunity from dictionary attacks, protection against active wiretapping and low error propogation. High speed encryption is possible because each plaintext symbol can be encrypted as soon as it is read. Because stream ciphers encrypt each successive character under a time-varying function of the key, the same plaintext character will not necessarily be encrypted to the same ciphertext character, so stream ciphers are not subject to dictionary attacks. Errors caused by insertion or deletion of bits of the ciphertext stream (whether accidental or malicious) are detected at the receiving end, as loss of synchronisation with the keystream results in the decrypted plaintext message being garbled from that point on. Bit flip errors will result in an error in the decryption of a single character, but there is no propogation of the error to the other characters in the message.

The synchronisation requirement of synchronous stream ciphers offers an advantage, in that active wiretapping is detected. However, it is also a disadvantage because whenever synchronisation is lost, we are unable to recover the original message from that point on. Retransmission of all or part of the message may be necessary. A further disadvantage of some stream ciphers is that all of the information of one plaintext symbol is contained in one ciphertext symbol, so there is low diffusion of the plaintext across the ciphertext.

In Section 2 the main cryptographic properties of stream ciphers are described, including methods of constructing Boolean functions for use in keystream generators. Section 3 contains descriptions of various models of keystream generators which have been proposed over the past twenty-five years, and an overview of their properties. Methods of attacking keystream generators are outlined in Section 4, and practical implementation issues for stream ciphers are considered in Section 5. In Section 6 the authors describe a new stream cipher, called LILI-II, which has been designed to be both secure against known attacks and efficient in hardware and software.

2. Fundamental Cryptographic Concepts

2.1. *Levels of security and types of attacks*

The level of security offered by a cipher is measured by the knowledge and facilities a cryptanalyst requires to attack the system. Common assumptions are that the cryptanalyst

(1) has complete knowledge of the structure of the cipher system,
(2) has access to a considerable amount of ciphertext, and
(3) knows the plaintext equivalent of some of the ciphertext.

If an attack is possible when only the first two assumptions are met, then the attack is described as a ciphertext-only attack. Ciphers which are susceptible to such attacks offer a very low level of security. If all three assumptions are required, then the attack is described as a known plaintext attack. The security offered by a stream cipher susceptible to known plaintext attacks is measured by the amount of keystream required and the amount of work the cryptanalyst must perform for the attack to be successful. Finally, if the cryptanalyst can obtain ciphertext corresponding to selected plaintext, the attack is termed a chosen plaintext attack. Ciphers which are secure from chosen plaintext attacks offer a high level of security. Note that, in general, for a stream cipher there is no difference between known plaintext and chosen plaintext attacks, as both provide the same information: knowledge of a segment of the keystream.

For a binary additive synchronous stream cipher, if some plaintext-ciphertext pairs are known, then the equivalent keystream is also revealed, as $z(t) = p(t) \oplus c(t)$. Therefore, the keystream generator is the critical component in the security of this type of stream cipher. The level of security provided by the cipher depends on the apparent randomness of the keystream. If it is possible to predict the entire keystream from a small known segment, then the cipher offers a low level of security.

For most stream ciphers, the keystream bit $z(t)$ is formed as a function of the internal state, and the next internal state is defined by the current internal state. If the output function is known, the goal of an attacker is to usually determine the internal state of the keystream generator at some time t, as then all following output bits of the keystream can be determined.

2.2. *Standard properties of keystream sequences*

The criteria used to evaluate a pseudorandom binary sequence for cryptographic application is that, given a section of the keystream, a cryptanalyst should not be able to predict the rest of the keystream any better than merely guessing. Three standard properties of pseudorandom binary sequences are used to measure the security of keystream generators: period, linear complexity and white noise statistics. Necessary, but not sufficient, requirements for suitable sequences include long period, large linear complexity and good statistical properties. These basic properties and their implications for the security of keystream generators are discussed in the remainder of this section.

2.2.1. *Period*

A deterministic pseudorandom number generator produces a sequence which is periodic, or ultimately periodic. The period of the keystream sequence produced should be substantially larger than the length of the message. If the keystream repeats itself within the same cryptogram, or if the same part of the keystream is used to encrypt two different messages, it may be possible for the cryptanalyst to recover a substantial part of the message using ciphertext alone. As described in [32], such an attack would consist of combining, using modulo two addition, the two sections of the ciphertext which are encrypted with the same part of the keystream. This results in the modulo two addition of the two binary plaintext messages.

Such a combination can be attacked using the redundancy of the plaintext, as described in [12].

2.2.2. *Linear complexity*

For any periodic sequence, there exists a linear feedback shift register (LFSR) which can produce the same sequence. The linear complexity of a binary sequence is defined to be the length of the shortest binary linear feedback shift register (LFSR) which can produce the sequence. If the linear complexity of a periodic binary sequence is denoted L, then the LFSR which produces the sequence can be constructed, provided $2L$ consecutive terms of the sequence are known, by application of the Berlekamp-Massey algorithm [34].

In order to avoid such a reconstruction, the linear complexity of the keystream should be sufficiently large. It may be possible for a keystream to have a high linear complexity, but have a "bad" linear complexity profile. The linear complexity profile measures the linear complexity of a sequence as it increases in length. A "good" keystream of increasing length i follows approximately the $i/2$ line for linear complexity profile [5].

2.2.3. *White noise characteristics*

The output of a pseudorandom number generator is not random. To provide security, a cryptanalyst should not be able to distinguish between the keystream segment and a random binary string of the same length. That is, the keystream should have white noise characteristics. Statistical properties required for a binary sequence to appear random are uniformly distributed single bits, pairs of bits, triples, and so on. The keystream should appear random in both the long and short term. If the keystream is sufficiently non-random, then a cryptanalyst may be able to combine knowledge of the keystream statistics with the redundancy of the plaintext to conduct a successful statistical attack. A set of statistical tests that can be used to measure the randomness of the sequence produced by the keystream generator is described in [28].

2.3. *Boolean functions*

Most keystream generators for stream ciphers use one or more Boolean functions as components. A function $f(x)$ is a Boolean function of n binary input variables, $x_1, x_2, \ldots x_n$, if $f(x) = f(x_1, x_2, \ldots x_n) : Z_2^n \to Z_2$. In this section, two well-known representations of Boolean functions are reviewed, and certain properties of Boolean functions based on these representations, which are important to keystream generators, such as nonlinearity, correlation, order of correlation immunity and balance are defined.

2.3.1. *Algebraic normal form*

The Algebraic Normal Form (ANF) of a Boolean function $f(x)$ is a representation of the function where the function is written as a modulo 2 sum of products:

$$f(x_1, x_2, \ldots, x_n = a_0 \oplus a_1 x_1 \oplus \ldots \oplus a_n x_n$$
$$\oplus a_{12} x_1 x_2 \oplus \ldots \oplus a_{n-1,n} x_{n-1} x_n$$
$$\oplus \ldots \oplus a_{12 \ldots n-1, n} x_1 x_2 \ldots x_n$$

for coefficients $a_i \in \{0, 1\}$.

The order of each product term in the function $f(x)$ is defined to be the number of variables the term contains. The algebraic order of a function provides a rough measure of the complexity of the function.

Definition 2.1: The algebraic order, $ord(f(x))$, of the function $f(x)$ is defined to be the maximum order of the product terms in the ANF for which the coefficient is 1.

Thus, the algebraic order of a Boolean function is an integer in the range $[0, n]$. The constant functions $f(x) = 0$ and $f(x) = 1$ are zero order functions. Affine functions, quadratic functions and cubic functions are functions of order one, two and three, respectively. All functions of order greater than one are known as nonlinear functions.

Definition 2.2: Let $x = (x_1, x_2, \ldots, x_n)$ and $\omega = (\omega_1, \omega_2, \ldots, \omega_n)$ be binary n-tuples. Define their dot product as $(\omega \cdot x) = \omega_1 x_1 \oplus \omega_2 x_2 \oplus \cdots \oplus \omega_n x_n$. Then the linear function $l_\omega(x)$ is defined as $l_\omega(x) = (\omega \cdot x) = \omega_1 x_1 \oplus \omega_2 x_2 \oplus \cdots \oplus \omega_n x_n$, where $\omega_i \in \{0, 1\}$ for $i = 1, 2, \ldots, n$.

The set of affine functions consists of the linear functions and their complements.

2.3.2. *Binary truth table*

The binary truth table for a Boolean function is a logic table listing the output of the function for every possible combination of input variables.

If $f(x) : Z_2^n \rightarrow Z_2$, then the binary truth table can be stored as an array of 2^n bits, with the integer representation of the input n-tuple used as the index to the array. There is a 1-1 correspondence between the ANF and the binary truth table of a function.

Definition 2.3: The Hamming weight of a Boolean function $f(x)$, denoted $wt_H(f(x))$, is defined to be the number of 1's in the binary truth table, and can be calculated as $wt_H(f(x)) = \sum_x f(x)$.

Definition 2.4: A Boolean function $f(x)$ is balanced if $wt_H(f(x)) = 2^{n-1}$. That is, the binary truth table of the function contains equal numbers of 0's and 1's.

Definition 2.5: The Hamming distance between two Boolean functions of the same number of variables, $f(x)$ and $g(x)$, denoted $dist_H(f, g)$, is defined to be the number truth table positions in which the functions differ, and can be calculated as

$$dist_H(f, g) = wt_H(f \oplus g) = \sum_x (f(x) \oplus g(x)).$$

Definition 2.6: The nonlinearity of a Boolean function, $f(x)$, denoted N_f is defined to be the minimum Hamming distance between $f(x)$ and any affine function. That is, $N_f = min_{\omega,c}\, dist_H(f(x), l_\omega(x)+c)$ where $c \in \{0,1\}$.

The nonlinearity of a Boolean function is a non-negative integer value. The minimum nonlinearity, zero, is attained by affine functions. The maximum nonlinearity value is known only for certain cases. For Boolean functions with even n, $n \geq 8$, the maximum nonlinearity is known to be $2^{n-1} - 2^{\frac{n}{2}-1}$ [41]. This value is attained by a class of Boolean functions known as bent functions. However, bent functions are of low algebraic order (they have order $ord(f_{bent}) \leq \frac{n}{2}, n > 2$) and they are not balanced (they have Hamming weight $wt_H(f_{bent}) = 2^{n-1} \pm 2^{\frac{n}{2}-1}$), which makes them unsuitable for use as combining functions in keystream generators in many cases. The maximum nonlinearity is unknown for Boolean functions with odd $n \geq 9$, and for balanced Boolean functions with even $n \geq 8$. However, an upper bound, which may not be tight, on the nonlinearity of such functions is given by $2^{n-1} - 2^{\lfloor \frac{n}{2} \rfloor} - 1$ [17].

If a Boolean function, selected for use as a combining function, has low nonlinearity, then the output of the keystream generator can be approximated by a linear function of the input values. This approach was used in the fast correlation attack on the nonlinear filter generator, presented in Section 4.3.2. Thus, for stream cipher applications, a Boolean function which is highly nonlinear is desirable, as it increases the resistance of the cipher to such attacks.

2.3.3. *Correlation*

The correlation between two Boolean functions of the same number of variables, $f(x)$ and $g(x)$, is a measure of the extent to which they approximate each other.

Definition 2.7: The coefficient of correlation, c, between two Boolean functions of the same number of variables, $f(x)$ and $g(x)$, is given by

$$c(f(x), g(x)) = 1 - \frac{dist_H(f(x), g(x))}{2^{n-1}}.$$

Hence the coefficient of correlation, c, is a rational number in the range $[-1, 1]$, where values of c close to 1 indicate that the functions are very similar, and values close to -1 indicate that the functions are substantially different. The coefficient of correlation can also be expressed as $c = 1 - 2p$, where $p = Pr(f(x) \neq g(x))$. Where $p \neq \frac{1}{2}$, if the value of $f(x)$ is known, then uncertainty about the value of $g(x)$ is reduced.

Definition 2.8: A nonlinear function $f(x)$ is described as having correlation immunity of order m, denoted $CI(m)$, if the coefficient of correlation between the function $f(x)$ and any Boolean function of order less than m is equal to zero.

Siegenthaler [48] showed that a tradeoff exists between the algebraic order, k, of a memoryless Boolean function and the order of correlation-immunity, m, that the function possesses, such that $k + m \leq n$.

2.4. *Constructing Boolean functions*

Having identified the properties required for Boolean functions, the next issue is how to obtain such a function. Possible methods include generating a function at random, using a direct construction method, using heuristic methods, or some combination of these.

2.4.1. *Random generation*

A naive method of obtaining a suitable Boolean function is to generate a function at random, and then test the function to determine the properties it possesses. If the function is unsuitable, it is discarded, and another generated. Where the desired properties are uncommon, such a method is not efficient. Where a combination of properties are required, obtaining a suitable function using random generation may be very difficult. For example, given that a function of $n = 7$ inputs is balanced, the probability that it is also $CI(1)$ is less than 0.0005. Additional requirements such as a high nonlinearity would further reduce the probability of obtaining a suitable function. Thus, more efficient methods should be considered.

2.4.2. *Direct algebraic methods*

Direct methods for constructing Boolean functions often construct functions to achieve a particular property. Since the introduction of the concept of correlation immunity by Siegenthaler in [47], several methods for constructing correlation immune Boolean functions have been proposed. In [46], a direct method for constructing balanced correlation immune functions is given, and in [45], Schneider shows how to construct every m^{th} order correlation immune function. In [17], a method of constructing highly nonlinear balanced Boolean functions is given. In [13], an exhaustive construction method based on linear codes is proposed, which can generate all correlation immune Boolean functions. As this is not efficient for large n, a construction method which can produce a higher-order correlation immune function from a lower-order function, using the generating matrix of a linear code, is also given. Many of the cryptographic properties of the lower-order function are inherited by the higher-order function, so a $CI(m)$ function can be constructed which will also be balanced and have known nonlinearity. An outline of this method, and the cryptographic properties of the produced function, follows.

Select a k-input Boolean function, $g(y)$, and the k by n generating matrix of an $[n, k, d]$ linear code, G, where n,k and d refer to the length, dimension and minimum nonzero weight, respectively, of the code. The n-input Boolean function $f(x)$ is obtained as $f(x) = g(xG^T)$. The properties of the constructed function $f(x)$ depend on both the properties of the Boolean function $g(y)$, and the matrix G. These properties are given in the following theorems from [13].

Theorem 2.1: *Let $g(y)$ be a k-input Boolean function. Let G be a generating matrix of an $[n, k, d]$ linear code, and $f(x) = g(xG^T)$. Then*

- *for any k-input Boolean function, $g(y)$, the correlation immunity of $f(x)$ is at least $d - 1$, and*

- *a necessary and sufficient condition for $f(x)$ to be $CI(m)$ is that for every $\alpha \epsilon Z_2^k$ with $d \leq wt_H(\alpha G) \leq m$, $g(y)$ is CI in $< \alpha, y >$, the inner product of α and y, where $< \alpha, y >= \alpha_1 y_1 \oplus \alpha_2 y_2 \oplus \ldots \oplus \alpha_k y_k$.*

Thus, for a given generating matrix G, a minimum value for the correlation immunity of $f(x)$ is is assured and conditions under which the correlation immunity may be higher are established.

Theorem 2.2: *Let $f(x) = g(xG^T)$, where g is a non-degenerate Boolean function of k variables and G is a generating matrix of an $[n, k, d]$ linear code. Then*

- *$f(x)$ is balanced if and only if $g(y)$ is balanced,*
- *$ord(f(x)) = ord(g(y))$, and*
- *$N_f = 2^{n-k} N_g$.*

2.4.3. *Heuristic design methods*

In contrast to direct construction methods, two heuristic techniques are described in [37, 38] in which quasi-random processes are used in an iterative manner to generate cryptographically strong balanced Boolean functions. The techniques are referred to as hill climbing and the genetic algorithm. Both methods are shown to be capable of generating highly nonlinear Boolean functions many times (hundreds and thousands of times) faster than exhaustive search, and a combination of both methods is shown to be more effective than either method alone [38]. For completeness, a brief outline of each method is given.

The hill climbing method is a recursive procedure involving complementing bits in the truth table of a balanced Boolean function, in order to improve nonlinearity while maintaining the balance of the function. The initial Boolean function is generated at random. The Walsh Hadamard Transform may be used to identify truth table positions for which modification will be effective. The recursion continues until a local maximum for nonlinearity is reached. A second heuristic technique is the so-called genetic algorithm. The genetic algorithm imitates an evolutionary process. The process starts with an initial pool of randomly generated functions (called the parents), and combines pairs of them in some manner (referred to as breeding), to obtain a new set of functions (the children). A new pool of functions is selected from the parents and children (culling) through the application of a suitable fitness function, and the process repeated. Genetic algorithms offer a flexible way of generating functions with desired properties, providing a suitable breeding scheme and fitness function are available. Genetic algorithms can also be adapted to search for functions with good performance with respect to combinations of criteria, such as high nonlinearity and balance, selected order of correlation immunity and balance, or selected combinations of properties. The basic outline of the genetic algorithm is as follows.

Genetic Algorithm:

- *Input:* i_{max}, the maximum number of iterations to perform; P, the size of the solution pool; *Breed*, the number of offspring produced by each parent pair;
- *Initialisation:* $i = 1$, where i is the iteration index, $j = 1$, where j is the pairing index.
- *Stopping Criterion:* The algorithm stops when i_{max} is reached.
- *Step* 1. Generate a pool of P random Boolean functions.
- *Step* 2. Calculate j_{max}, the number of possible pairings in the solution pool.
- *Step* 3. For the pairing with index j, perform the breeding operation to produce a number of offspring equal to *Breed*.
- *Step* 4. If $j < j_{max}$, increment j by 1 and go to Step 3.
- *Step* 5. Select the best P solutions from the list of children and the current pool of solutions.
- *Step* 6. Reset j to 1. If $i < i_{max}$, increment i by 1 and go to Step 2.
- *Step* 7. Stop the procedure.
- *Output:* Pool of suitable Boolean functions.

A method of combining hill climbing and the genetic algorithm, in which hill climbing is applied to each new Boolean function in a genetic algorithm, increasing the nonlinearity, is shown in [38] to be more effective than either technique alone. The combined method was found to be effective in finding Boolean functions which were highly nonlinear, balanced and satisfied $CI(1)$.

3. Types of Synchronous Stream Ciphers

3.1. *One-time pad*

The Vernam one-time pad is a binary additive stream cipher which uses a truly random binary sequence, equal in length to the message length, as the keystream. This is the only cipher which offers perfect secrecy; that is, it is unconditionally secure against ciphertext-only attacks. However, high volume use of such ciphers is not practical, due to the severe key management problems the generation and distribution of truly random sequences involves. For practical security, pseudorandom sequences produced by deterministic algorithms are used in place of truly random sequences. The remainder of this section describes keystream generators used to produce such pseudorandom keystream sequences.

3.2. Block ciphers in OFB mode

A symmetric block cipher, such as the Data Encryption Algorithm, used in Output Feedback (OFB) mode may provide the keystream for a stream cipher. When a block cipher is used in OFB mode, then a fixed subset of L of the lower-order (rightmost) bits from each output block of the cipher is used in forming the keystream of a stream cipher, the remaining $n - L$ bits being discarded. This subset of L bits is added bitwise modulo-two to the next L bits of the message (plaintext) to produce the corresponding L bits of the cryptogram (ciphertext). The same L bits are also fed back into the L lower-order positions of the input block using a left-shift, to be used in generating the next output block. In general, the OFB mode of a block cipher produces a good pseudorandom keystream, provided the block cipher in electronic codebook (ECB) mode is secure from attack. Due to the complex nature of block ciphers, it is possible only to estimate the standard cryptographic properties for the produced binary sequence. The period of the block cipher can be estimated as $P = 2^{n-1}$ if $n = L$, and $P = 2^{n/2}$ if $n < L$ [10, 32]. Note that in many commercial applications, DES in OFB mode, with either one-bit or eight-bit feedback is used, despite the shorter expected period. Generally, the diffusion properties of a block cipher produce a keystream which is secure from statistical attack.

3.3. Table-based stream ciphers

Several recently published keystream generators have been table-based. These include the keystream generator RC4 [39]. Although the statistical properties of keystreams produced by these generators appear to be good, other properties of the keystream sequences are more difficult to determine. In this section a brief description of RC4 will be given.

Although no explicit key length is given for RC4, a 256 byte key is implied. However, key lengths of up to 256 bytes are acceptable. The key is used to generate a random permutation, denoted $P = \{p_i\}_{i=0}^{255}$, of the integers $0, \dots, 255$, to be used as the initial values in the table. The initialisation process is as follows.

Initialisation Process:

- *Step 1*. For $i = 0, \dots, 255$ set $p_i = i$.
- *Step 2*. Set $i_1 = 0$ and $i_2 = 0$.
- *Step 3*. For $j = 0$ to 255 do

  (1) $i_2 = (k_{i_1} + p_j + i_2) \bmod 256$
  (2) Swap p_j with p_{i_2}.
  (3) $i_1 = (i_1 + 1) \bmod 256$.

The permutation table P is then used to produce the keystream. The contents of the table vary over time. The algorithm for both producing a keystream of length N bytes and varying the permutation table P is as follows.

Keystream Production:

- *Step 1*. Set $x = 0$ and $y = 0$.
- *Step 2*. For $j = 1$ to N do

  (1) $x = (x + 1) \bmod 256$
  (2) $y = (p_x + y) \bmod 256$.
  (3) Swap p_x with p_y.
  (4) $Index = (p_x + p_y) \bmod 256$.
  (5) $z_j = p_{Index}$.

To date, no attack faster than exhaustive search has been found for the RC4 algorithm. The only types of attacks proposed are methods which allow one to distinguish RC4 from other keystream generators provided a large amount of keystream is known [18, 25], and possible rekeying attacks described in [19] (see Section 5.2 for a description of rekeying).

3.4. *Shift register based keystream generators*

Many designs for keystream generators for stream ciphers are based on shift registers, particularly linear feedback shift registers (LFSRs), because LFSRs are easily implemented in hardware and they allow for high encryption and decryption rates. Also, sequences produced by LFSRs with primitive characteristic polynomials possess certain important cryptographic properties: large periods and good long-term and short-term statistical properties. However, due to the linearity, LFSR sequences are easily predicted from a short segment [34], which makes LFSRs unsuitable for use as keystream generators on their own. Instead they are used as components in keystream generators. Nonlinearity is introduced into the keystream either explicitly or implicitly. Descriptions of four common methods for constructing LFSR-based keystream generators follow.

3.4.1. *Nonlinear combination of LFSRs*

One method for forming keystream sequences involves combining the outputs of several, say n, regularly clocked LFSRs by a nonlinear Boolean function, f, as shown in Figure 2. Such keystream generators are known as nonlinear combination generators and the nonlinear Boolean function f is called a combining function.

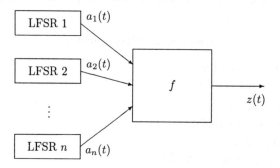

Fig. 2.   Nonlinear combination generator.

Where the combining function is memoryless, a nonlinear Boolean function f of n binary random variables is used. The keystream bit produced at time instant t is the output of the Boolean function f when the inputs to f are the outputs of the n LFSRs at time t, that is, $z(t) = f(a_1(t), a_2(t), \dots, a_n(t))$. The cryptographic properties of sequences produced in this manner have been thoroughly investigated in [42]. If the outputs of n LFSRs with primitive feedback polynomials whose lengths $L_1, L_2, \dots, L_n$ are pairwise relatively prime are combined using a nonlinear Boolean function $f(x_1, x_2, \dots, x_n)$, then the period, P, and the linear complexity, L, of $z(t)$ are given by $P = \prod_{i=1}^{n}(2^{L_i} - 1)$ and $L = f(L_1, L_2, \dots, L_n)$, respectively, with the latter Boolean function transformed by evaluating the addition and multiplication operations in the function over the integers, rather than over $GF(2)$. The period of keystream sequences produced by such generators can be increased by increasing the lengths of the component LFSRs, and the linear complexity can be increased by increasing both the lengths of the component LFSRs and the algebraic order of the Boolean function. However, for memoryless combining functions an increase in the algebraic order of a function is achieved at the cost of decreasing the correlation immunity of the function (see Section 2.3.3), which increases the susceptibility of the generator to correlation attacks as will be described in Section 4.3.

3.4.2. *Nonlinear filter generator*

A second method for forming keystream sequences involves using a nonlinear Boolean function to combine the outputs of several stages of a single regularly clocked LFSR, as shown in Figure 3. Such keystream generators are described as nonlinear filter generators. A nonlinear filter generator can also be represented as a nonlinear combination generator where all of the n LFSRs have the same feedback function and the n initial states are phase shifts of the nonlinear filter generator initial state.

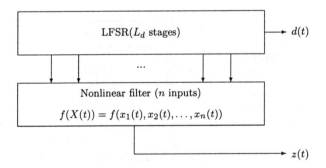

Fig. 3.   Nonlinear filter generator.

The cryptographic properties of a sequence produced by such a generator are partially described in [42], where it is noted that the period of $z(t)$ is $2^L - 1$ or a divisor of $2^L - 1$, and the linear complexity of $z(t)$ is less than or equal to $\sum_{i=1}^{k} \binom{L}{i}$, where k is the algebraic order of the Boolean function defining the sequence.

A basic condition to be satisfied for cryptographic applications is that the output sequence is balanced and independent, provided the input sequences are assumed to be so. This is true only if the combining Boolean function f is balanced. Thus, for most nonlinear filter generators f is a balanced Boolean function. If the feedback function for $LFSR_d$ is primitive of degree L_d, and f is balanced, then the period of z is exactly $2^{L_d} - 1$ [50].

3.4.3. *Generators with memory*

For a nonlinear combination generator with M bits of memory, the combining function consists of two subfunctions: a nonlinear Boolean function, f, of $n + M$ binary random variables used to produce the keystream output, and a next state memory function, F, of $n + M$ binary variables used to update the memory, as shown in Figure 4. If the state of the memory at time t is denoted $S(t) = (s_1(t), s_2(t) \ldots s_M(t))$, then $z(t)$, the keystream bit produced at time instant t, and $S(t + 1)$, the next state of the memory, are given by the following equations:

$$z(t) = f(a_1(t), a_2(t), \ldots, a_n(t), s_1(t), \ldots s_M(t)), \quad t \geq 0,$$

$$S(t + 1) = F(a_1(t), a_2(t), \ldots, a_n(t), s_1(t), \ldots s_M(t)), \quad t \geq 0,$$

where $S(0)$ is the initial state of the memory.

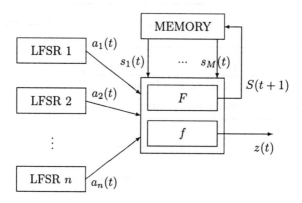

Fig. 4.   Nonlinear combination generator with memory.

The cryptographic properties of sequences produced by nonlinear combiners using functions with memory were investigated in [42], where it is shown that, apart from some initial nonperiodic terms, z is ultimately periodic with period P given by $P = \prod_{i=1}^{n}(2^{L_i} - 1)$ and linear complexity approximately equal to period length for appropriately chosen functions f and F. The linear complexity can be much greater than that obtained by linear combination generators using memoryless combining functions, as the tradeoff between algebraic order and correlation immunity which exists for nonlinear combiner generators with memoryless combining functions can be overcome by the inclusion of only one bit of memory [42]. However, weaknesses in such generators still exist, as noted in [36], and demonstrated by successful fast correlation attacks on the summation generator [27], see Section 4.2 for other attacks on summation generator.

3.4.4. *Clock-controlled LFSRs*

If an LFSR-based keystream generator is regularly clocked, then a keystream bit is produced every time the underlying LFSRs are clocked. For irregularly clocked keystream generators, keystream bits are not produced every time the underlying data shift registers are clocked, but at irregular intervals. Irregular clocking is an implicit source of nonlinearity.

Many recent stream cipher proposals use irregularly clocked keystream generators because some regularly clocked keystream generators have been shown to be vulnerable to correlation attacks [48] and fast correlation attacks [35, 43] which use a correlation measure based on the Hamming distance. For irregularly clocked keystream generators, this correlation measure cannot be applied. However, these generators may be susceptible to other divide and conquer attacks, including correlation attacks, see Section 4.3.3.

For irregularly clocked LFSR-based keystream generators, the output of one LFSR may be used to control the clock of another. The simplest keystream generator designs use only two binary LFSRs; one of these is a control register, $LFSR_C$, and the other is the data generating register, $LFSR_D$, as illustrated in Figure 5. Usually $LFSR_C$ is regularly clocked, and the output of this register, c, controls the clocking of $LFSR_D$. $LFSR_D$ is described as a clock-controlled LFSR (CCLFSR). The output of $LFSR_D$ then forms the keystream sequence, z. The clocking of $LFSR_D$ is termed constrained if there exists some fixed maximum number of times $LFSR_D$

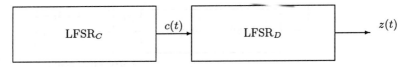

Fig. 5. A simple clock-controlled LFSR scheme.

may be clocked before a keystream bit is produced. Otherwise, it is termed unconstrained.

A simple example of a clock-controlled keystream generator for which the clocking is constrained is the step1-step2 generator, where LFSR_C is regularly clocked, and controls the output of LFSR_D in the following manner. At time instant t, LFSR_C is stepped. If $c(t)$, the output of LFSR_C at time t, is 0, then LFSR_D is stepped once, and if $c(t)$ is 1, then LFSR_D is stepped twice. The output of LFSR_D forms the keystream bit, $z(t)$. The step1-step2 generator is vulnerable to a correlation attack: in [54], a constrained embedding attack on LFSR_D is proposed, which can be used as a first stage in a divide and conquer attack on a step1-step2 generator. A more general attack, which can be applied to other constrained clock-controlled generators, is proposed in [20].

The shrinking generator [7] is another example of a keystream generator with irregular output. The shrinking generator consists of two regularly clocked binary LFSRs. Denote these as LFSR_A and LFSR_S, as shown in Figure 6, and denote the lengths of these LFSRs as L_A and L_S, respectively. The shrinking generator output is a "shrunken" version or subsequence of the output of LFSR_A, with the subsequence elements selected according to the position of 1's in the output sequence of LFSR_S: the keystream sequence z consists of those bits in the sequence a for which the corresponding bit in the sequence s is a 1. The other bits of a, for which the corresponding bit of s is a 0, are deleted. If the LFSR feedback polynomials are primitive, then a and s are maximum-length sequences with periods $2^{L_A} - 1$ and $2^{L_S} - 1$, respectively. If, in addition, L_A and L_S are relatively prime, then it is shown in [7] that the period of z is $(2^{L_A} - 1)2^{L_S-1}$ and that the linear complexity, L, of z satisfies $L_A 2^{L_S-2} < L \leq L_A 2^{L_S-1}$. The shrinking generator is vulnerable to the correlation attack described in Section 4.3.3.

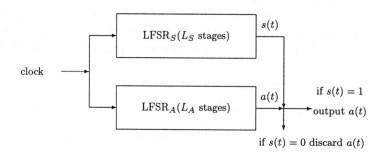

Fig. 6.   The shrinking generator.

4. Techniques for Analysis of Stream Ciphers

Several methods for attacking LFSR-based stream ciphers are well known. These include time/memory/data tradeoff attacks, correlation attacks, and divide and conquer attacks. In this section, these methods are outlined. Some of these attacks are possible against the different models of keystream described in Section 3, although in this section examples will only be selected from LFSR based generators.

4.1. *Time/memory/data tradeoff attacks*

Time/memory tradeoff attacks on stream ciphers (see [3] and [26]) are known plaintext attacks, conducted in two phases. The first phase is a preprocessing phase, during which the attacker tries different keys and stores the outcome in a table. In the second (realtime) phase, the attacker has a segment of known keystream and uses the precomputed table with the objective of recovering the internal state at a known time.

Let S, M, T, P, and D denote the cardinality of the internal state space, the memory (in binary words of size equal to $\log_2 S$), the computational time (in table lookups), the precomputational time (in table lookups), and the amount of data (without re-keying, this is the keystream length), respectively.

The preprocessing phase is as follows. Select M random internal initial states, $\{x_i\}_{i=1}^{M}$. For each of the x_i, compute the output string y_i of length $log(N)$, and store the (x_i, y_i) pair in increasing order of y_i in RAM. In the realtime phase, given a known keystream of $D + logN - 1$ bits, a sliding window is used to produce all D possible strings of length $log(N)$. These strings are then used to look up the precomputed table. If a match is found with string y_i, then the initial internal state was x_i. The time-memory tradeoffs satisfy $T \cdot M = S$, $P = M$ and $D = T$.

There is also a more general time-memory-data tradeoff [4] which asserts that $T \cdot M^2 \cdot D^2 = S^2$, $P = S/D$, $D^2 \leq T$. This decreases D at the cost of increasing P.

In Section 6.3.1, the application of such attacks is considered against the LILI-II keystream generator.

4.2. *Divide and conquer attacks*

Divide and conquer attacks on keystream generators work on each component of the keystream generator separately, and sequentially solve for individual subkeys. The attack is generally performed as a known plaintext attack, although sufficient redundancy in the plaintext may permit a ciphertext only attack.

For LFSR-based keystream generators, the objective of the attack is to recover the initial contents of a subset of the component LFSRs from a known segment of the keystream sequence. These attacks require knowledge of the structure of the keystream generator. If the entire structure of the generator is known, and the secret key is only the LFSR initial states, for a keystream generator consisting of n LFSRs, the total number of keys to be searched in a brute force attack is $\prod_{i=1}^{n}(2^{L_i} - 1)$, where L_i is the length of the ith LFSR. Using a divide and conquer attack that determines individual LFSR states sequentially can reduce the total number of keys to be searched to $\sum_{i=1}^{n}(2^{L_i} - 1)$.

Although divide and conquer attacks were devised for regularly clocked keystream generators, the approach can be extended to generators which incorporate irregular clocking. We provide an example of a divide and conquer attack on the summation generator below, which is a memory-based keystream generator.

The summation generator with two inputs from [42] can be described by the Boolean equations:

$$z(t) = a(t) + b(t) + s(t-1),$$

$$s(t) = a(t)b(t) + (a(t) + b(t))s(t-1),$$

where $z(t)$, $s(t)$, $a(t)$, $b(t)$ represent the keystream, memory bit and output of two LFSRs, respectively. The lengths of the LFSRs are L_1 and L_2. As mentioned in Section 3.4.3, if L_1 and L_2 are relatively prime and the LFSRs are defined by primitive polynomials, then the period and linear complexity of $z(t)$ are both approximately $(2^{L_1} - 1)(2^{L_2} - 1)$. In [11] a divide and conquer attack was described on the summation generator under the assumptions that $L_1 < L_2$, and at least $L_2 + 1$ keystream bits are known, say $z(0), \dots, z(L_2)$. The attack procedure is as follows.

Attack Procedure:

- *Step* 1. Select $a(0), b(0)$
  - derive $s(-1)$ to match $z(0)$,
  - derive $s(0)$.
- *Step* 2. Select $a(1), \dots, a(L_1 - 1)$
  - derive $b(1), \dots, bL_1 - 1$,
  - derive $s(1), \dots, s(L_1 - 1)$.
- *Step* 3. Determine $a(L_1), \dots, a(L_2)$ by LFSR$_1$.

- *Step* 4. Substitute $a(i)$ values to determine the values of $b(i)$.
- *Step* 5. Test additional values.

The search space for this attack is $2(2^{L_1} - 1)$ which is much reduced from the exhaustive key attack complexity of $(2^{L_1} - 1)(2^{L_2} - 1)$.

4.3. *Correlation attacks*

Correlation between two binary segments is a measure of the extent to which they approximate each other. Correlation attacks on LFSR-based keystream generators are based on statistical dependencies between the observed keystream sequence and underlying shift register sequences. The objective of the attacks is to use the correlation between the known keystream segment and the underlying LFSR sequences to sequentially recover the initial contents of each of the component LFSRs.

4.3.1. *Correlation attacks on regularly clocked generators*

The first correlation attack, proposed by Siegenthaler [48], is a divide and conquer attack on a regularly clocked LFSR-based keystream generator; the nonlinear combination generator (shown in Figure 2). For this attack the keystream is viewed as a noisy version of an underlying LFSR sequence, where the noise is additive and independent of the underlying LFSR sequence. The problem can be viewed as one of decoding a linear block code over a binary symmetric channel. As the clocking is regular, both the observed keystream segment and segments of the underlying LFSR sequences have the same length and the Hamming distance is used as a measure of correlation.

To identify the actual initial state of each component LFSR, the correlation between the keystream and the LFSR sequence is calculated for each possible LFSR initial state. The correct initial state is assumed to be that for which the correlation is highest. The correlation attack procedure is implemented by the following algorithm, where $a_i^r = \{a_i(t)\}_{t=1}^r$ denotes the output of the ith LFSR, the LFSR under attack.

Basic Correlation Attack Procedure:

- *Input:* structure of keystream generator and observed keystream sequence of length r, $z^r = \{z(t)\}_{t=1}^r$.
- For each possible subkey a_i:
  (1) generate the sequence $a_i^r = \{a_i(t)\}_{t=1}^r$,

(2) compute the correlation between a_i^r and z^r,

(3) if correlation between a_i^r and z^r is high, preserve the subkey as a possible candidate.

- *Output:* set of candidate subkeys (initial states for the ith LFSR).

The attack is considered successful if there are only a few possible sub-keys. This requires exhaustive search of the state of each LFSR to determine which is highly correlated to the keystream. If such a procedure can be conducted successfully for each of the input LFSR's, then the number of keys to be tested is reduced from $\prod_{i=1}^{n}(2^{L_i} - 1)$ to $\sum_{i=1}^{n}(2^{L_i} - 1)$.

4.3.2. *Fast correlation attack*

The concept of a fast correlation attack, which outperforms exhaustive search over the initial states of the individual shift registers, was first introduced in [35] as an attack on the nonlinear combiner generator. The keystream sequence $z = \{z(t)\}_{t=1}^{\infty}$ is regarded as a noisy version of an unknown underlying LFSR sequence $a_i = \{a_i(t)\}_{t=1}^{\infty}$ for some known i, where $1 \leq i \leq n$. Parity checks based on the underlying LFSR feedback polynomial are used in a bit-by bit reconstruction procedure, in which the observed keystream segment $z^N = \{z(t)\}_{t=1}^{N}$ is modified to reconstruct the underlying LFSR segment $a_i^N = \{a_i(t)\}_{t=1}^{N}$.

Fast correlation attacks for other regularly clocked LFSR-based keystream generators based on iterative error-correction algorithms have also been proposed, including an attack on the nonlinear filter generator [43], and a more recent fast correlation attack in [30]. We provide a summary of results from [43] below.

When applied to the nonlinear filter generator, the aim of the fast correlation attack based on iterative error-correction is to iteratively modify the observed keystream sequence into a linear transform of the underlying LFSR sequence corresponding to any linear approximation to the filter function with a nonzero correlation coefficient. If the attack is successful, then such a linear transform is identified by exhaustive search over all such linear approximations. That is, for any assumed linear approximation to the filter function, the unknown LFSR initial state is recovered by solving the corresponding set of linear equations and, as the filter function is known, the candidate LFSR initial state is then tested for correctness by producing a candidate keystream for comparison with the known keystream sequence.

To apply the fast correlation attack procedure to the nonlinear filter generator, the observed segment of the keystream sequence $z^N = \{z(t)\}_{t=1}^{N}$

is regarded as a noisy version of a segment of *a nonzero linear transform of* the unknown underlying LFSR sequence $d^N = \{d(t)\}_{t=1}^N$, rather than merely a noisy version of d^N. That is, $z(t) = l(X(t)) \oplus e(t)$, where $l(X(t)) = \sum_{j=1}^n c_j x_j(t)$, with $c_j \in \{0,1\}$, and where not all c_j are equal to zero. The corresponding correlation coefficient is defined as $c = (1 - 2p) > 0$ and is equal to the correlation coefficient between f and l, that is, $p = \Pr(f \neq l)$. Note that any (feedforward) linear transform of d is also a linear recurring sequence satisfying the same feedback polynomial as d.

There are three stages in applying the fast correlation attack to the nonlinear filter generator. The first stage, which can be precomputed as it is independent of the keystream, determines the correlation between the filter function, f, and linear approximations of the function. In the second stage, parity-checks are used to iteratively modify the keystream. The final stage involves recovering the LFSR initial state from the modified keystream. The relationship between the various stages of the fast correlation attack on the nonlinear filter generator is shown in Figure 7.

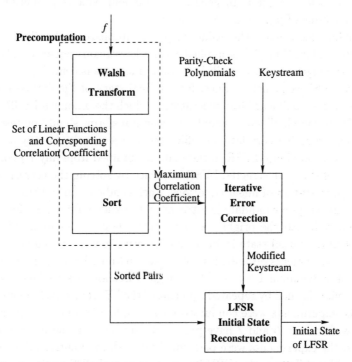

Fig. 7.   The fast correlation attack procedure on nonlinear filter generator.

The linear function l specifying a linear transform of d is to be determined by the attack. In principle, it can be any linear function of n variables with a positive correlation coefficient to f. The correlation coefficients between f and all the linear functions of the same n variables are precomputed by the Walsh transform technique [42], of complexity $O(n2^n)$. Let $\mathcal{L}^+$ denote the set of linear functions with a positive correlation coefficient to f, and let $\mathcal{L}^-$ denote the set of linear functions with negative correlation coefficients. It is expected that the fast correlation attack will, if successful, very likely converge to a linear transform with the maximum correlation coefficient, and the success of the attack predominantly depends on how large c is. Consequently, c is assumed to be the maximum absolute value of the correlation coefficient corresponding to the best affine approximation to f. If the value of c is obtained for $l \in \mathcal{L}^-$, then the binary complement of the keystream sequence is used. The probability p used in the attack is assumed to be given as $(1-c)/2$, where c is the maximum (positive) correlation coefficient corresponding to the best linear approximation to f. For a fixed value of n, the value of p varies depending on a particular Boolean function used as a filter. In general, for random balanced functions, the expected value of p increases as n increases.

Having determined the value of p, the iterative probabilistic error-correction algorithm [23] is applied to the keystream sequence. The output of the algorithm is very likely to be a linear transform, $l \in \mathcal{L}^+$, of the original LFSR sequence d. Thus, for the final stage of the fast correlation attack on the nonlinear filter generator, in which the unknown LFSR initial state is recovered, all such candidate linear transforms should be tested, in order of decreasing correlation coefficients. Consequently, for any assumed $l \in \mathcal{L}^+$, the candidate LFSR initial state is obtained from any L_d consecutive bits of the reconstructed LFSR sequence by solving the corresponding nonsingular system of linear equations. Each candidate initial state is then used to generate a candidate keystream by the known filter function which is compared with the original keystream. If they are identical, then the correct LFSR initial state is found and is very likely to be unique.

The experiments described in [43] used a 63-bit LFSR with a primitive feedback polynomial $1 + x + x^{63}$ of weight three and a set of parity-check polynomials formed by repeated squaring of the LFSR feedback polynomial.

The experiments were conducted for $n = 5, 6, 7, 8$ and 9, as filter functions with $n = 5, 6$ and 7 have previously appeared in the literature. For each n, the experiments were performed for both consecutive (CONS) and full positive difference set (FPDS) tap settings, as FPDS settings are much more resistant to several other attacks: the conditional correlation attack [1] and the inversion attack [24]. Also, two sets of parity-check polynomials were used: a set of 6 parity-check polynomials requiring an observed keystream length of 2500 bits, and a set of 9 parity-check polynomials requiring an observed keystream length of 20000 bits. For each of these four sets of conditions and each n, ten different seeds were used for the filter function and twenty different seeds for the LFSR initial state. Thus, for each value of n and for each set of the conditions, the fast correlation attack was performed two hundred times.

The resetting and stopping criteria for the fast correlation attack, selected through preliminary experimentation and with respect to the experimental results outlined in [44], are as follows: the attack was reset when the average residual error probability dropped below 0.01, when the number of satisfied parity-checks remained unchanged for four iterations or when the cumulative number of complementations in a round reached 1000; the attack stopped when 90% of the parity-checks were satisfied, 35 rounds were completed or if an error-free sliding window of length 63 was found at the end of any round. The search for a sliding window began once the level of satisfied parity-checks exceeded 70%. The attack was considered successful if the correct LFSR initial state was obtained.

The outcomes of the experiments were considered with respect to the number of inputs to the filter function, the theoretical probability of correlation noise, $p = (1 - c)/2$, where c was the maximum absolute value of the correlation coefficients of the filter function to linear functions, and the empirical probability of noise defined as the relative Hamming distance between the keystream segment and the best affine transform of the LFSR segment (if the attack was successful, instead of the best affine transform we took the linear transform the attack converged to). The main observation is that the iterative error-correction algorithm, if successful, mainly converges to one of the best linear transforms of the original LFSR sequence or to a linear transform that is close to being the best with respect to the correlation coefficient magnitude.

For a fixed number n of inputs to the filter, and a fixed set of conditions for tap setting and the number of parity-check polynomials, two hundred correlation attacks were carried out. Table 1 shows the success rate as the average percentage of successful attacks for each value of n, under each set of conditions. It can be seen that the percentage of successful attacks is significantly affected by the number of parity-checks used. As n increases, increasing the number of parity-check polynomials and consequently the length of the output sequence will increase the probability of successful recovery of the LFSR initial state. Table 1 also shows that the tap setting has little effect on the success of the attack, with the proportion of successful attacks being much the same for both the tap settings used.

Table 1. Success rate versus n.

n	6 parity-checks		9 parity-checks	
	CONS	FPDS	CONS	FPDS
5	86	84	100	100
6	72	74	100	100
7	23	8	100	100
8	0	1	94	96
9	0	0	35	28

4.3.3. *Correlation attacks on clock-controlled generators*

More recently, correlation attacks which can be applied to keystream generators based on irregularly clocked LFSRs have been proposed [54, 20, 21, 22, 29]. For irregularly clocked keystream generators, a correlation attack may be used as the first stage in a divide and conquer attack. The attack targets the component LFSRs involved in producing the keystream bits. Once the correct initial states of these LFSRs are identified, the clock control component can be attacked.

The correlation attack on irregularly clocked keystream generators follows the same procedure as the basic correlation attack on regularly clocked keystream generators as described in Section 4.3.1.

The major point of difference between the basic correlation attacks on regularly clocked and irregularly clocked keystream generators is in the measure of correlation. For an irregularly clocked keystream generator, the keystream may be viewed as an irregularly decimated version of some longer underlying sequence produced by a regularly clocked keystream generator. As the known keystream segment and the underlying segment are of different lengths, the Hamming distance is no longer useful as a basis of the measure of correlation. Instead, a different measure of correlation is required.

A commonly used measure for the distance between two strings of different lengths is the Levenshtein distance.

Definition 4.1: The Levenshtein distance, $LD(a, b)$, between two binary strings a and b is defined to be the minimum number of edit operations (deletion, insertion, substitution) needed to transform a, of length N_1, into b, of length N_2.

Note that for $N_1 \geq N_2$, in obtaining the minimum distance, the insertion operation is not required, and that the minimum distance cannot be less than $N_1 - N_2$. Where clocking is constrained, the distance measure may be based on the constrained Levenshtein distance [20], which is similar to the Levenshtein distance, but with a constraint on the maximum number of consecutive deletions which can occur in any edit sequence.

Definition 4.2: The constrained Levenshtein distance, $CLD(a, b)$, between two binary strings a and b is defined to be the minimum number of edit operations (deletion, insertion, substitution) needed to transform a, of length N_1, into b, of length N_2, subject to the constraint that no more than d consecutive bits of a may be deleted.

All correlation attacks involve a process of hypothesis testing. For a sequence a^m produced by the regularly clocked keystream generator, and the observed keystream sequence z^n, the hypothesis to be tested is

$$H_0 : z^n \text{ and } a^m \text{ are independent, versus}$$

$$H_1 : z^n \text{ and } a^m \text{ are correlated.}$$

As with any hypothesis testing, there are two types of errors which can occur. Let the error made by deciding that the two strings are independent when, in fact, they are correlated be described as "missing the event", and the error made by deciding that the two strings are correlated when, in fact, they are independent be described as "a false alarm". Denote the probabilities of these events by P_m and P_f, respectively. A relationship between these probabilities exists, and is dependent on the lengths of the sequences, n and m. For a given value of n, m should be chosen as a function of n and P_m. The recovery will be successful if, for fixed P_m, increasing n will result in a decrease of P_f.

In [22], several correlation attacks using measures of correlation which are similar to, or based on the Levenshtein distance are analysed, and the conditions under which these attacks can be successful are determined. The so-called embedding and probabilistic correlation attacks on a CCLFSR are examined, for the constrained and unconstrained cases. Constrained embedding attacks can be successful if enough keystream is available, whereas unconstrained embedding attacks are successful only if the probability of bit deletion is less than 0.5. In that case, a probabilistic correlation attack may be successful. Such an attack has been conducted against the shrinking generator [49], see Section 3.4.4 for a description of this generator.

5. Implementation Issues

In designing a keystream generator there are many practical design considerations. These may vary with the actual communications network which the keystream generator is being intended for use. In general we must balance security against efficiency. The goal in designing a stream cipher is to have the actual key length to be a true measure of the security of the algorithm. As was described in Section 4, this is not a simple task with the development of more sophisticated attacks.

There are several factors in relation to efficiency which may need to be considered when designing a stream cipher. One important consideration is to design an algorithm which is efficient in software or hardware as well. There may be a need to rekey the stream cipher in order to maintain synchronisation. In some communications network, such as a packet based system, frequent resynchronisation may be required. The goal here is to design an efficient procedure that does not compromise the security. In this section we discuss these practical considerations.

5.1. *Efficient implementation*

In general the fastest implementation of an encryption algorithm is in special purpose hardware. A hardware implementation may offer an order of magnitude or more improvement over a software version of the algorithm. Hardware also offers a more secure environment for key storage and generation. We shall mention briefly some of the issues in designing LFSR-based keystream generators.

5.1.1. *LFSR configuration*

The Fibonacci configuration and the Galois configuration (see Figure 8) are two common methods of configuring an LFSR. The Fibonacci configuration is efficient in hardware as it only requires a single shift register N bits long and a few XOR gates, however it is inefficient in software as the individual bits must be collected by a sequence of shifts and a mask then XORed together. The Galois configuration is more efficient in software as it applies

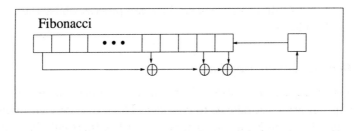

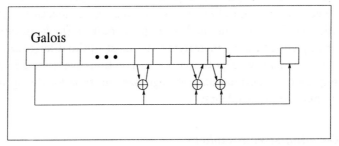

Fig. 8.   Fibonacci and Galois styles of LFSR implementation.

an XOR as a single operation on the register where the XOR value is the primitive polynomial. By initialising a Galois or Fibonacci register using reverse processing, the same output sequence can be generated by different states although the final state of the register will differ.

In software, when using more than one processor word for an LFSR register, a possible optimisation technique for a Galois configuration involves reversing the order of every second word value and the feedback polynomial associated with that word. Other speed gains can be created by using the word size of the processor, unsigned values where possible, and left shifting of the words was found to involve less instructions than right shifting.

5.1.2. *Boolean function implementation*

Many keystream generators involve the explicit use of Boolean functions. It is generally more efficient for programs to make use of table look-ups for Boolean functions in preference to processing the Boolean function in algebraic form using multiple instructions. Table look-ups, however, produce overhead in the form of memory management. For example, if the table becomes larger than the cache memory allowance (and is therefore unable to be stored in cache), its use is inefficient due to extra moving from storage into cache. If possible, table look-ups should be computed as far ahead as possible. This is so that other instructions can process whilst the program is finding the address.

5.1.3. Generic software guidelines

In implementing an encryption algorithm in software, there are several generic optimisation techniques to consider. These include identifying and streamlining bottlenecks, avoiding pipeline flushes (particularly in loops), reducing cache misses, unrolling loops, comparing bit masks to typecasting, identifying alternative methods of implementation (look-up tables versus direct implementation), identifying the most optimal language for the application (by identifying potential for instruction-level parallelism), and identifying the most optimal data storage size and type for the application (generally platform dependent).

5.2. Re-keying stream ciphers

In general a re-keying of a stream cipher should include a method for reinitialisation using both the secret key and an additional initialisation variable which is sent in the clear. Let k denote the secret key and v_i denote the initialisation vector for the ith re-keying. Let z denote the keystream, so that $z = F(k, v_i)$. Now F can be considered as the composition of two internal operations, G and R, where G is the keystream generator and R is the re-keying operation. Assuming that the action of the secret key is only to determine the initial state, let S_0 denote the initial state of G, generated by the re-keying function R, so that $S_0 = R(k, v_i)$. Then $z = G(S_0) = G(R(k, v_i))$. Most LFSR-based stream ciphers in the open literature implicitly have $v_i = 0$ for all i, and $R(k, 0) = k$ so that the secret key is used directly as the LFSR initial contents.

To break a particular instance of the cipher G, the cryptanalyst must recover the initial state S_0, using knowledge of the structure of G and some amount of the keystream z. With resynchronisation, the cryptanalyst has access to related keystreams, produced under the same k and for different but known v_i's, typically sequential or differing only a few bits. The cryptanalyst's task is to recover k, given a set of (v_i, z) pairs.

5.2.1. Re-keying models

The number of (v_i, z) pairs and the lengths of the known keystream z in these pairs available to an attacker are dependent on the application. Most practical applications are covered by one of two different models. As our models correspond to the resync models in [8], we use the same model names: fixed resync and requested resync.

Fixed Resync. Fixed resync applies to the situation where messages are transmitted as a series of frames, with each frame encrypted separately using the secret key and the known frame number as an initialisation vector. For example, this model applies to mobile telecommunications and internet transmissions. The v_i's are transmitted with each packet in the clear, so that the packets can be routed separately and reassembled in the correct order at the receiver. Typically, the v_i are the output of a counter and are strongly related. If the re-keying operation is not sufficiently strong, then these relationships may be exploited in order to extract the secret key, k.

Under this model, the task for the cryptanalyst is restricted in the following ways. The length of the known keystream sequences is limited to the size of the frames (typically 64 to 1500 bytes per frame). The number of (v_i, z) pairs is the same as the number of frames in the message. Under this model, the v_i are selected by routing protocols, so are known but not chosen by an attacker.

Requested Resync. Requested resync applies to the situation where messages are transmitted across an error prone channel, which can cause a loss of synchronisation. Following this, the receiver requests retransmission, using a different initialisation vector. Typically, the number of retransmissions required to complete the communication is expected to be small.

In this case, the v_i's are sent in the clear, but need not be sequential and may be chosen. Under this model, the task for the cryptanalyst has the following features. The length of the known keystream sequences is equivalent to the message length. The number of (v_i, z) pairs is equal to the number of requests for retransmission, which is expected to be limited. A notable difference in the cryptanalyst's task between the two models is that in this case the v_i's may be chosen (possibly adaptively).

5.2.2. *Security requirements*

Designing for $R(k, v_i)$ together with $G(S_0)$ must satisfy the following basic security requirements for F: *Given multiple output keystreams resulting from different v_i values, it is infeasible to recover the secret key k.* However, to avoid security weaknesses due to re-keying such as those discussed in [2] and [19], it is reasonable, but not necessary, to impose a stronger security requirement for R only: *Given multiple initial states resulting from different v_i values, it is infeasible to recover the secret key k.*

Attacks on k can be categorised by the required amount of work (time, memory, processor availability tradeoff), the number of (v_i, z) pairs and the amount of known keystream. Under fixed resync, attacks which require an amount of keystream in excess of the frame length cannot be successful. Under the requested resync model, the attack scenario is more powerful, as the v_i may be chosen, thus presenting the possibility of differential (or other correlation) cryptanalysis of F. The amount of keystream available is bounded by the message length, which may be quite large, but the number of chosen v_i should be limited by the transmission protocol. The worst case scenario is where requested resync is used with an unlimited number of requests permitted per message.

In the resynchronisation scenario, the secret key reconstruction attacks on F are at least as difficult as the attacks (on F) targeting the corresponding initial states. However, it is not true that the secret key reconstruction attacks on F in the resynchronisation scenario are at least as difficult as the initial state reconstruction attacks on G in the usual scenario without resynchronisation (as this scenario does not assume a set of known keystreams obtained from unknown, but related initial states). Ideally, we would like a re-keying operation that does not reduce the practical security provided by G against attacks in the scenario without resynchronisation. As the v_i can be chosen or are strongly related, and $S_0 = R(k, v_i)$, a reasonable condition for security is that simple changes in the value of v_i result in complex, unpredictable changes in the value of S_0. This method is applied in Section 6.1.3 to the LILI-II keystream generator.

6. LILI-II Keystream Generator

In this section we describe a new keystream generator called LILI-II [6] which has been designed to be efficient and prevent attacks described in Section 4. LILI-II is a specific cipher from the LILI (see [51]) family of keystream generators. The development of LILI-II was motivated by the response to the LILI-128 keystream generator, included as a stream cipher candidate for NESSIE [14]. Although the design for the LILI keystream generators is conceptually simple, it produces output sequences with provable properties with respect to basic cryptographic security requirements. Hypothesised attacks on LILI-128 [2, 31] and the request for a rekeying proposal prompted a review of the LILI-128 parameters to ensure provable security properties could be maintained while achieving an effective key size of 128 bits.

6.1. *Description of LILI-II keystream generator*

The LILI-II keystream generator is a simple and fast keystream generator using two binary LFSRs and two functions to generate a pseudorandom binary keystream sequence. The structure of the LILI keystream generators is illustrated in Figure 9. The components of the keystream generator can be grouped into two subsystems based on the functions they perform: clock control and data generation. The LFSR for the clock-control subsystem is regularly clocked. The output of this subsystem is an integer sequence which controls the clocking of the LFSR within the data-generation subsystem. If regularly clocked, the data-generation subsystem is a simple nonlinearly filtered LFSR (see Section 3.4.2).

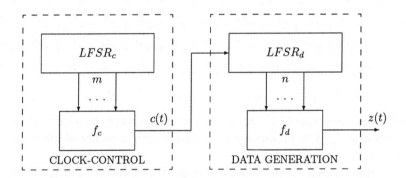

Fig. 9.   Overview of LILI keystream generators.

The state of LILI-II is defined to be the contents of the two LFSRs. The functions f_c and f_d are evaluated on the current state data, and the feedback bits are calculated. Then the LFSRs are clocked and the keystream bit is output. At initialisation the 128 bit key and a publicly known 128 bit initialisation vector are combined to form the initial values of the two shift registers. This initialisation process uses the LILI-II structure itself, and can also be used for rekeying. All valid keys produce a different keystream and there are no known weak keys.

The LILI keystream generators may be viewed as clock-controlled non-linear filter generators. Such a system, with the clock control provided by a stop-and-go generator, was examined in [15]. However, the use of stop-and-go clocking produces repetition of the nonlinear filter generator output in the keystream, which may permit attacks. This system is an improvement on that proposal, as stop-and-go clocking is avoided. For LILI-II keystream generators, $LFSR_d$ is clocked at least once and at most four times between the production of consecutive keystream bits.

6.1.1. Clock control subsystem

The clock-control subsystem of LILI-II uses a pseudorandom binary sequence produced by a regularly clocked LFSR, $LFSR_c$, of length 128 and a function, f_c, operating on the contents of $m = 2$ stages of $LFSR_c$ to produce a pseudorandom integer sequence, $c = \{c(t)\}_{t=1}^{\infty}$. The feedback polynomial of $LFSR_c$ is chosen to be the primitive polynomial

$$x^{128} + x^{126} + x^{125} + x^{124} + x^{123} + x^{122} + x^{119} + x^{117} + x^{115} + x^{111} + x^{108}$$
$$+x^{106} + x^{105} + x^{104} + x^{103} + x^{102} + x^{96} + x^{94} + x^{90} + x^{87} + x^{82} + x^{81}$$
$$+x^{80} + x^{79} + x^{77} + x^{74} + x^{73} + x^{72} + x^{71} + x^{70} + x^{67} + x^{66} + x^{65}$$
$$+x^{61} + x^{60} + x^{58} + x^{57} + x^{56} + x^{55} + x^{53} + x^{52} + x^{51} + x^{50} + x^{49}$$
$$+x^{47} + x^{44} + x^{43} + x^{40} + x^{39} + x^{36} + x^{35} + x^{30} + x^{29} + x^{25} + x^{23}$$
$$+x^{18} + x^{17} + x^{16} + x^{15} + x^{14} + x^{11} + x^9 + x^8 + x^7 + x^6 + x^1 + 1$$

and the initial state of $LFSR_c$ can be any state except the all zero state. It follows that $LFSR_c$ produces a maximum-length sequence of period $P_c = 2^{128} - 1$.

At time instant t, the contents of stages 0 and 126 of $LFSR_c$ are input to the function f_c and the output of f_c is an integer $c(t)$, such that $c(t) \in \{1, 2, 3, 4\}$. The function f_c is given by

$$f_c(x_0, x_{126}) = 2(x_0) + x_{126} + 1.$$

This function was chosen to be a bijective mapping so that the distribution of integers $c(t)$ is close to uniform. Thus $c = \{c(t)\}_{t=1}^{\infty}$ is a periodic integer sequence with period equal to $P_c = 2^{128} - 1$.

6.1.2. Data generation subsystem

The data-generation subsystem of LILI-II uses the integer sequence c produced by the clock-control subsystem to control the clocking of a binary LFSR, $LFSR_d$, of length $L_d = 127$. The contents of a fixed set of $n = 12$

stages of $LFSR_d$ are input to a specially constructed Boolean function, f_d. The binary output of f_d is the keystream bit $z(t)$. After $z(t)$ is produced, the two LFSRs are clocked and the process repeated to form the keystream $z = \{z(t)\}_{t=1}^{\infty}$.

The feedback polynomial of $LFSR_d$ is chosen to be the primitive polynomial

$$x^{127} + x^{121} + x^{120} + x^{114} + x^{107} + x^{106} + x^{103} + x^{101} + x^{97} + x^{96} + x^{94}$$

$$+x^{92} + x^{89} + x^{87} + x^{84} + x^{83} + x^{81} + x^{76} + x^{75} + x^{74} + x^{72} + x^{69}$$

$$+x^{68} + x^{65} + x^{64} + x^{62} + x^{59} + x^{57} + x^{56} + x^{54} + x^{52} + x^{50} + x^{48}$$

$$+x^{46} + x^{45} + x^{43} + x^{40} + x^{39} + x^{37} + x^{36} + x^{35} + x^{30} + x^{29} + x^{28}$$

$$+x^{27} + x^{25} + x^{23} + x^{22} + x^{21} + x^{20} + x^{19} + x^{18} + x^{14} + x^{10} + x^8$$

$$+x^7 + x^6 + x^4 + x^3 + x^2 + x + 1$$

and the initial state of $LFSR_d$ is not permitted to be the all zero state. Then $LFSR_d$ produces a maximum-length sequence of period $P_d = 2^{127} - 1$, which is a Mersenne Prime.

The 12 inputs to f_d are taken either from $LFSR_d$ positions (0, 1, 3, 7, 12, 20, 30, 44, 65, 80, 96, 122) which form a full positive difference set (see [24]). The function selected for f_d is balanced, highly nonlinear and has first order correlation immunity relative to the positions of 12 stages used as inputs to f_d. The function f_d has nonlinearity of 1992, and an algebraic order of 10. The function f_d was created using the heuristic techniques outlined in Section 2.4.3.

6.1.3. *Key loading and re-keying*

In this section we describe the method for initial key loading and for rekeying of the LILI-II keystream generator, using the secret key and an additional initialisation (or reinitialisation) vector. This process satisfies requirements for re-keying as set out in Section 5.2. Let k denote the 128-bit secret key and v_i denote the 128-bit initialisation vector for the ith rekeying. If the initialisation vector is of shorter length, then multiple copies of the vector will be concatenated to form a 128-bit vector.

The key loading or rekeying process for the LILI-II keystream generator involves the use of two applications of the LILI-II keystream generator, with the output from one application used to form the LFSR initial states for the next application. Specifically, the process is as follows. The initial state of $LFSR_c$ is obtained by XORing the two 128-bit binary strings k and v_i.

The initial state of $LFSR_d$ is obtained by deleting the first bit of k and the last bit of v_i, and XORing the two resulting 127-bit binary strings. Now the cipher is run to produce an output string of length 255 bits. For the second application of the cipher, the first 128 bits of this output string are used to form the initial state of $LFSR_c$, and the remaining 127 bits are used to form the initial state of $LFSR_d$. The cipher is run again to produce an output string of length 255 bits. The output from this second application is used to form the initial state of the keystream generator when we begin keystream production. As previously, the first 128 bits form the initial state of $LFSR_c$, and the remaining 127 bits form the initial state of $LFSR_d$.

6.2. Keystream properties

It can be shown (see [51]) that for LILI-II:

- the period of the keystream is $(2^{128-1})(2^{127} - 1)$,
- the linear complexity of the keystream is at least 2^{175},
- a period of a keystream is approximately balanced in the distribution of ones and zeros.

6.3. Possible attacks

From Section 4, a number of attacks should be considered with respect to the LILI-II keystream generator. These are known-plaintext attacks conducted under the standard assumption that the cryptanalyst knows the complete structure of the generator, and the secret key is only the initial states of the component shift registers.

6.3.1. Time/memory/data tradeoff attacks

For the time-memory attacks described in Section 4.1, a $2^{128} \cdot 2^{127} = 2^{255}$ tradeoff could be used although, as this requires time equivalent to exhaustive key search and an excessive amount of memory and known keystream, such an attack is certainly not feasible. The more general time-memory-data tradeoff [4] asserts that $T \cdot M^2 \cdot D^2 = S^2$, $P = S/D$, $D^2 \leq T$. This decreases D at the cost of increasing P. For example, one may choose $M = D = S^{1/3}$ and $T = P = S^{2/3}$, but for LILI-II, with $S = 255$, this gives $M = D = 2^{85}$ and $T = P = 2^{170}$, clearly worse than exhaustive key search. Alternatively, to reduce the time required (with a corresponding reduction in D), we can increase the amount of memory required, and obtain, for example, $M = 2^{127}$, $D = 2^{64}$ and $T = 2^{128}$, although this is still no better than exhaustive search, and requires an excessive amount of memory. The tradeoffs permitted by this attack result in either M or T being in excess of 2^{128}, when applied to LILI-II.

In any case, the use of the initialisation scheme (the key-loading/rekeying algorithm) to expand the 128-bit secret key into a 255 bit initial state renders the time-memory attacks on LILI-II infeasible, as their performance is at best, no better than exhaustive key search.

6.3.2. Attacks on irregularly clocked $LFSR_d$

Suppose a keystream segment of length N is known, say $\{z(t)\}_{t=1}^{N}$. This is a decimated version of a segment of length M of the underlying regularly clocked nonlinearly filtered $LFSR_d$ sequence, $g = \{g(i)\}_{i=1}^{M}$, where $M \geq N$. The objective of correlation attacks targeting $LFSR_d$ is to recover the initial state of $LFSR_d$ by identifying the segment $\{g(i)\}_{i=1}^{M}$ that $\{z(t)\}_{t=1}^{N}$ was obtained from through decimation, using the correlation between the regularly clocked sequence and the keystream, without knowing the decimating sequence.

The complexity of such an attack on LILI-II is $(2^{127} - 1)$ multiplied by the complexity of computing the correlation measure, and then add on the complexity of recovering the corresponding $LFSR_c$ state (see [51]). That is, much worse than exhaustive key search.

6.3.3. Attacks targeting $LFSR_c$

A possible approach to attacking the proposed generator is by targeting the clock-control sequence produced by $LFSR_c$. Guess an initial state of $LFSR_c$, say $\{\hat{a}(t)\}_{t=1}^{L_c}$. Use the known $LFSR_c$ feedback function and the function f_c to generate the decimating sequence $\{\hat{c}(t)\}_{t=1}^{N}$ for some $N \geq L_c$. Then position the known keystream bits $\{z(t)\}_{t=1}^{N}$ in the corresponding positions of $\{\hat{g}(i)\}_{i=1}^{\infty}$, the nonlinear filter generator output when regularly clocked. At this point we have some (not all consecutive) terms in the nonlinear filter generator output sequence and are trying to reconstruct a candidate initial state for $LFSR_d$.

Note that the amount of trial and error involved in guessing the initial state of $LFSR_c$ is the same as for guessing the secret key. Thus, the performance of any divide and conquer attack targeting $LFSR_c$ will be worse than exhaustive key search.

6.4. *Implementation*

The current software implementation of LILI-II runs at 6.07 Mbps on a 300MHz Pentium III using Borland C++ 5.5, and 3.75Mbps on a 450MHz SPARCv9 using gcc 2.95.3. In comparison the LILI-128 stream cipher, which offered lower levels of security, achieved speeds of 6.65Mbps and 4.12 Mbps under the respective conditions. As has recently been shown in [33], that by using parallel copies of the data register, it is possible for a LILI-II implementation to achieve throughput equal to the underlying clock speed, the same as LILI-128. With current microprocessor clock speeds exceeding 1GHz, this implies LILI-II in hardware can have throughput of over 125 Megabytes/second.

7. Conclusion

Synchronous stream ciphers are the standard method of encryption to provide confidentiality of messages. In this paper we have examined various models of stream ciphers. Particular emphasis has been placed on LFSR based designs. Such designs can offer high security in relation to the standard cryptographic properties of large period, large linear complexity and white noise characteristics. Despite these properties, as was outlined in Section 4, many of these keystream generators can be attacked. Such attacks highlight the flaw in designing an encryption algorithm to satisfy various properties. In designing an encryption algorithm one must satisfy the standard properties as well as search for new flaws that may have been introduced in a particular design. This may require thorough examination by independent third parties.

In addition to security there are many efficiency considerations to be considered in both designing and implementing encryption algorithms. It is a relatively simple task to design a highly secure encryption algorithm. It is much more difficult to design an algorithm which is both secure and highly efficient. Various concepts related to efficient implementation of LFSR based stream ciphers were discussed in Section 5. A generic secure and efficient method of rekeying keystream generators was presented. Such a rekeying procedure is required in many applications.

In Section 6 a new keystream generator which has been designed by the authors, LILI-II, is described. This algorithm has been designed to be especially efficient in hardware as well as offering suitable speed in software for most applications. LILI-II has been designed to satisfy the standard cryptographic properties as well as being secure from the attacks described in Section 4. As was discussed there does not seem to be any attack on LILI-II more efficient than exhaustive key search. However, further analysis of this algorithm by other researchers is required to measure its security.

The keystream generators described in Section 3 except for the OFB mode of a block cipher and RC4 algorithm were bit based. As in the case of LILI-II algorithm, which is over 100 times slower in software, there are some limitations in efficiency in software in many bit based designs. Over the past few years several word based keystream generators have emerged which allow for greater efficiency in software than most bit based stream ciphers including SNOW [16], PANAMA [9], SOBER [40] and ORYX [52]. Besides ORYX which was shown to be a seriously flawed design, [53], little is known about the security of these other stream ciphers. Much greater analysis of each of these algorithms is required. Future word based stream ciphers should be designed to take advantage of the inherent parallelism of microprocessors and build on the extensive analysis of bit based keystream generators.

References

1. R. Anderson, *Searching for optimum correlation attack*, Proceedings of Fast Software Encryption '94, Lecture Notes in Computer Science, **1008** (1995), 137–143.
2. S. Babbage, *Cryptanalysis of LILI-128*, https://www.cosic.esat.kuleuven.ac.be/nessie/reports/
3. S. Babbage, *A space/time tradeoff in exhaustive search attacks on stream ciphers*, European Convention on Security and Detection, IEE Conference Publication, **408** (1995).
4. A. Biryukov and A. Shamir, *Cryptanalytic time/memory/data tradeoffs for stream ciphers*, Advances in Cryptology – ASIACRYPT 2000, Lecture Notes in Computer Science, **1976** (2000), 1–13.
5. G. Carter, *Aspects of Linear Complexity*, PhD thesis, University of London, (1989).
6. A. Clark, E. Dawson, J. Fuller, J. Dj. Golić, H.-J. Lee, S.-J. Moon, W. Millan and L. Simpson. *The LILI-II keystream generator*, to appear in the Proceedings of the 7th Australasian Conference on Information Security and Privacy – ACISP 2002.
7. D. Coppersmith, H. Krawczyk and Y. Mansour, *The shrinking generator*, Advances in Cryptology – CRYPTO'93, Lecture Notes in Computer Science, **773** (1993), 22–39.
8. J. Daemen, R. Govaerts and J. Vanderwalle, *Resynchronization weaknesses in synchronous stream ciphers*, Advances in Cryptology – CRYPTO'93, Lecture Notes in Computer Science, **765** (1994), 159–167.
9. J. Daemen and C. Clapp, *Fast hashing and stream encryption with PANAMA*,

Fast Software Encryption, Lecture Notes in Computer Science, **1372** (1998), 60–74.

10. D.W. Davies and G.I.P. Parkin, *The average size of the keystream in output feedback encipherment*, Cryptography, Proceedings of the Workshop on Cryptography 1982, Springer-Verlag, Berlin, 1983, 263–279.

11. E. Dawson and A. Clark, *Divide and conquer attacks on certain classes of stream ciphers*, Cryptologia, **18(1)** (1994), 25–40.

12. E. Dawson and L. Nielsen, *Automated cryptanalysis of XOR plaintext strings*, Cryptologia, **10(2)** (1996).

13. E. Dawson and C. Wu, *Constuction of correlation immune boolean functions*, ICICS'97, Lecture Notes in Computer Science, **1334** (1997), 170–180.

14. E. Dawson, A. Clark, J. Golić, W.Millan, L. Penna and L. Simpson, *The LILI-128 keystream generator*, https://www.cosic.esat.kuleuven.ac.be/nessie/workshop/submissions.html.

15. C. Ding, G. Xiao and W. Shan, *The Stability Theory of Stream Ciphers*, Lecture Notes in Computer Science, **561** (1991).

16. P. Ekdahl and T. Johansson, *SNOW-a new stream cipher*, in Proceedings of the first NESSIE Workshop, Belgium, 2000.

17. E. Filiol and C. Fontaine, *Highly nonlinear balanced booleans functions with a good correlation-immunity*, Advances in Cryptology – EUROCRYPT '98, Lecture Notes in Computer Science, **1403** (1998), 475–488.

18. S. Fluhrer and D. McGrew, *Statistical analysis of the alleged RC4 keystream generator*, Proceedings of Fast Software Encryption 2000, Lecture Notes in Computer Science, **1978** (2000), 19–30.

19. S. Fluhrer, I. Mantin and A. Shamir, *Weaknesses in the key scheduling algorithm of RC4.*, Proceedings of the Eighth Annual Workshop on Selected Areas in Cryptography -SAC'2001, Lecture Notes in Computer Science, **2259** (2001), 1–24.

20. J. Dj. Golić and M. J. Mihaljević, *A generalized correlation attack on a class of stream ciphers based on the Levenshtein distance*, Journal of Cryptology, **3(3)** (1991), 201–212.

21. J. Dj. Golić and S. Petrović, *A generalized correlation attack with a probabilistic constrained edit distance*, Advances in Cryptology – EUROCRYPT'92, Lecture Notes in Computer Science, **658** (1993), 472–476.

22. J. Dj. Golić and L. O'Connor, *Embedding and probabilistic correlation attacks on clock controlled shift registers*, Advances in Cryptology – EUROCRYPT'94, Lecture Notes in Computer Science, **950** (1994), 230–243.

23. J. Dj. Golić, M. Salmasizadeh, A. Clark, A. Khodkar and E. Dawson, *Discrete optimisation and fast correlation attacks*, Cryptography: Policy and Algorithms, Lecture Notes in Computer Science, **1029** (1996), 186–200.

24. J. Dj. Golić, *On the security of nonlinear filter generators*, Fast Software Encryption – Cambridge'96, Lecture Notes in Computer Science, **1039** (1996), 173–188.

25. J. Dj. Golić, *Linear statistical weakness of alleged RC4 keystream generator*, Advances in Cryptology – EUROCRYPT'97, Lecture Notes in Computer Science, **1233** (1997), 226–238.

26. J. Dj. Golić, *Cryptanalysis of alleged A5 stream cipher*, Advances in Cryptology – EUROCRYPT'97, Lecture Notes in Computer Science, **1233** (1997), 239–255.
27. J. Dj. Golić, M. Salmasizadeh and E. Dawson, *Fast correlation attacks on the summation generator*, Journal of Cryptology, **13(2)** (2000), 245–262.
28. H. Gustafson, *Statistical Analysis of Symmetric Ciphers*, PhD thesis, Information Security Research Centre, Queensland University of Technology, Brisbane, Australia, July 1996.
29. T. Johansson, *Reduced complexity correlation attacks on two clock-controlled generators*, Advances in Cryptology – ASIACRYPT'98, Lecture Notes in Computer Science, **1514** (1998), 342–356.
30. T. Johansson and F. Jönsson, *Improved fast correlation attacks on stream ciphers via convolutional codes*, Advances in Cryptology – EUROCRYPT'99, Lecture Notes in Computer Science, **1592** (1999), 347–362.
31. F. Jönsson and F. Johansson, *A fast correlation attack on LILI-128*, http://www.it.lth.se/thomas/papers/paper140.ps
32. R.R. Junemann, *Analysis of certain aspects of output-feedback mode*, Advances in Cryptology – Proceedings of CRYPTO'82, Plenum, (1983), 99–127.
33. H. Lee and S. Moon, *High-speed implementation LILI-II keystream generator*, Private Communication, 2002.
34. J. L. Massey, *Shift register synthesis and BCH decoding*, IEEE Trans. Inform. Theory, **15** (1969), 122–127.
35. W. Meier and O. Staffelbach, *Fast correlation attacks on certain stream ciphers*, Journal of Cryptology, **1(3)** (1989), 159–167.
36. W. Meier and O. Staffelbach, *Correlation properties of combiners with memory in stream ciphers*, Journal of Cryptology, **5(1)** (1992), 67–88.
37. W. Millan, *Analysis and Design of Boolean Functions for Cryptographic Applications*, PhD thesis, Information Security Research Centre, Queensland University of Technology, Brisbane, Australia, January 1998.
38. W. Millan, A. Clark and E. Dawson, *Heuristic design of cryptographically strong balanced boolean functions*, Advances in Cryptology – EUROCRYPT'98, Lecture Notes in Computer Science, **1403** (1998), 489–499.
39. R. Rivest, *The RC4 encryption algorithm*, RSA Data Security Inc., March 1992.
40. G. Rose, *A stream cipher based on linear feedback over $GF(2^8)$*, Information Security and Privacy – ACISP'98, Lecture Notes in Computer Science, **1438** (1998), 135–146.
41. O. Rothaus, *On bent functions*, Journal of Combinatorial Theory(A), **20** (1976), 300–305.
42. R. Rueppel, *Analysis and Design of Stream Ciphers*, Springer-Verlag, Berlin, 1986.
43. M. Salmasizadeh, L. Simpson, J. Dj. Golić and E. Dawson, *Fast correlation attacks and multiple linear approximations*, Information Security and Privacy – ACISP'97, Lecture Notes in Computer Science, **1270** (1997), 228–239.
44. M. Salmasizadeh, *Statistical and Correlation Analysis of Certain Shift Register Based Stream Ciphers*, PhD thesis, Information Security Research Centre, Queensland University of Technology, Brisbane, Australia, June 1997.

45. M. Schneider, *On the construction and upper bounds of balanced and correlation-immune functions*, Selected Areas in Cryptography – SAC'97, (1997), 73–87.

46. J. Seberry, X. Zhang and Y. Zheng, *On constructions and nonlinearity of correlation immune functions*, Advances in Cryptology – EUROCRYPT'93, Lecture Notes in Computer Science, **765** (1994), 181–199.

47. T. Siegenthaler, *Correlation immunity of nonlinear combining functions for cryptographic applications*, IEEE Information Theory, **IT-30** (1984), 776–780.

48. T. Siegenthaler, *Decrypting a class of stream ciphers using ciphertext only*, IEEE Transactions on Computers, **34(1)** (1985), 81–85.

49. L. Simpson, J. Dj. Golić and E. Dawson, *A probabilistic correlation attack on the shrinking generator*, Information Security and Privacy – ACISP'98, Lecture Notes in Computer Science, **1438** (1998), 147–158.

50. L. Simpson, *Divide and Conquer Attacks on Shift Register Based Stream Ciphers*, PhD thesis, Information Security Research Centre, Queensland University of Technology, Brisbane, Australia, November 1999.

51. L. Simpson, E. Dawson, J. Dj. Golić and W. Millan, *LILI keystream generator*, Proceedings of the Seventh Annual Workshop on Selected Areas in Cryptology – SAC'2000, Lecture Notes in Computer Science, **2012** (2001), 248–261.

52. T I A TR45.0.A, *Common Cryptographic Algorithms*, Telecommunications Industry Association, Vienna VA., USA, June 1995, Rev B.

53. D. Wagner, L. Simpson, E. Dawson, J. Kelsey, W. Millan and B. Schneier, *Cryptanalysis of ORYX*, Proceedings of the Fifth Annual Workshop on Selected Areas in Cryptography - SAC'98, Lecture Notes in Computer Science, **1556** (1998), 296–305.

54. M. Živković, *An algorithm for the initial state reconstruction of the clock-controlled shift register*, IEEE Transactions on Information Theory, **37** (1991), 1488–1490.

QUANTUM ERROR-CORRECTING CODES

Keqin Feng

Department of Mathematical Sciences, Tsinghua University
Beijing, 100084, China
E-mail: kfeng@math.tsinghua.edu.cn

Quantum error correction is one of the important requirements in both quantum computing and quantum communication. The mathematical theory of quantum codes (QC) has been formulated in 1996-1998 and developed rapidly since then. In these lecture notes we introduce the basic concepts on QC, mathematical methods to construct good QC and some related topics.

1. Introduction

In the past half century, classical error-correcting codes have been developed, widely and deeply, into a huge body, effectively used in digital and network communication to detect and correct errors in transmission channels, and demonstrated remarkable applications of pure mathematical branches, such as number theory, algebra and algebraic geometry, in practical engineering fields. In the quantum case, error correction is also one of the important requirements not only for quantum communication, but also for quantum computing. There are at least the following reasons to make error correction in the quantum case more critical than in the classical case.

(i) Classical information can be read, transcribed (into any medium) and duplicated at will. Quantum information, on the other hand, cannot be read or duplicated in general without being disturbed. Thus, for example, we cannot use "repetition codes" to correct quantum information.

(ii) An interaction of a quantum system with its environment can lead to the so-called decoherence effects and can greatly influence, or even completely destroy, subtle quantum mechanisms.

(iii) In classical information processing, measurements usually remain original information. But in the quantum case, any measurement is like a kind of noise in some sense since quantum information would change its state after being measured.

(iv) Digital information has discrete and finite states (finite field $\mathbb{F}_q$), but quantum information occupies continuous states (complex number field $\mathbb{C}$).

These appear to make long reliable quantum computation or communication practically impossible and hopeless. A breakthrough came in 1995-96 with the pioneering papers of Shor [38] and Steane [40] who reduced very complicated entangled errors into simple types of errors (Pauli operations σ_x, σ_z and σ_y) acting on quantum information bit-by-bit. Shor made the first quantum error-correcting code in a physical way by using 9 quantum bits (qubits) to correct one-bit error. Soon after, Steane was able to reduce encoding message into 7 qubits. An optimal way to correct one qubit, by an entangled state of 5 qubits, was first shown by Laflamme et al. [24] and Bennett et al. [4] in 1996. The first attempts to develop a methodology to design quantum codes in a systematic and mathematical way were due to Calderbank and Shor [7], Steane [41], Gottesman [15] and Calderbank et al. [5, 6] in 1996-1998. Particularly in [6], a neat and rigorous formalization of quantum codes was established and an effective mathematical method to construct nice quantum codes was presented by using character theory of finite abelian groups and symplectic geometry over finite fields which made it possible to construct quantum codes from classical codes over $\mathbb{F}_2$ or $\mathbb{F}_4$. Since then the parameters of constructed quantum codes improved rapidly by using classical Reed-Muller codes [43], Reed-Solomon codes [10], BCH codes [18] and algebraic-geometry codes [9, 8]. At the same time, non-binary quantum codes were formulated [1, 21, 28, 37, 31]. It is to a large extent due to these remarkable works that the vision of real quantum computing and communication is much closer.

In this article, we introduce the basic mathematical concepts on quantum error-correcting codes, main methods to construct quantum codes and some theoretical results in quantum code theory, a theory with a six-year history.

2. Classical Error-Correcting Codes

3. Quantum Error-Correcting Codes

4. Basic Methods to Construct Quantum Codes

5. Weight Enumerator and Singleton Bound

6. Nonbinary Case

2. Classical Error-Correcting Codes

We start from a quick view of classical error-correcting codes since quantum codes have similar formulations as classical codes in many places and can be effectively constructed from classical codes. We just describe basic facts on classical codes we need in these notes. For more details see the books [26] or [27].

2.1. *Three basic parameters of a code*

Definition 2.1: Let $\mathbb{F}_q$ be the finite field with q elements. A *code* C is defined just as a subset of $\mathbb{F}_q^n$. A vector v in C is called a *codeword*. The number n is called *length*, $K = |C|$ (the *size* of C), $k = \log_q K$ is called the *number of information digits* in C. The number $\frac{k}{n} (\leq 1)$ is called the *transmission ratio* which measures the effectiveness of the code C.

We have one more important parameter — the minimal distance which measures the error-correcting ability of a code C.

Definition 2.2: For a vector $v = (v_1, \ldots, v_n)$ in $\mathbb{F}_q^n$ we define the *Hamming weight* of v by

$$w_H(v) = \sharp\{1 \leq i \leq n \,|v_i \neq 0\}.$$

The *Hamming distance* of two vectors $u = (u_1, \ldots, u_n)$ and $v = (v_1, \ldots, v_n) \in \mathbb{F}_q^n$ is defined by

$$d_H(u, v) = \sharp\{1 \leq i \leq n \,|u_i \neq v_i\} = w_H(u - v).$$

In this section we simply denote d_H and w_H by d and w, but after Section 3 we should use d_H and w_H in order to distinguish them from the quantum weight w_Q.

The Hamming distance is really a "mathematical" distance since it satisfies the following basic properties. For $u, v, w \in \mathbb{F}_q^n$,

(H1) $d(u, v) > 0$; and $d(u, v) = 0 \iff u = v$.
(H2) $d(u, v) = d(v, u)$.
(H3) (The triangle inequality) $d(u, w) \leq d(u, v) + d(v, w)$.

Definition 2.3: Let C be a code over $\mathbb{F}_q$ with length n ($C \subseteq \mathbb{F}_q^n$). The *minimal distance* of C is defined by

$$d = d(C) = \min\{d(c, c')|c, c' \in C, \; c \neq c'\}.$$

The following result shows that the minimal distance $d(C)$ is the proper parameter to judge the ability of a code C to detect and correct digital errors.

Theorem 2.4: (*Basic theorem*) *Let C be a code with minimal distance $d = d(C)$. The code C can detect $\leq d-1$ digital errors and correct $\leq [\frac{d-1}{2}]$ digital errors.*

Proof: Alice sends a codeword $c \in C$ to Bob, but has error vector $e \in \mathbb{F}_q^n$ with $1 \leq w(e) \leq d-1$, so that Bob receives the vector $y = c + e$, $y \neq c$. By definition of $d = d(C)$ we know that $d(c, c') \geq d$ for each codeword $c' \in C$, $c' \neq c$. Now $d(c, y) = w(y - c) = w(e) \leq d-1$. Thus, y is not a codeword, so that Bob finds y has errors.

Moreover, if $w(e) \leq [\frac{d-1}{2}]$, then $d(c, y) = w(e) \leq [\frac{d-1}{2}]$ and $d(c', y) \geq d(c', c) - d(c, y) \geq d - [\frac{d-1}{2}] > [\frac{d-1}{2}]$ for any $c' \in C$, $c' \neq c$. This means that c is the unique codeword closest to the received y. In this way y can be decoded to c correctly.   □

Now we have introduced the basic parameters of a code C: n, k (or K) and d. We denote this code C over $\mathbb{F}_q$ by $(n, K, d)_q$ or $[n, k, d]_q$. A good code should have big values of $\frac{k}{n}$ and d. But there are some restrictions, or bounds between parameters.

Theorem 2.5: (*Hamming bound*) *Let C be a code $(n, K, d)_q$, and $l = [\frac{d-1}{2}]$. Then*

$$q^n \geq K \sum_{i=0}^{l} (q-1)^i \binom{n}{i}. \tag{1}$$

Proof: For $v \in \mathbb{F}_q^n$ and integer r $(0 \leq r \leq n)$, let $S_r(v)$ be the (closed) ball with center v and radius r,

$$S_r(v) = \{u \in \mathbb{F}_q^n \mid d(u, v) \leq r\}.$$

The size of the ball is $|S_r(v)| = \sum_{i=0}^{r} (q-1)^i \binom{n}{i}$. By the triangle inequality and $2l = 2[\frac{d-1}{2}] < d$, we know that all balls $\{S_l(c) | c \in C\}$ are disjoint. Thus,

$$q^n = |\mathbb{F}_q^n| \geq \sum_{c \in C} |S_l(c)| = K \cdot \sum_{i=0}^{l} (q-1)^i \binom{n}{i}.$$

□

Definition 2.6: A code $(n, K, d)_q$ is called *perfect* if (1) becomes equality

$$q^n = K \sum_{i=0}^{l} (q - 1)^i \binom{n}{i}.$$

Example 2.7: Consider the following binary code C in $\mathbb{F}_2^7$

$$C = \left\{ \begin{array}{ll} (0010111), & (1101000) \\ (1001011), & (0110100) \\ (1100101), & (0011010) \\ (1110010), & (0001101) \\ (0111001), & (1000110) \\ (1011100), & (0100011) \\ (0101110), & (1010001) \\ (0000000), & (1111111) \end{array} \right\}.$$

This is a code $(7, 16, 3)_2$ or $[7, 4, 3]_2$. Since $l = 1$ and $2^7 = 2^4(1 + \binom{7}{1})$, C is a perfect code.

In order to construct good codes and to have practical decoding algorithms, we need to consider codes with more algebraic structure.

2.2. Linear codes

Definition 2.8: A code C in $\mathbb{F}_q^n$ is called *linear* if C is an $\mathbb{F}_q$-subspace of the $\mathbb{F}_q$-vector space $\mathbb{F}_q^n$.

Let $k = \dim_{\mathbb{F}_q} C$. Then $K = |C| = q^k$ and $k = \log_q K$, so that $\dim_{\mathbb{F}_q} C$ is the number of information digits of C. For the minimal distance of a linear code we have the following result.

Lemma 2.9: *For a linear code C,*

$$d(C) = \min\{w(c) | c \in C, \ c \neq 0\}.$$

Proof: We denote the right-hand side by d'. Then there exists $0 \neq c \in C$ such that $w(c) = d'$. Since C is linear so that $0 \in C$, $d' = w(c) = d(c, 0 \geq d(C)$. On the other hand, we have $c, c' \in C$, $c \neq c'$, such that $d(C) = d(c, c') = w(c - c')$. Since C is linear, we have $0 \neq c - c' \in C$. Thus $d(C) = w(c - c') \geq d'$. Therefore $d(C) = d'$. $\qquad\square$

Let C be a linear code $[n, k, d]_q$. Then C is an $\mathbb{F}_q$-subspace of $\mathbb{F}_q^n$ with $\dim_{\mathbb{F}_q} C = k$. We may choose an $\mathbb{F}_q$-basis $\{u_1, \ldots, u_k\}$ of C,

$$u_i = (a_{i1}, a_{i2}, \ldots, a_{in}) \quad (1 \le i \le k) \, (a_{ij} \in \mathbb{F}_q).$$

Each codeword $c \in C$ can be expressed uniquely by

$$c = b_1 u_1 + b_2 u_2 + \cdots + b_k u_k \quad (b_i \in \mathbb{F}_q)$$
$$= (b_1, \ldots, b_k)G,$$

where

$$G = \begin{bmatrix} u_1 \\ u_2 \\ \vdots \\ u_k \end{bmatrix} = \begin{vmatrix} a_{11} & a_{12} & \cdots & a_{1n} \\ a_{21} & a_{22} & \cdots & a_{2n} \\ \multicolumn{4}{c}{\cdots\cdots\cdots\cdots\cdots} \\ a_{k1} & a_{k2} & \cdots & a_{kn} \end{vmatrix}$$

is called a *generating matrix* of the linear code C. Note that G is a $k \times n$ matrix over $\mathbb{F}_q$ with $\text{rank}_{\mathbb{F}_q} G = k$. On the other hand, by linear algebra, any subspace C of $\mathbb{F}_q^n$ with dimension k is the intersection of $n - k$ hyperplanes in $\mathbb{F}_q^n$. Namely, there exists an $(n - k) \times n$ matrix over $\mathbb{F}_q$ with rank $= n - k$

$$H = \begin{bmatrix} v_1 \\ v_2 \\ \vdots \\ v_{n-k} \end{bmatrix} = \begin{vmatrix} b_{11} & b_{12} & \cdots & b_{1n} \\ b_{21} & b_{22} & \cdots & b_{2n} \\ \multicolumn{4}{c}{\cdots\cdots\cdots\cdots\cdots\cdots} \\ b_{n-k,1} & b_{n-k,2} & \cdots & b_{n-k,n} \end{vmatrix} \quad (b_{ij} \in \mathbb{F}_q),$$

such that for $c = (c_1, c_2, \ldots, c_n) \in \mathbb{F}_q^n$,

$$c \in C \Leftrightarrow v_i c^\top = \sum_{\lambda=1}^{n} b_{i\lambda} c_\lambda = 0 \in \mathbb{F}_q \quad (1 \le i \le n - k)$$

$$\Leftrightarrow Hc^\top = 0 \in \mathbb{F}_q^{n-k}.$$

H is called a *parity-check matrix* of the linear code C since for any vector c in $\mathbb{F}_q^n$ one can check if c is a codeword in the linear code C by computing $Hc^\top$. Moreover, the parity-check matrix can be used to determine (or design) the minimal distance of a linear code.

Lemma 2.10: *Let C be an $[n, k, d]$ $(d \ge 2)$ linear code over $\mathbb{F}_q$, H be a parity-check $(n - k) \times n$ matrix of C. Then:*

(1) *Any $d - 1$ distinct column vectors of H are $\mathbb{F}_q$-linearly independent.*
(2) *There are d distinct column vectors of H which are $\mathbb{F}_q$-linearly dependent.*

Proof: Let $H = [w_1, w_2, \ldots, w_n]$, where $w_1, \ldots, w_n$ are column vectors in $\mathbb{F}_q^{n-k}$. For each vector $c = (c_1, c_2, \ldots, c_n) \in \mathbb{F}_q^n$, $w_H(c) = l$, $c_{j_1} \cdot c_{j_2} \cdots c_{j_l} \neq 0$ and $c_j = 0$ for $j \neq j_1, \ldots, j_l$. Then we have

$$c \in C \Leftrightarrow 0 = Hc^\mathsf{T} = [w_1, w_2, \ldots, w_n] \begin{pmatrix} c_1 \\ c_2 \\ \vdots \\ c_n \end{pmatrix} = c_{j_1} w_{j_1} + \cdots + c_{j_l} w_{j_l}.$$

Therefore a codeword in C with weight l corresponds to l linearly dependent columns in H. The conclusion of Lemma 2.10 follows from this. □

Example 2.11: (Hamming codes) Consider the following matrix over $\mathbb{F}_2$

$$H_3 = \begin{bmatrix} 1001011 \\ 0101110 \\ 0010111 \end{bmatrix}.$$

The 7 columns of H_3 are just all non-zero vectors in $\mathbb{F}_2^3$. Let C_3 be the binary linear code with parity-check matrix H_3,

$$C_3 = \{c \in \mathbb{F}_2^7 | \, H_3 c^\mathsf{T} = 0\}.$$

The code length of C_3 is $n = 7$ and $k = \dim_{\mathbb{F}_2} C_3 = n - \mathrm{rank} H_3 = 7 - 3 = 4$. Since the columns of H_3 are distinct, any two columns of H_3 are linearly independent. There are three columns in H_3 which are linearly dependent. By Lemma 2.10, $d = d(C_3) = 3$. Therefore C_3 is a binary $[7, 4, 3]$ linear code. In fact, C_3 is just the binary code in Example 2.7. Similarly, for each integer $t \geq 3$, let H_t be the $t \times (2^t - 1)$ matrix over $\mathbb{F}_2$ such that the $2^t - 1$ columns of H_t are all non-zero vectors in $\mathbb{F}_2^t$. Let C_t be the binary linear code with parity-check matrix H_t. Then C_t is a binary code $[n, k, d] = [2^t - 1, 2^t - 1 - t, 3]$. All binary codes C_t ($t \geq 3$) are perfect, called *binary Hamming codes*.

For linear codes we have the following bound.

Theorem 2.12: (*Singleton bound for linear codes*) *For a linear code* $[n, k, d]_q$ *we have* $d \leq n - k + 1$.

Proof: Let H be the parity-check matrix of a linear code $[n, k, d]_q$. The size of H is $(n - k) \times n$ and $\mathrm{rank}_{\mathbb{F}_q} H = n - k$. Thus, there exist $n - k \mathbb{F}_q$-linearly independent columns in H. On the other hand, any $n - k + 1$ distinct columns of H are $\mathbb{F}_q$-linearly dependent. By Lemma 2.10 we know that $d = d(C) \leq n - k + 1$. □

A linear code which reaches the Singleton bound $(d = n - k + 1)$ is called an *MDS (maximal distance separable) code*. It is easy to see that a linear code C is an MDS code if and only if any $n - k$ distinct columns of its parity-check matrix are linearly independent.

Definition 2.13: Let C be a linear code $[n, k, d]$ over $\mathbb{F}_q$. The subset of $\mathbb{F}_q^n$ given by

$$C^\perp = \{v \in \mathbb{F}_q^n | (c, v) = 0 \text{ for all } c \in C\}$$

is also a linear code over $\mathbb{F}_q$ where, for $c = (c_1, c_2, \ldots, c_n)$ and $v = (v_1, v_2, \ldots, v_n) \in \mathbb{F}_q^n$, (c, v) is the *inner product*

$$(c, v) = cv^\top = \sum_{i=1}^n c_i v_i \in \mathbb{F}_q.$$

$C^\perp$ is called the *dual code* of C. If $C \subseteq C^\perp$, then C is called *self-orthogonal*. Let $\{v_1, \ldots, v_k\}$ be a basis of a linear code C. Then C is self-orthogonal if and only if v_i $(1 \le i \le k)$ are orthogonal to each other: $v_i v_j^\top = 0 \in \mathbb{F}_q$ $(1 \le i, j \le k)$.

Lemma 2.14: *Let C be an $[n, k]_q$ linear code. Then:*

(1) $C^\perp$ *is an $[n, n - k]_q$ linear code:* $\dim_{\mathbb{F}_q} C^\perp = n - \dim_{\mathbb{F}_q} C$.
(2) $(C^\perp)^\perp = C$.
(3) *A generating (parity-check) matrix of C is a parity-check (generating) matrix of $C^\perp$.*

2.3. Weight enumerator and MacWilliams identity

Definition 2.15: Let C be an $[n, k]_q$ linear code. For each i $(0 \le i \le n)$, let A_i be the number of $c \in C$ with $w(c) = i$. The *weight enumerator* of C is defined by

$$f_C(x, y) = \sum_{i=0}^n A_i x^{n-i} y^i \in \mathbb{Z}[x, y].$$

It is easy to see that $A_0 = 1$, $A_0 + A_1 + \cdots + A_n = |C| = q^k$ and $d(C)$ is the smallest positive integer d such that $A_d \ge 1$.

Theorem 2.16: *(MacWilliams Identity)* Let C be an $[n, k]_q$ linear code and $C^\perp$ be the dual code of C. Then

$$f_{C^\perp}(x, y) = q^{-k} f_C(x + (q - 1)y, x - y).$$

There are many proofs of Theorem 2.16. We choose one proof by using the additive character of $\mathbb{F}_q^n$ since these characters are also used to construct quantum codes in Section 4. We list basic definitions and facts on character theory of finite abelian groups. For proofs and more details see [26] or [25].

Definition 2.17: A *character* of a finite abelian group $G = (G, +)$ is a homomorphism $\chi \colon G \text{ to } \mathbb{C}^*$ from G to the multiplicative group $\mathbb{C}^* = \mathbb{C} \backslash \{0\}$ of non-zero complex numbers. Namely, for $g_1, g_2 \in G$, we have $\chi(g_1 + g_2) = \chi(g_1)\chi(g_2)$.

Let $m = |G|$. Then for each $g \in G$, $g^m = 1$ and $\chi(g)^m = \chi(g^m) = \chi(1) = 1$. Thus, all $\chi(g)$ $(g \in G)$ are m-th roots of 1. We denote the trivial character (mapping all $g \in G$ to 1) by 1. Let $\hat{G}$ be the set of all characters of G. For $\chi_1, \chi_2 \in \hat{G}$, the mapping $\chi_1\chi_2$ defined by $(\chi_1\chi_2)(g) = \chi_1(g)\chi_2(g)$, and $\overline{\chi_1}$ defined by $\overline{\chi_1}(g) = \overline{\chi_1(g)} = \chi_1(g)^{-1}$ are also characters of G and $\overline{\chi_1}\chi_1 = 1$. In this way, we are making $\hat{G}$ into an abelian group, called the *character group* of G.

It is easy to see by the structure theorem of finite abelian groups that the two groups G and $\hat{G}$ are isomorphic. Particularly, $|G| = |\hat{G}|$. One more thing we need is the orthogonal relations between characters.

Lemma 2.18: *Let G be a finite abelian group. Then:*

(1) *For $\chi \in \hat{G}$,*

$$\sum_{g \in G} \chi(g) = \begin{cases} |G|, & \text{if } \chi = 1, \\ 0, & \text{otherwise.} \end{cases}$$

(2) *For $g \in G$,*

$$\sum_{\chi \in \hat{G}} \chi(g) = \begin{cases} |G|, & \text{if } g = 1, \\ 0, & \text{otherwise.} \end{cases}$$

Now we determine the character group $\widehat{\mathbb{F}_q^n}$ of the additive group $\mathbb{F}_q^n$. Let $q = p^m$ where p is a prime number and $m \geq 1$. We introduce the *trace mapping*

$$T : \mathbb{F}_q \to \mathbb{F}_p, \quad T(\alpha) = \alpha + \alpha^p + \alpha^{p^2} + \cdots + \alpha^{p^{m-1}} \quad (\alpha \in \mathbb{F}_q).$$

It is well known that T is a surjective homomorphism from the additive group $\mathbb{F}_q$ to the additive group $\mathbb{F}_p$.

Theorem 2.19: *Let* $\zeta = e^{\frac{2\pi i}{p}}$ *and* $T\colon \mathbb{F}_q \to \mathbb{F}_p$ *be the trace mapping* $(q = p^m)$*. For each* $a = (a_1, a_2, \ldots, a_n) \in \mathbb{F}_q^n$ $(a_i \in \mathbb{F}_q)$*, the mapping*

$$\lambda_a \colon \ \mathbb{F}_q^n \to \{1, \zeta, \ldots, \zeta^{p-1}\} \subseteq \mathbb{C}^*$$

$$\lambda_a(b) = \lambda_a(b_1, \ldots, b_n) = \zeta^{T(a \cdot b)}$$

$$= \zeta^{T(a_1 b_1 + \cdots + a_n b_n)} \quad (b = (b_1, \ldots, b_n) \in \mathbb{F}_q^n)$$

is a character of the additive group $\mathbb{F}_q^n$*. Also* $\widehat{\mathbb{F}_q^n} = \{\lambda_a | a \in \mathbb{F}_q^n\}$*.*

Lemma 2.20: *Let* C *be an* $\mathbb{F}_q$*-linear subspace of* $\mathbb{F}_q^n$*. For* $v \in \mathbb{F}_q^n$*,*

$$\sum_{u \in C} \lambda_u(v) = \begin{cases} |C|, & \text{if } v \in C^{\perp}, \\ 0, & \text{otherwise.} \end{cases}$$

Now we can prove Theorem 2.16.

Proof: For each $u = (u_1, u_2, \ldots, u_n) \in \mathbb{F}_q^n$ $(u_i \in \mathbb{F}_q)$, we define a polynomial

$$g_u(z) = \sum_{v \in \mathbb{F}_q^n} \lambda_u(v) z^{w(v)} \in \mathbb{C}[z],$$

where $w(v) = w_H(v)$ is the Hamming weight of v and $\lambda_u(v) = \zeta^{T(u \cdot v)}$ is an additive character of $\mathbb{F}_q^n$. Then

$$g_u(z) = \sum_{v_1, \ldots, v_n \in \mathbb{F}_q} \zeta^{T(u_1 v_1 + \cdots + u_n v_n)} z^{w(v_1) + \cdots + w(v_n)}$$

$$\left(w(v_i) = \begin{cases} 0, & \text{if } v_i = 0, \\ 1, & \text{if } v_i \in \mathbb{F}_q^*. \end{cases} \right)$$

$$= \left(\sum_{v_1 \in \mathbb{F}_q} \zeta^{T(u_1 v_1)} z^{w(v_1)} \right) \cdots \left(\sum_{v_n \in \mathbb{F}_q} \zeta^{T(u_n v_n)} z^{w(v_n)} \right).$$

For $u = 0$ $(w(u) = 0)$ we have

$$\sum_{v \in \mathbb{F}_q} \zeta^{T(uv)} z^{w(v)} = \sum_{v \in \mathbb{F}_q} z^{w(v)} = 1 + (q-1)z, \tag{2}$$

and for $u \neq 0$ $(w(u) = 1)$ we have

$$\sum_{v \in \mathbb{F}_q} \zeta^{T(uv)} z^{w(v)} = 1 + \sum_{v \in \mathbb{F}_q^*} \zeta^{T(uv)} z^{w(v)} = 1 + \sum_{v \in \mathbb{F}_q^*} \zeta^{T(uv)} z$$

$$= 1 + \left(\sum_{v \in \mathbb{F}_q^*} \zeta^{T(v)} \right) z = 1 + \left(\sum_{v \in \mathbb{F}_q^*} \lambda_1(v) \right) z = 1 - z. \tag{3}$$

(2) and (3) can be put together as

$$g_u(z) = (1-z)^{w(u)}(1+(q-1)z)^{n-w(u)}.$$

Thus

$$\sum_{u \in C} g_u(z) = \sum_{u \in C} (1-z)^{w(u)}(1+(q-1)z)^{n-w(u)}$$

$$= \sum_{i=0}^{n} A_i (1+(q-1)z)^{n-i}(1-z)^i. \tag{4}$$

On the other hand,

$$\sum_{u \in C} g_u(z) = \sum_{u \in C} \sum_{v \in \mathbb{F}_q^n} \lambda_u(v) z^{w(v)} = \sum_{v \in \mathbb{F}_q^n} z^{w(v)} \sum_{u \in C} \lambda_u(v)$$

$$= q^k \sum_{v \in C^\perp} z^{w(v)} \quad \text{(by Lemma 2.20)}$$

$$= q^k \sum_{i=0}^{n} A_i^\perp z^i.$$

Therefore

$$f_{C^\perp}(x,y) = x^n \sum_{i=0}^{n} A_i^\perp \left(\frac{y}{x}\right)^i = x^n q^{-k} \sum_{u \in C} g_u\left(\frac{y}{x}\right)$$

$$= x^n q^{-k} \sum_{i=0}^{n} A_i (1+(q-1)\frac{y}{x})^{n-i}\left(1-\frac{y}{x}\right)^i \quad \text{(by (4))}$$

$$= q^{-k} \sum_{i=0}^{n} A_i (x+(q-1)y)^{n-i}(x-y)^i$$

$$= q^{-k} f_C(x+(q-1)y, x-y).$$

This completes the proof of Theorem 2.16. $\qquad\square$

3. Quantum Error-Correcting Codes

3.1. *Quantum bits and quantum codes*

In digital communication a bit is an element of a finite field $\mathbb{F}_q$. An n-bit is a vector in $\mathbb{F}_q^n$. In quantum communication a quantum bit (qubit) and n-qubit are objects in vector spaces over the complex number field $\mathbb{C}$.

Definition 3.1: A *qubit* is a non-zero vector in $\mathbb{C}^2$. Physicists usually denote a $\mathbb{C}$-basis of $\mathbb{C}^2$ by $|0>$ and $|1>$. Thus, a qubit $|v>$ can be expressed by

$$|v> = \alpha|0> + \beta|1> \quad (\alpha, \beta \in \mathbb{C},\ (\alpha, \beta) \neq (0,0)).$$

An *n-qubit* is a non-zero vector in the tensor product space $(\mathbb{C}^2)^{\otimes n} = \mathbb{C}^{2^n}$. We usually choose the following $\mathbb{C}$-basis of $\mathbb{C}^{2^n}$:

$$\{|a_0 a_1 \cdots a_{n-1}> = |a_0> \otimes |a_1> \otimes \cdots \otimes |a_{n-1}> : (a_0, \ldots, a_{n-1}) \in \mathbb{F}_2^n\}.$$

Thus, an n-qubit can be expressed by

$$(0 \neq)|v> = \sum_{(a_0, \ldots, a_{n-1}) \in \mathbb{F}_2^n} c_{a_0, \ldots, a_{n-1}} |a_0, \ldots, a_{n-1}> = \sum_{a \in \mathbb{F}_2^n} c_a |a>,$$

where $c_a \in \mathbb{C}$. We have the Hermitian inner product in $\mathbb{C}^{2^n}$: for

$$|v> = \sum_{a \in \mathbb{F}_2^n} c_a |a> \quad \text{and} \quad |u> = \sum_{a \in \mathbb{F}_2^n} b_a |a> \quad (c_a, b_a \in \mathbb{C}),$$

the *Hermitian inner product* of $|v>$ and $|u>$ is defined by

$$< u|v> = \sum_{a \in \mathbb{F}_2^n} \overline{b_a} c_a \in \mathbb{C}.$$

If $< u|v> = 0$, then $|v>$ and $|u>$ are called *orthogonal*.

There is one thing in quantum physics which is different from the classical case. In digital communication, two n-digits $u = (u_0, \ldots, u_{n-1})$ and $v = (v_0, \ldots, v_{n-1}) \in \mathbb{F}_q^n$ are *distinct* if $u \neq v$ (namely, if there is $i, 1 \leq i \leq n$, such that $u_i \neq v_i$). In the quantum case, for two n-qubits $|u>$ and $|v>$,

(I) If $|u> = \alpha|v>$ for some $\alpha \in \mathbb{C}^* = \mathbb{C} \backslash \{0\}$, $|u>$ and $|v>$ are considered as the *same quantum state*.

(II) $|u>$ and $|v>$ are *distinguishable* (= totally distinct) if $< u|v> = 0$.

Definition 3.2: A *quantum code* Q is just a subspace of $(\mathbb{C}^2)^{\otimes n} = \mathbb{C}^{2^n}$. The number n is called the *length* of the code Q. We denote $K = \dim_{\mathbb{C}} Q$ and $k = \log_2 K$, so that $k \leq n$.

3.2. *Quantum error group*

A quantum error on $\mathbb{C}^{2^n}$ is a unitary $\mathbb{C}$-linear operator acting on $\mathbb{C}^{2^n}$ bit by bit. Since each qubit $|v> = \alpha|0> + \beta|1>$ is a vector in $\mathbb{C}^2$, a quantum error can be expressed by a unitary matrix with respect to the basis $|0>$ and $|1>$. There are three basic errors (Pauli matrices) acting on a qubit:

$$\sigma_x = \begin{pmatrix} 0 & 1 \\ 1 & 0 \end{pmatrix}, \quad \sigma_x|v> = \beta|0> + \alpha|1>, \quad \sigma_x\begin{pmatrix} \alpha \\ \beta \end{pmatrix} = \begin{pmatrix} \beta \\ \alpha \end{pmatrix}, \quad (bit \ error)$$

$$\sigma_z = \begin{pmatrix} 1 & 0 \\ 0 & -1 \end{pmatrix}, \quad \sigma_z|v> = \alpha|0> - \beta|1>,$$

$$\sigma_z\begin{pmatrix} \alpha \\ \beta \end{pmatrix} = \begin{pmatrix} \alpha \\ -\beta \end{pmatrix}, \quad (phase \ error)$$

$$\sigma_y = i\sigma_x\sigma_z = \begin{pmatrix} 0 & -i \\ i & 0 \end{pmatrix}, \quad \sigma_y|v> = -i\beta|0> + i\alpha|1>, \quad \sigma_y\begin{pmatrix} \alpha \\ \beta \end{pmatrix} = \begin{pmatrix} -i\beta \\ i\alpha \end{pmatrix}$$

$$\left(I = \begin{pmatrix} 1 & 0 \\ 0 & 1 \end{pmatrix}, \quad I|v> = |v>, \quad no \ error \right).$$

Among them there are the following relations:

$$\sigma_x^2 = \sigma_z^2 = \sigma_y^2 = I, \quad \sigma_x\sigma_z = -\sigma_z\sigma_x.$$

A quantum error operation acting on $(\mathbb{C}^2)^{\otimes n}$ has the following form:

$$e = i^\lambda w_0 \otimes w_1 \otimes \cdots \otimes w_{n-1},$$

where $i = \sqrt{-1}$, $\lambda \in \{0, 1, 2, 3\}$, $w_j \in \{I, \sigma_x, \sigma_y, \sigma_z\}$, $(0 \le j \le n-1)$ is the $\mathbb{C}$-linear unitary operator acting on a basis element $|a> = |a_0> \otimes |a_1> \otimes \cdots \otimes |a_{n-1}>$ $(a = (a_0, \ldots, a_{n-1}) \in \mathbb{F}_2^n)$ of $\mathbb{C}^{2^n}$ by

$$e|a> = i^\lambda(w_0|a_0>) \otimes (w_1|a_1>) \otimes \cdots \otimes (w_{n-1}|a_{n-1}>).$$

The set of quantum error operators

$$E_n = \{i^\lambda w_0 \otimes w_1 \otimes \cdots \otimes w_{n-1} | 0 \le \lambda \le 3,$$

$$w_j \in \{I, \sigma_x, \sigma_y, \sigma_z\}, \ 0 \le j \le n-1\}$$

is a finite non-abelian group ($|E_n| = 4^{n+1}$), where for $e = i^\lambda w_0 \otimes w_1 \otimes \cdots \otimes w_{n-1}$ and $e' = i^{\lambda'} w_0' \otimes w_1' \otimes \cdots \otimes w_{n-1}'$,

$$ee' = i^{\lambda\lambda'}(w_0w_0') \otimes (w_1w_1') \otimes \cdots \otimes (w_{n-1}w_{n-1}').$$

Example 3.3: $(n = 2)$ For $e = I_2 \otimes \sigma_x$, $e' = \sigma_y \otimes \sigma_z$,

$$ee' = \sigma_y \otimes (\sigma_x\sigma_z) = i^3\sigma_y \otimes \sigma_y, \quad e'e = \sigma_y \otimes (\sigma_z\sigma_x) = i\sigma_y \otimes \sigma_y.$$

It is easy to see that the center of the group E_n is

$$C(E_n) = \{i^\lambda = i^\lambda(I_2 \otimes \cdots \otimes I_2) | 0 \leq \lambda \leq 3\}.$$

Let $\overline{E_n} = E_n/C(E_n)$. Then $|\overline{E_n}| = 4^n$. Since $ee' = \pm e'e$ for any $e, e' \in E_n$, we know that the quotient group $\overline{E_n}$ is an abelian 2-elementary group ($\bar{e}^2 = 1$ for each $\bar{e} \in \overline{E_n}$). Therefore $\overline{E_n}$ will be isomorphic to the additive group $\mathbb{F}_2^{2n}$.

Now we give $e = i^\lambda w_0 \otimes w_1 \otimes \cdots \otimes w_{n-1}$ a new expression

$$e = i^{\lambda+l} X_{(a)} Z_{(b)}, \quad (a = (a_0, \ldots, a_{n-1}), \ b = (b_0, \ldots, b_{n-1}) \in \mathbb{F}_2^n)$$

where

$$(a_i, b_i) = \begin{cases} (0,0), & \text{if } w_i = I_2, \\ (1,0), & \text{if } w_i = \sigma_x, \\ (0,1), & \text{if } w_i = \sigma_z, \\ (1,1), & \text{if } w_i = \sigma_y \quad (= i\sigma_x\sigma_z), \end{cases}$$

and $l = \#\{0 \leq i \leq n - 1 | w_i = \sigma_y\}$. Therefore

$$X_{(a)} = w_0' \otimes w_1' \otimes \cdots \otimes w_{n-1}', \quad w_i' = \begin{cases} I_2, & \text{if } a_i = 0, \\ \sigma_x, & \text{if } a_i = 1. \end{cases}$$

$$Z_{(b)} = w_0'' \otimes w_1'' \otimes \cdots \otimes w_{n-1}'', \quad w_i'' = \begin{cases} I_2, & \text{if } b_i = 0, \\ \sigma_z, & \text{if } b_i = 1. \end{cases}$$

From $\sigma_y = i\sigma_x\sigma_z$ we know that $i^\lambda w_0 \otimes w_1 \otimes \cdots \otimes w_{n-1} = i^{\lambda+l} X_{(a)} Z_{(b)}$. The following result shows the advantage of this new notation.

Lemma 3.4: (1) *The action of $X_{(a)}$ and $Z_{(b)}$ $(a, b \in \mathbb{F}_2^n)$ on the basis $|v> = |v_0 v_1 \cdots v_{n-1}>$ $(v \in \mathbb{F}_2^n)$ of $\mathbb{C}^{2^n}$ is given by*

$$X_{(a)}|v> = |a + v>, \quad Z_{(b)}|v> = (-1)^{b \cdot v}|v>.$$

(2) *For $e = i^\lambda X_{(a)} Z_{(b)}$ and $e' = i^{\lambda'} X_{(a')} Z_{(b')}$ in E_n,*

$$ee' = (-1)^{a \cdot b' + a' \cdot b} e'e.$$

Proof: (1) By definition $\sigma_x|0> = |1>$ and $\sigma_x|1> = |0>$. Since $X_{(a)} = w_0 \otimes w_1 \otimes \cdots \otimes w_{n-1}$, where

$$w_i = \begin{cases} I_2, & \text{if } a_i = 0, \\ \sigma_x, & \text{if } a_i = 1, \end{cases} \quad (0 \leq i \leq n - 1)$$

we know that $w_i|v_i> = |v_i + a_i>$ and $X_{(a)}|v> = |a + v>$.

Similarly $Z_{(b)} = w_0 \otimes w_1 \otimes \cdots \otimes w_{n-1}$, where

$$w_i = \begin{cases} I_2, & \text{if } b_i = 0, \\ \sigma_z, & \text{if } b_i = 1. \end{cases}$$

From $\sigma_z |0> = |0>$ and $\sigma_z |1> = -|1>$ we know that $w_i |v_i> = (-1)^{b_i v_i} |v_i>$ and $Z_{(b)} |v> = (-1)^{b \cdot v} |v>$.

(2) For each basis element $|v> (v \in \mathbb{F}_2^n)$,

$$\begin{aligned} ee'|v> &= i^{\lambda + \lambda'} X_{(a)} Z_{(b)} X_{(a')} Z_{(b')} |v> \\ &= (-1)^{b' \cdot v} i^{\lambda + \lambda'} X_{(a)} Z_{(b)} X_{(a')} |v> \\ &= (-1)^{b' \cdot v} i^{\lambda + \lambda'} X_{(a)} Z_{(b)} |v + a'> \\ &= (-1)^{b' \cdot v + b \cdot (v + a')} i^{\lambda + \lambda'} X_{(a)} |v + a'> \\ &= (-1)^{b' \cdot v + b \cdot (v + a')} i^{\lambda + \lambda'} |v + a + a'> . \end{aligned}$$

Similarly

$$e'e|v> = (-1)^{b \cdot v + b' \cdot (v + a)} i^{\lambda + \lambda'} |v + a + a'> .$$

Therefore

$$ee' = (-1)^{a \cdot b' + a' \cdot b} e'e. \qquad \square$$

For $e = i^{\lambda} X_{(a)} Z_{(b)} \in E_n$, the image of e in $\overline{E_n} = E_n / C(E_n)$ is denoted by

$$\bar{e} = (a|b) \quad (a, b \in \mathbb{F}_2^n).$$

Then for $e' = i^{\lambda'} X_{(a')} Z_{(b')}$, $\bar{e}' = (a'|b')$,

$$\bar{e}\bar{e}' = \overline{ee'} = \overline{i^{\lambda + \lambda'} (-1)^{a \cdot b' + a' \cdot b} X(a + a') Z(b + b')} = (a + a'|b + b')(= \bar{e}'\bar{e}).$$

This shows that $\overline{E_n}$ is naturally isomorphic to the additive group $\mathbb{F}_2^{2n}$. Moreover, by Lemma 3.4(2) we know that

$$ee' = e'e \Leftrightarrow (-1)^{a \cdot b' + a' \cdot b} = 1 \Leftrightarrow a \cdot b' + a' \cdot b = 0 \in \mathbb{F}_2. \tag{5}$$

We introduce the *symplectic inner product* $(,)_s$ in $\mathbb{F}_2^{2n}$: for $(a|b)$ and $(a'|b') \in \mathbb{F}_2^{2n}$,

$$((a|b), (a'|b'))_s = a \cdot b' + a' \cdot b = (a, b) \begin{pmatrix} 0 & I_n \\ I_n & 0 \end{pmatrix} \begin{pmatrix} a' \\ b' \end{pmatrix} \in \mathbb{F}_2.$$

For a subspace C of $\mathbb{F}_2^{2n}$, the *symplectic dual* of C is defined as the subspace

$$C_s^{\perp} = \{v \in \mathbb{F}_2^{2n} | (v, c)_s = 0 \text{ for all } c \in C\}.$$

We have $\dim_{\mathbb{F}_2} C_s^{\perp} + \dim_{\mathbb{F}_2} C = 2n$ and $(C_s^{\perp})^{\perp} = C$. If $C \subseteq C_s^{\perp}$, this subspace is called *symplectic self-orthogonal.* From formula (5) we have the following result.

Lemma 3.5: *Let S be a subgroup of E_n. Then S is abelian if and only if $\bar{S}$ is a symplectic self-orthogonal subspace of $\overline{E_n} = \mathbb{F}_2^{2n}$.*

3.3. *Quantum error correction*

Definition 3.6: The *quantum weight* of an error

$$e = i^{\lambda} w_0 \otimes w_1 \otimes \cdots \otimes w_{n-1} = i^{\lambda'} X_{(a)} Z_{(b)}, \quad \bar{e} = (a|b),$$

$$(w_i \in \{I, \sigma_x, \sigma_y, \sigma_z\}, \ a = (a_0, \ldots, a_{n-1}), \ b = (b_0, \ldots, b_{n-1}) \in \mathbb{F}_2^n)$$

is defined by

$$w_Q(e) = w_Q(\bar{e}) = \sharp\{0 \leq i \leq n - 1 | a_i = 1 \text{ or } b_i = 1\}$$

$$= \sharp\{0 \leq i \leq n - 1 | w_i \neq I_2\},$$

which is just the number of qubits having errors. For $0 \leq l \leq n$, let

$$E_n(l) = \{e \in E_n | W_Q(e) \leq l\}, \quad \overline{E_n}(l) = \{\bar{e} \in \overline{E_n} | W_Q(\bar{e}) \leq l\}.$$

Then $|E_n(l)| = 4 \cdot \sum_{i=0}^{l} 3^i \binom{n}{i}$, $|\overline{E_n}(l)| = \sum_{i=0}^{l} 3^i \binom{n}{i}$.

Now we are ready to describe the ability to correct quantum errors for a quantum code $Q \subseteq \mathbb{C}^{2^n}$. In the classical case, both the error $e = (e_1, \ldots, e_n)$ and the codeword $c = (c_1, \ldots, c_n)$ are objects in $\mathbb{F}_q^n$; an error e acts on a codeword by addition $e + a \in \mathbb{F}_q^n$. For a code C in $\mathbb{F}_q^n$, C can correct l errors means that C satisfies the following condition:

If $c_1, c_2 \in C, c_1 \neq c_2$ (distinct), and $e_1, e_2 \in \mathbb{F}_q^n, w_H(e_1) \leq l, w_H(e_2) \leq l$, then $c_1 + e_1 \neq c_2 + e_2$ (distinct).

For the quantum case, the quantum code $Q \subseteq \mathbb{C}^{2^n}$ and the quantum error $e \in E_n$ are different types of objects. Note that e is a unitary $\mathbb{C}$-linear operator acting on $|v> \in Q$ by $e|v> \in \mathbb{C}^{2^n}$. Remember that the concept "distinct" in the classical case corresponds to the concept "distinguishable" in the quantum case. We have the following definition.

Definition 3.7: A quantum code $Q \subseteq \mathbb{C}^{2^n}$ can *correct* $\leq l$ *quantum errors* $(0 \leq l \leq n)$ means that Q satisfies the following condition:

If $|v_1>, |v_2> \in Q$, $< v_1|v_2 > = 0$ (distinguishable), and $e_1, e_2 \in E_n(l)$, then $< v_1|e_1e_2|v_2 > = 0$ ($e_1|v_1 >$ and $e_2|v_2 >$ are distinguishable).

A quantum code Q is said to have *minimal distance* d if Q satisfies the following condition:

If $|v_1>, |v_2> \in Q$, $< v_1|v_2 > = 0$, and $e \in E_n(d-1)$, then $< v_1|e|v_2 > = 0$.

Since $w_Q(ee') \le w_Q(e) + w_Q(e')$ for any $e, e' \in E_n$, it is obvious from the definitions that we have the following result.

Lemma 3.8: *Let d be the minimal distance of a quantum code Q, $l = [\frac{d-1}{2}]$. Then Q can correct $\le l$ quantum errors.*

Similar to the classical case, a quantum code Q in $\mathbb{C}^{2^n}$ has three basic parameters: the length n, dimension $K = \dim_{\mathbb{C}} Q$ (or $k = \log_2 K$) and the minimal distance d. We denote this quantum code Q by $((n, K, d))$ or $[[n, k, d]]$. We also have bounds related to the parameters of a quantum code Q which are analogues of the classical Hamming bound and Singleton bound.

Before introducing the quantum Hamming bound, one difference between classical and quantum cases should be indicated. Let C be a classical code $(n, K, d)_q$. Namely, C is a subset of $\mathbb{F}_q^n$ with $|C| = K$ and minimal distance d. For $l = [\frac{d-1}{2}]$, the sets $e + C$ $(w_H(e) \le l)$ are disjoint. Therefore we obtain the Hamming bound

$$q^n = |\mathbb{F}_q^n| \ge \sum_{\substack{e \\ w_H(e) \le l}} |e + C| = |C| \cdot \sum_{\substack{e \\ w_H(e) \le l}} 1 = K \cdot \sum_{i=0}^{l} (q-1)^i \binom{n}{i}.$$

But in the quantum case, for a quantum code Q with parameters $((n, K, d))$, eQ is a subspace of $\mathbb{C}^{2^n}$ for any quantum error $e \in E_n$ since Q is a subspace and e is a linear operator of $\mathbb{C}^{2^n}$. Now Q can correct $\le l$ quantum errors, but this does not imply that the subspaces $\{\bar{e}Q\}(w_H(\bar{e}) \le l)$ are orthogonal to each other. The reason is that even for $e \ne 1$ $(w_Q(e) \ge 1)$ a codeword $|v> \in Q$ can be fixed by e: $e|v> = \alpha|v>$ $(\alpha \in \mathbb{C}^*)$ ($|v>$ is an eigenvector of e) so that $eQ \cap Q \ne (0)$.

Definition 3.9: Let Q be a quantum code with minimal distance d and put $l = [\frac{d-1}{2}]$. If the subspaces $\bar{e}Q$ $(w_H(\bar{e}) \le l)$ of $\mathbb{C}^{2^n}$ are orthogonal to each other, then Q is called *pure*.

Lemma 3.10: (*Quantum Hamming bound*) *For a pure quantum code* $((n, K, d))$, $l = [\frac{d-1}{2}]$, *we have*

$$2^n \geq K \sum_{i=0}^{l} 3^i \binom{n}{i}.$$

Proof:

$$2^n = \dim_{\mathbb{C}} \mathbb{C}^{2^n} \geq \sum_{\bar{e} \in \overline{E_n}(l)} \dim(\bar{e}Q) = \dim Q \cdot |\overline{E_n}(l)| = K \cdot \sum_{i=0}^{l} 3^i \binom{n}{i}. \quad \square$$

A quantum code is called *perfect* if it meets the quantum Hamming bound. We will introduce the quantum Singleton bound in Section 5.

4. Basic Methods to Construct Quantum Codes

4.1. *Stabilizer quantum codes*

In this section we describe the mathematical method to construct quantum codes introduced in [6] and its variations. The idea of the method is as follows. Firstly we choose an abelian subgroup G of E_n. Such an abelian subgroup G comes from a symplectic self-orthogonal code C in $\mathbb{F}_2^{2n}$ ($C \subseteq C_s^\perp$) by Lemma 3.5. Then the action of G on $\mathbb{C}^{2^n}$ makes $\mathbb{C}^{2^n}$ a direct sum of χ-eigenspaces $Q(\chi)$ ($\chi \in \hat{G}$). For each subspace $Q = Q(\chi)$, $k = n - \log_2 |G|$ and its minimal distance can be determined by certain properties of C and $C_s^\perp$. In this way, a nice quantum code Q can be obtained from a classical code C.

We need a little representation theory of a finite abelian group. Let G be a finite abelian group acting on a finite-dimensional $\mathbb{C}$-vector space V. Each $g \in G$ is a $\mathbb{C}$-linear operator of V and for $g, g' \in G$,

$$(gg')(v) = g(g'v), \quad gg^{-1}(v) = 1(v) \quad (v \in V),$$

where $1 \in G$ is the identity operator on V.

Let $\hat{G}$ be the character group of G. For each character $\chi \in \hat{G}$, we define a linear operator over V by

$$e_\chi = \frac{1}{|G|} \sum_{g \in G} \bar{\chi}(g) g.$$

Usually $\{e_\chi \mid \chi \in \hat{G}\}$ is called the *system of orthogonal primitive idempotent operators*. Their orthogonal and idempotent properties can be justified by the following lemma.

Lemma 4.1: (1) $e_\chi^2 = e_\chi$ and $e_\chi e_{\chi'} = 0$ if $\chi \neq \chi'$.

(2) $1 = \sum\limits_{\chi \in \hat{G}} e_\chi$.

(3) *For any* $g \in G$, $g e_\chi = \chi(g) e_\chi$.

Proof: (1) By the orthogonal property of characters and $\bar{\chi} = \chi^{-1}$,

$$e_\chi e_{\chi'} = \frac{1}{|G|^2} \sum_{g \in G} \bar{\chi}(g)g \cdot \sum_{h \in G} \overline{\chi'}(h)h \quad (\text{let } gh = a)$$

$$= \frac{1}{|G|^2} \sum_{a,g \in G} \bar{\chi}(g)\overline{\chi'}(ag^{-1})a$$

$$= \frac{1}{|G|^2} \sum_{a \in G} \overline{\chi'}(a)a \cdot \sum_{g \in G} (\bar{\chi}\chi')(g)$$

$$= \begin{cases} 0, & \text{if } \bar{\chi}\chi' \neq \chi_0 (\Leftrightarrow \chi \neq \chi'), \\ \frac{1}{|G|} \sum_{a \in G} \bar{\chi}(a)a = e_\chi, & \text{if } \chi = \chi'. \end{cases}$$

(2) $\sum\limits_{\chi \in \hat{G}} e_\chi = \frac{1}{|G|} \sum\limits_{\chi \in \hat{G}} \sum\limits_{g \in G} \bar{\chi}(g)g = \frac{1}{|G|} \sum\limits_{g \in G} g \sum\limits_{\chi \in \hat{G}} \bar{\chi}(g) = 1$.

(3) $g e_\chi = \sum\limits_{h \in G} \bar{\chi}(h)gh = \sum\limits_{a \in G} \bar{\chi}(ag^{-1})a = \chi(g) \sum\limits_{a \in G} \bar{\chi}(a)a = \chi(g)e_\chi$. $\qquad\square$

Lemma 4.2: *Suppose that a finite abelian group G acts on a finite-dimensional complex vector space V such that each $y \in G$ is a Hermitian linear operator. For each $\chi \in \hat{G}$ we denote*

$$V(\chi) = e_\chi V = \{e_\chi(v) | v \in V\}.$$

(1) *For $v \in V(\chi)$, $g \in G$ we have*

$$gv = \chi(g)v.$$

Therefore $V(\chi)$ is a common eigenspace for all Hermitian operators $g \in G$.

(?) *We have the following direct decomposition of V:*

$$V = \bigoplus_{\chi \in \hat{G}} V(\chi).$$

This means that each $v \in V$ has unique expression $v = \sum\limits_{\chi \in \hat{G}} v(\chi)$, $v(\chi) \in V(\chi)$. Moreover, if all $g \in G$ are unitary linear operators on V, then $V(\chi)$ and $V(\chi')$ are hermitian-orthogonal for all $\chi, \chi' \in \hat{G}$, $\chi \neq \chi'$.

Proof: (1) For $v \in V(\chi)$, $g \in G$, we have

$$gv = ge_\chi(w) \quad \text{(for some } w \in V\text{)}$$
$$= \chi(g)e_\chi(w) = \chi(g)v.$$

(2) For each $v \in V$,

$$v = \left(\sum_{\chi \in \hat{G}} e_\chi \right) v = \sum_{\chi \in \hat{G}} v_\chi,$$

where $v_\chi = e_\chi v \in V(\chi)$. On the other hand, if we have $v = \sum_{\chi \in \hat{G}} u_\chi$, $u_\chi \in V(\chi)$, then $u_\chi = e_\chi w_\chi$ for some $w_\chi \in V$. By the orthogonal property of $\{e_\chi\}$ we know that for each $\chi \in \hat{G}$,

$$v_\chi = e_\chi v = e_\chi \left(\sum_{\chi' \in \hat{G}} u_{\chi'} \right) = e_\chi \left(\sum_{\chi' \in \hat{G}} e_{\chi'} w_{\chi'} \right)$$

$$= \sum_{\chi' \in \hat{G}} (e_\chi e_{\chi'} w_{\chi'}) = e_\chi w_\chi = u_\chi.$$

Therefore we proved that $V = \bigoplus_{\chi \in \hat{G}} V(\chi)$. The last statement is a well-known fact in linear algebra. □

Let us go back to the case that G is an abelian subgroup of E and $G \cong (\mathbb{F}_2^k, +)$. Then G has a basis:

$$\{g_j = i^{\lambda_j} X(a_j) Z(b_j) | 1 \leq j \leq k\} \quad (a_j, b_j \in \mathbb{F}_2^n),$$

where $\lambda_j \in \mathbb{Z}_4$ such that $\lambda_j \equiv a_j b_j \pmod 2$. Since $\sigma_x = \begin{pmatrix} 0 & 1 \\ 1 & 0 \end{pmatrix}$ and $\sigma_z = \begin{pmatrix} 1 & 0 \\ 0 & -1 \end{pmatrix}$ are Hermitian, we know that $X(a_j)$ and $Z(b_j)$ are Hermitian, so that

$$g_j^\dagger \triangleq \bar{g}_j^T = (-i)^{\lambda_j} Z(b_j)^T X(a_j)^T = (-i)^{\lambda_j} Z(b_j) X(a_j)$$

$$= i^{\lambda_j} (-1)^{a_j b_j} \cdot (-1)^{a_j b_j} X(a_j) Z(b_j)$$

$$= i^{\lambda_j} X(a_j) Z(b_j) = g_j.$$

This means that g_j $(1 \leq j \leq k)$ are Hermitian. Since G is an abelian group generated by g_j $(1 \leq j \leq k)$, we know that all $g \in G$ are Hermitian linear operators acting on $\mathbb{C}^{2^n}$. By Lemma 4.2, we have the direct decomposition

$$\mathbb{C}^{2^n} = \bigoplus_{\chi \in \hat{G}} Q(\chi),$$

where $Q(\chi)$ is the χ-eigenspace of G:

$$Q(\chi) = e_\chi \mathbb{C}^{2^n} = \{v \in \mathbb{C}^{2^n} | gv = \chi(g)v \text{ for all } g \in G\}.$$

After these preparations, we can now state and prove the most important result in these lecture notes.

Theorem 4.3: *Suppose that C is a self-orthogonal subspace of $\mathbb{F}_2^{2n}$ with respect to the symplectic inner product $(\ ,\ )_s$, $\dim C = k$. Let*

$$C_s^\perp = \{u \in \mathbb{F}_2^{2n} | <u, v> = 0 \quad \text{for all } v \in C\},$$

$$d = w_Q(C_s^\perp \backslash C) = \min\{w_Q(v)|v \in C_s^\perp \backslash C\}.$$

Then there exists an $[[n, n-k, d]]$ quantum code.

Proof: As we have shown before, $\bar{G} = C$ can be lifted to an abelian subgroup G of E and $G \cong \mathbb{F}_2^k$. Then $\mathbb{C}^{2^n} = \bigoplus_{\chi \in \hat{G}} Q(\chi)$, where

$$Q(\chi) = e_\chi \mathbb{C}^{2^n} = \{v \in \mathbb{C}^{2^n} | gv = \chi(g)v \quad \text{for all } g \in G\}.$$

We want to show that each $Q(\chi)$ is an $[[n, n-k, d]]$ quantum code.

First of all, we show $\dim_\mathbb{C} Q(\chi) = 2^{n-k}$ for any $\chi \in \hat{G}$. For doing this, we consider the actions of E on the set $\{Q(\chi)|\chi \in \hat{G}\}$. For any $|v> \in Q(\chi)$ and $g \in G$, $g|v> = \chi(g)|v>$, $(\chi(g) \in C)$. Thus for any $e \in E$,

$$g|e|v> = ge|v> = (-1)^{(\bar{g}, \bar{e})_s} eg|v> = (-1)^{(\bar{g}, \bar{e})_s} \chi(g)e|v \quad (\forall g \in G).$$

It is easy to see that

$$\chi_{\bar{e}} : G \rightarrow \{\pm 1\} \subseteq \mathbb{C}^*, \quad \chi_{\bar{e}}(g) = (-1)^{(\bar{g}, \bar{e})_s}$$

is a character of G. Thus we have

$$g|e|v> = \chi_{\bar{e}}(g)\chi(g)e|v> = \chi'(g)e|v> \quad (\forall g \in G),$$

which means that $e|v> \in Q(\chi')$ and $e: Q(\chi) \rightarrow Q(\chi')$, where $\chi' = \chi_{\bar{e}} \cdot \chi$. Since E is a group, we know that e is bijective, so that $\dim_\mathbb{C} Q(\chi) = \dim_\mathbb{C} Q(\chi')$ When e runs through E, $\chi_{\bar{e}}$ can take all characters of G. This implies that all $\dim_\mathbb{C} Q(\chi)$ $(\chi \in \hat{G})$ are the same. Let N be this common dimension, then

$$2^n = \dim \mathbb{C}^{2^n} = \sum_{\chi \in \hat{G}} \dim Q(\chi) = \sum_{\chi \in \hat{G}} N = N \cdot |\hat{G}| = N|G| = N \cdot 2^k,$$

therefore $\dim Q(\chi) = N = 2^{n-k}$ for all $\chi \in \hat{G}$.

Next we show $d(Q(\chi)) \geq w_Q(C_s^\perp \setminus C)$. For doing this, we need to show that if $e \in E_{d-1}$, $v_1, v_2 \in Q(\chi)$, $<v_1|v_2> = 0$, then $<v_1|e|v_2> = 0$.

(1) If $\bar{e} \in \bar{G} = C$, then $<v_1|e|v_2> = \chi(\bar{e})<v_1|v_2> = 0$.

(2) Otherwise, $\bar{e} \notin C$. From $w_Q(\bar{e}) = w_Q(e) \leq d-1$ and the assumption $(C_s^\perp \setminus C) \cap \overline{E_{d-1}} = \emptyset$, we know that $\bar{e} \notin C_s^\perp$, which means that there exists $\bar{e}' \in \bar{G}$ such that $ee' = -e'e$. Then for $v_2 \in Q(\chi)$,

$$e'e|v_2> = -ee'|v_2> = -\chi(e')e|v_2>, \quad -\chi(e') \neq \chi(e').$$

Therefore $e|v_2 > \in Q(\chi')$, where $\chi' \neq \chi$. Since $v_1 \in Q(\chi)$ and $Q(\chi)$ is orthogonal to $Q(\chi')$, we have $<v_1|e|v_2> = 0$. This completes the proof of the theorem. $\qquad \square$

Definition 4.4: The quantum codes constructed by Theorem 4.3 are called *stabilizer* or *additive quantum codes*.

It is easy to see that such a quantum code is pure (Definition 3.9) if $C \cap \overline{E_{d-1}} = (0)$.

Example 4.5: For $n = 5$, $k = 1$, $d = 3$, we have $2^{n-k} = 16 = (1 + 3 \cdot 5) = \sum_{i=0}^{1} 3^i \binom{5}{i}$. Therefore a $[[5,1,3]]$ additive quantum code is perfect if it exists. Such a quantum code has been found by Calderbank and Shor [7], Gottesman [15] and Laflamme et al. [24] as the first perfect quantum code in the world.

Consider a subspace (= classical binary linear code) C of $\mathbb{F}_2^{10}$ with generating matrix:

$$
\begin{bmatrix}
1 & 1 & 0 & 0 & 0 & 0 & 0 & 1 & 0 & 1 \\
0 & 1 & 1 & 0 & 0 & 1 & 0 & 0 & 1 & 0 \\
0 & 0 & 1 & 1 & 0 & 0 & 1 & 0 & 0 & 1 \\
0 & 0 & 0 & 1 & 1 & 1 & 0 & 1 & 0 & 0
\end{bmatrix}
=
\begin{bmatrix}
v_1 \\
v_2 \\
v_3 \\
v_4
\end{bmatrix},
\tag{6}
$$

where, for example, the first row vector v_1 means $(a|b)$ with $a = (11000)$, $b = (00101)$. It can be checked that v_1, v_2, v_3, v_4 are linearly independent over $\mathbb{F}_2$ and orthogonal with respect to the symplectic inner product $(\ ,\ )_s$. Therefore $\dim C = 4$ and C is self-orthogonal. We have $C_s^\perp \supseteq C$ and $\dim C_s^\perp = 10 - 4 = 6$, so we need two more vectors in $C_s^\perp$ to make a basis of $C_s^\perp$ with v_1, v_2, v_3, v_4. In this way we find

$$C_s^\perp = C \oplus (00000|11111)\mathbb{F}_2 \oplus (11111|00000)\mathbb{F}_2.$$

It can be calculated that $w_Q(C) = 4$ and $w_Q(C_s^\perp) = 3$ ($(11000|00101) + (11111|00000) = (00111|00101) \in C_s^\perp$ and $w_Q(00111|00101) = 3$). Therefore $w_Q(C_s^\perp \backslash C) = 3$ and the quantum code $Q(\chi)$ constructed by C is pure. Since $2n = 10$, $k = n - \dim C = 5 - 4 = 1$, $d = 3$. We know that $Q(\chi)$ is an $[[n, k, d]] = [[5, 1, 3]]$ quantum code.

Let us write down the code $Q = Q(\chi_0)$ explicitly, where χ_0 is the trivial character of G. Recall that G is generated by

$$g_j = i^{\lambda_j} X(a_j) Z(b_j) \quad (1 \leq j \leq 4).$$

While $v_j = (a_j|b_j)$, $\lambda_j \in \mathbb{Z}_4$ and $\lambda_j \equiv a_j \cdot b_j \pmod{2}$. It is easy to see from (6) that $a_j \cdot b_j = 0 \in \mathbb{F}_2$ ($1 \leq j \leq 4$). So we may choose $\lambda_j = 0$ for all j, $1 \leq j \leq 4$. Therefore

$$G = \langle g_1, g_2, g_3, g_4 \rangle$$

$$= \{1, g_2, g_3, g_4, g_1 g_2, \cdots, g_3 g_4, g_1 g_2 g_3, \cdots, g_2 g_3 g_4, g_1 g_2 g_3 g_4\}, \quad |G| = 16.$$

$$g_1 = X(11000) Z(00101) = \sigma_x \otimes \sigma_x \otimes \sigma_z \otimes I_2 \otimes \sigma_z.$$

$$g_2 = X(01100) Z(10010) = \sigma_z \otimes \sigma_x \otimes \sigma_x \otimes \sigma_z \otimes I_2.$$

$$g_3 = \cdots, \quad g_4 = \cdots.$$

Since $k = 1$ and $\dim_\mathbb{C} Q = 2^k = 2$, we need to find two vectors in $\mathbb{C}^{2^n} = \mathbb{C}^{32}$ as a basis of Q. We have

$$Q = Q(\chi_0) = \{v \in \mathbb{C}^{32} | gv = \chi_0(g)v = v \text{ for all } g \in G\},$$

which means that Q consists of the fixed vector by all $g \in G$. It is natural to consider the "average" of a vector $|v\rangle \in \mathbb{C}^{32}$,

$$|\overline{v}\rangle = \sum_{h \in G} h|v\rangle .$$

Since for each $g \in G$, $g|\overline{v}\rangle = g \sum_{h \in G} h|v\rangle = \sum_{h \in G} (gh)|v\rangle = \sum_{h \in G} h|v\rangle = |\overline{v}\rangle$, thus $|\overline{v}\rangle \in Q$. Then we try $|v\rangle = |00000\rangle$ and $|11111\rangle$ to compute

$$|v_0\rangle = \sum_{h \in G} h|00000\rangle$$

$$= X(00000) Z(00000)|00000\rangle + \sum_{j=1}^{4} g_j|00000\rangle + \sum_{1 \leq i < j \leq 4} g_i g_j|00000\rangle$$

$$+ \sum_{1 \leq i < j < k \leq 4} g_i g_j g_k|00000\rangle + g_1 g_2 g_3 g_4|00000\rangle$$

$$= |00000> + (|11000> + |01100> + |00110> + |00011> + |10001>)$$
$$- (|10100> + |01010> + |00101> + |10010> + |01001>)$$
$$+ (|11110> + |01111> + |10111> + |11011> + |11101>),$$

by using $X(a)|v> = |v+a>$ and $Z(b)|v> = (-1)^{b \cdot v}|v>$. For example:

$$g_1 g_2 = X(11000)Z(00101)X(01100)Z(10010)$$
$$= X(11000)X(01100)Z(00101)Z(10010) \cdot (-1)^{(00101) \cdot (01100)}$$
$$= -X(10100)Z(10111).$$

Therefore

$$g_1 g_2 |00000> = -X(10100)Z(10111)|00000> = -X(10100)|00000>$$
$$= -|10100>.$$

$$|v_1> = \sum_{h \in G} h|11111>$$
$$= |11111> + (|00111> + |10011> + |11001> + |11100> + |01110>)$$
$$- (|01011> + |10101> + |11010> + |01101> + |10110>)$$
$$- (|00001> + |10000> + |01000> + |00100> + |00010>).$$

$|v_0>$ and $|v_1>$ are independent, therefore the quantum code Q generated by $|v_0>$ and $|v_1>$ is 2-dimensional and can correct any error (σ_x, σ_z, or σ_y) in any one qubit of its five qubits.

4.2. *From classical codes to quantum codes*

Theorem 4.3 presents a general principle to construct quantum codes from χ-eigenspaces of an abelian subgroup G of E acting on $\mathbb{C}^{2^n}$. Such a group G comes from a binary self-orthogonal linear code C of the symplectic space $\mathbb{F}_2^{2n}$. Therefore Theorem 4.3 establishes a connection between classical binary codes and quantum codes. In this subsection we specify this general connection to give several concrete methods which make it possible to construct quantum codes from known classical binary codes.

Theorem 4.6: ([6]) *Suppose there exists a binary $[n, k, d]$ linear code C with $C \supseteq C^\perp$, where $C^\perp$ is the dual code of C with respect to the ordinary inner product in $\mathbb{F}_2^n$. Then we get an additive $[[n, 2k-n, d]]$ quantum code.*

Proof: From $\dim_{\mathbb{F}_2} C = k$, $\dim_{\mathbb{F}_2} C^\perp = n - k$ and $C \supseteq C^\perp$, we know that $k \geq \frac{n}{2}$. Consider the subspace $\bar{G}$ of $\mathbb{F}_2^{2n}$ defined by

$$\bar{G} = \{(v_1|v_2) \in \bar{E} = \mathbb{F}_2^{2n} | v_1, v_2 \in C^\perp\}$$
$$= C^\perp \oplus C^\perp.$$

So $\dim_{\mathbb{F}_2} \bar{G} = 2(n - k)$.

Then it is easy to see that

$$\bar{G}_s^\perp = \{(v_1, v_2) \in \bar{E} | v_1, v_2 \in C\}, \quad \dim_{\mathbb{F}_2} \bar{G}_s^\perp = 2k,$$

and $\bar{G} \subseteq \bar{G}_s^\perp$. From Theorem 4.3 we can construct an $[[n, k', d']]$ quantum code, where $k' = n - 2(n-k) = 2k - n$ and $d' \geq w_Q(\bar{G}_s^\perp \backslash \bar{G}) = w_H(C \backslash C^\perp) \geq d$. $\qquad \square$

Example 4.7: (Quantum Hamming codes) Consider the binary Hamming code C_3 in Section 2, Example 2.11. The generating matrix of C_3 can be rewritten as

$$M = \begin{bmatrix} 1110100 \\ 0111010 \\ 0011101 \end{bmatrix}.$$

C_3 is a binary $[7, 3, 4]$ linear code. The binary linear code C_3' having generating matrix

$$M' = \begin{bmatrix} 0 & \\ 0 & M \\ 0 & \\ 1 & 1111111 \end{bmatrix}$$

is called an *extended Hamming code*. C_3' is an $[8, 4, 4]$ self-dual code ($C_3' = C_3'^\perp$). By Theorem 4.6 we get an $[[8, 0, 4]]$ quantum code Q. Since $\dim_{\mathbb{C}} Q = 2^k = 1$, the quantum code Q has essentially only one quantum code word in $\mathbb{C}^{2^8} = \mathbb{C}^{256}$ which can correct one quantum error (σ_x, σ_z, or σ_y) in any qubit of its eight qubits. This quantum code Q is not good. We modify it by choosing the linear code $\bar{G}$ in $\mathbb{F}_2^{16}$ with generating matrix

$$\begin{bmatrix} 01110100 & 00111010 \\ 00111010 & 00011101 \\ 00011101 & 01001110 \\ 11111111 & 00000000 \\ 00000000 & 11111111 \end{bmatrix}.$$

This time we have $n = 8$, $k = \dim_{\mathbb{F}_2} \bar{G} = 5$, $\bar{G} \subseteq \bar{G}_s^{\perp}$ and $w_Q(\bar{G}_s^{\perp} \backslash \bar{G}) = w_Q(\bar{G}_s^{\perp}) = 3$. By Theorem 4.3 we get a pure $[[8, 3, 3]]$ quantum code Q'. The quantum code Q' is good because we have no $[[7, 3, 3]]$ quantum code by the quantum Hamming bound.

Theorem 4.8: ([6]) *If we have binary linear codes* $C_1 = [n, k_1, d_1]$, $C_2 = [n, k_2, d_2]$ *and* $C_1^{\perp} \subseteq C_2$ *(so that* $n \leq k_1 + k_2$*), then we get an additive quantum code* $[[n, k_1 + k_2 - n, \min\{d_1, d_2\}]]$.

Proof: Let G_i and H_i be generating and parity-check matrices of the linear code C_i $(i = 1, 2)$. Consider the linear code C of $\mathbb{F}_2^{2n}$ with generating matrix

$$\begin{pmatrix} H_1 & 0 \\ 0 & H_2 \end{pmatrix} \begin{matrix} n - k_1 \\ n - k_2 \end{matrix}.$$

The assumption $C_1^{\perp} \subseteq C_2$ implies that $H_1 H_2^{\top} = 0$, and $C_s^{\perp}$ has parity-check matrix $\begin{pmatrix} H_2 & 0 \\ 0 & H_1 \end{pmatrix}$ and generating matrix $\begin{pmatrix} G_2 & 0 \\ 0 & G_1 \end{pmatrix}$. From these we know that $C \subseteq C_s^{\perp}$ and $w_Q(C_s^{\perp} \backslash C) \geq \min\{d_1, d_2\}$. By Theorem 4.3 we get an $[[n, k', \min\{d_1, d_2\}]]$ quantum code, where $k' = n - \dim C = n - (n - k_1 + n - k_2) = k_1 + k_2 - n$. $\qquad \square$

Using more technical skills, Steane [41] improved Theorem 4.8 to obtain better quantum codes. Based on Theorem 4.8 and Steane's improvement, several authors, using classical Reed-Muller codes [43], BCH codes [18] and algebraic-geometry codes [9], constructed a series of quantum codes with better parameters than some quantum codes listed in the table of [6].

4.3. *Classical codes over* $\mathbb{F}_4$ *and quantum codes*

In this subsection we introduce how to construct quantum codes from classical codes over $\mathbb{F}_4$.

The finite field $\mathbb{F}_4$ has four elements: $0, 1, \omega$, and $\bar{\omega} = \omega^2 = \omega + 1$. Consider the following mapping:

$$\Phi : \bar{E} = \mathbb{F}_2^{2n} \rightarrow \mathbb{F}_4^n,$$

$$v = (a|b) = (a_0, \ldots, a_{n-1}|b_0, \ldots, b_{n-1}) \longmapsto$$

$$\Phi(v) = \omega a + \bar{\omega} b = (\omega a_0 + \bar{\omega} b_0, \ldots, \omega a_{n-1} + \bar{\omega} b_{n-1}).$$

Since $\{\omega, \bar{\omega}\}$ is a basis of $\mathbb{F}_4$ over $\mathbb{F}_2$, we know that Φ is an isomorphism of $\mathbb{F}_2$-vector spaces.

Lemma 4.9: (1) Φ *keeps weights:* $w_Q(v) = w_H(\Phi(v))$ *for* $v \in \mathbb{F}_2^{2n} = \bar{E}$.

(2) Φ *keeps inner products:* $(v, v')_s =< \Phi(v), \Phi(v') >_{th}$ *for all* $v, v' \in \bar{E}$, *where for* $\alpha = (\alpha_0, \dots, \alpha_{n-1})$, $\beta = (\beta_0, \dots, \beta_{n-1}) \in \mathbb{F}_4^n$, $< \alpha, \beta >_{th}$ *is the trace-hermitian inner product defined by*

$$< \alpha, \beta >_{th}= T_r \left(\sum_{i=0}^{n-1} \alpha_i \bar{\beta}_i \right) = T_r \left(\sum_{i=0}^{n-1} \alpha_i \beta_i^2 \right) = \sum_{i=0}^{n-1} (\alpha_i \beta_i^2 + \alpha_i^2 \beta_i).$$

Proof: (1) Let $v = (a|b)$, $v' = (a'|b')$, $a = (a_0, \dots, a_{n-1})$, $b = (b_0, \dots, b_{n-1})$, $a' = (a'_0, \dots, a'_{n-1})$, $b' = (b'_0, \dots, b'_{n-1}) \in \mathbb{F}_2^n$. Then $\Phi(v) = (\omega a_0 + \bar{\omega} b_0, \dots, \omega a_{n-1} + \bar{\omega} b_{n-1})$. Since $\{\omega, \bar{\omega}\}$ is a basis of $\mathbb{F}_4$ over $\mathbb{F}_2$, we know that

$$w_Q(a_i|b_i) = 0 \Leftrightarrow (a_i, b_i) = (0,0) \Leftrightarrow \omega a_i + \bar{\omega} b_i = 0 \Leftrightarrow w_H(\omega a_i + \bar{\omega} b_i) = 0.$$

Therefore $w_Q(v) = w_H(\Phi(v))$.

(2)

$$< \Phi(v), \Phi(v') >_{th} = T_r \left(\sum_{i=0}^{n-1} (\omega a_i + \omega^2 b_i)(\omega^2 a'_i + \omega b'_i) \right)$$

$$= \sum_{i=0}^{n-1} T_r(a_i a'_i + \omega^2 a_i b'_i + \omega a'_i b_i + b_i b'_i)$$

$$= \left(\sum_{i=0}^{n-1} a_i b'_i \right) T_r(\omega^2) + \left(\sum_{i=0}^{n-1} a'_i b_i \right) T_r(\omega)$$

$$= \sum_{i=0}^{n-1} a_i b'_i + \sum_{i=0}^{n-1} a'_i b_i$$

$$= (v, v')_s.$$

Since $T_r(1) = 0$, $T_r(\omega) = T_r(\omega^2) = \omega + \omega^2 = 1$. $\qquad\square$

By Lemma 4.9 we know that
C is a linear code of $\mathbb{F}_2^{2n} \Leftrightarrow \Phi(C)$ is an additive code of $\mathbb{F}_4^n$,
$C \subseteq C_s^\perp \Leftrightarrow \Phi(C) \subseteq \Phi(C)_{th}^\perp$, and $d_s(C) = d_H(\Phi(C))$, $|\Phi(C)| = |C| = 2^{\dim_{\mathbb{F}_2} C}$, where

$$\Phi(C)_{th}^\perp = \{\alpha \in \mathbb{F}_4^n| < \alpha, \beta >_{th}= 0 \text{ for all } \beta \in \Phi(C)\},$$

$$d_H(\Phi(C)) = \min\{w_H(\alpha)|\alpha \in \Phi(C), \quad \alpha \neq 0\}.$$

From these facts, Theorem 4.3 can be transformed into the following form.

Theorem 4.10: ([6]) *Let C be an additive code of $\mathbb{F}_4^n$ (which means that C is an additive subgroup of $\mathbb{F}_4^n$), $C \subseteq C_{th}^\perp$, $|C| = 2^{n-k}$. Then we get an $[[n,k,d]]$ quantum code, where $d = w_H(C_{th}^\perp \backslash C) = \min\{w_H(v) | v \in C_{th}^\perp \backslash C\}$.*

Theorem 4.10 justifies the name "additive quantum codes" given by Calderbank et al.

The next result shows that if C is a linear code of $\mathbb{F}_4^n$, which means that C is an $\mathbb{F}_4$-linear subspace of $\mathbb{F}_4^n$, then the trace-hermitian product on $\mathbb{F}_4^n$ can be replaced by the hermitian product. In this case, the quantum code Q derived from C is called a *linear quantum code*.

Theorem 4.11: *Suppose that C is a linear code of $\mathbb{F}_4^n$. Then*

$$C \subseteq C_{th}^\perp \Leftrightarrow C \subseteq C_h^\perp,$$

where

$$C_h^\perp = \{v \in \mathbb{F}_4^n | <v, u>_h = 0 \text{ for all } u \in C\},$$

and for $u = (u_0, \ldots, u_{n-1})$, $v = (v_0, \ldots, v_{n-1}) \in \mathbb{F}_4^n$, $<u, v>_h = \sum_{i=0}^{n-1} u_i v_i^2 \in \mathbb{F}_4$.

Proof: ($\Leftarrow$) Let $u, v \in C$. If $<u, v>_h = 0$, then $<u, v>_{th} = T_r <u, v>_h = 0$.

($\Rightarrow$) Suppose that $C \subseteq C_{th}^\perp$. For any $c \in C$, we have $<c, v>_{th} = 0$ for each $v \in C$. Let $<c, v>_h = a + b\omega$ ($a, b \in \mathbb{F}_2$, since $\{1, \omega\}$ is a basis of $\mathbb{F}_4$ over $\mathbb{F}_2$). Then $0 = <c, v>_{th} = T_r(a + b\omega) = b$, so that $<c, v>_h = a$. Moreover, since C is a linear code of $\mathbb{F}_4^n$, $\omega v \in C$ so that

$$0 = <c, \omega v>_{th} = T_r(<c, \omega v>_h) = T_r((\omega^2(a + b\omega)) = T_r(a\omega^2 + b) = a.$$

Therefore $<c, v>_h = a + b\omega = 0$ for each $v \in C$, which means that $c \in C_h^\perp$ for any $c \in C$. Thus $C \subseteq C_h^\perp$.   □

Definition 4.12: A subset C of $\mathbb{F}_4^n$ is called an *even code* if for each $c \in C$, $w_H(c)$ is an even integer.

Lemma 4.13: *Suppose that $C \subseteq \mathbb{F}_4^n$.*

(1) *If C is an even and additive code, then C is self-orthogonal with respect to $< , >_{th}$ ($C \subseteq C_{th}^\perp$).*

(2) *If $C \subseteq C_{th}^\perp$ and C is an $\mathbb{F}_4$-linear code, then C is an even code.*

Proof: (1) Let $u = (u_0, \ldots, u_{n-1})$, $v = (v_0, \ldots, v_{n-1})$, $u_i, v_i \in \mathbb{F}_4$. Suppose that $u_i = a + b\omega$, $v_i = c + d\omega$, where $a, b, c, d, \in \mathbb{F}_2$. Then

$$T_r(u_i \overline{v_i}) = T_r((a + b\omega)\overline{(c + d\omega)}) = T_r((a + b\omega)(c + d\omega^2)) = ad + bc.$$

On the other hand, $w_H(v_i) = 0 \Leftrightarrow a = b = 0$. Thus

$$w_H(u_i) \equiv ab + a + b \pmod{2}.$$

Similarly we have

$$w_H(v_i) \equiv cd + c + d,$$
$$w_H(u_i + v_i) \equiv (a + c)(b + d) + a + b + c + d \pmod{2}.$$

Therefore, for each i $(0 \leq i \leq n - 1)$,

$$w_H(u_i + v_i) + w_H(u_i) + w_H(v_i) \equiv ad + bc = T_r(u_i \overline{v_i}) \pmod{2},$$

which implies that $w_H(u+v) + w_H(u) + w_H(v) \equiv <u, v>_{th} \pmod{2}$. From this the condition (1) can be obtained.

(2) For each $u = (u_0, \ldots, u_{n-1}) \in C$, $u_i \in \mathbb{F}_4$, by assumption we have $\omega u \in C$ and

$$0 = <u, \omega u>_{th} = T_r \left(\sum_{i=0}^{n-1} u_i \omega^2 u_i^2 \right) = T_r \left(\sum_{i=0}^{n-1} \omega^2 u_i^3 \right)$$
$$\equiv w_H(u) T_r(\omega^2) \equiv w_H(u) \pmod{2},$$

since for $u_i \in \mathbb{F}_4$,

$$u_i^3 = \begin{cases} 1, & \text{if } u_i \neq 0, \\ 0, & \text{if } u_i = 0. \end{cases}$$

Therefore C is an even code. $\qquad\square$

Example 4.14: (Quantum Hamming codes via $\mathbb{F}_4$) Let $m \geq 2$ be an integer and α be a primitive element of $\mathbb{F}_{4^m}$, which means $\alpha \in \mathbb{F}_{4^m}^*$ and the multiplicative order of α is $s = 4^m - 1$. By choosing a basis $\{v_1, \ldots, v_m\}$ of $\mathbb{F}_{4^m}$ over $\mathbb{F}_4$ (for example, $\{v_1, \ldots, v_m\} = \{1, \alpha, \ldots, \alpha^{m-1}\}$), each element $c \in \mathbb{F}_{4^m}$ can be uniquely expressed by $c = c_1 v_1 + \cdots + c_m v_m$ ($c_i \in \mathbb{F}_4$) or by the vector $c = (c_1, \ldots, c_m) \in \mathbb{F}_4^m$. In this way we view $\mathbb{F}_{4^m}$ as an $\mathbb{F}_4$-vector space $\mathbb{F}_4^m$.

Consider the $\mathbb{F}_4$-linear code C in $\mathbb{F}_4^n$ $(n = \frac{s}{3} = \frac{4^m-1}{3})$ with parity-check matrix

$$H = [1^\top, \alpha^\top, (\alpha^2)^\top, \ldots, (\alpha^{n-1})^\top],$$

where $(\alpha^i)^\top$ means the column vector of α^i. Since the order of α is $s = 4^m - 1$, we know that $\{\alpha^n, \alpha^{2n}\} = \{\omega, \omega^2\}$. Hence $\{\alpha^i, \alpha^{i+n}, \alpha^{i+2n}\} = \{\alpha^i, \omega\alpha^i, \omega^2\alpha^i\}$ for each i ($0 \le i \le n-1$). Therefore the n column vectors in H are just the projective points in the projective space $\mathbb{P}^{m-1}(\mathbb{F}_4)$. Therefore C is the Hamming code over $\mathbb{F}_4$ with $[n, k, d] = [\frac{4^m-1}{3}, \frac{4^m-1}{3}-m, 3]$. It is well known that $C^\perp$ is a $[\frac{4^m-1}{3}, m, 4^{m-1}]$ linear code over $\mathbb{F}_4$ and each non-zero codeword in $C^\perp$ has Hamming weight 4^{m-1}. Therefore $C^\perp$ is an even code. By Lemma 4.13 we know that $C^\perp$ is self-orthogonal with respect to $<, >_{th}$ and it is easy to see that $C^\perp \subseteq (C^\perp)^\perp_h = (C^\perp)^\perp = C$. From Theorem 4.10 and $|C^\perp| = 2^{2m}$ we know that there exist $[[n, k, d]] = [[\frac{4^m-1}{3}, \frac{4^m-1}{3}-2m, 3]]$ quantum codes for all $m \ge 2$. All the quantum codes in this series are perfect since they are pure codes and

$$2^{n-k} = 2^{2m} = 1 + 3\left(\frac{4^m-1}{3}\right) = 1 + 3n.$$

For $m = 2$ we obtain a $[[5, 1, 3]]$ quantum code again. For more quantum codes constructed from classical codes over $\mathbb{F}_4$, see [6] and [45].

5. Weight Enumerator and Singleton Bound

Besides constructing good quantum codes, there are other topics on quantum codes being investigated. In this section we show one of these topics: the weight enumerator.

5.1. *MacWilliams identity for quantum codes*

Let Q be an $((n, K))$ quantum code, which means that Q is a subspace of $\mathbb{C}^{2^n}$ with $\dim_\mathbb{C} Q = K$. Let $P\colon \mathbb{C}^{2^n} \to Q$ be the orthogonal projection.

Definition 5.1: Let

$$B_i = \frac{1}{K^2} \sum_e T_r^2(eP), \quad B_i^\perp = \frac{1}{K} \sum_e T_r(ePeP),$$

where $e = \sigma_0 \otimes \sigma_1 \otimes \cdots \otimes \sigma_{n-1}$ and $w_Q(e) = i$, $\sigma_i \in \{I_2, \sigma_x, \sigma_y, \sigma_z\}$. Since e is a Hermitian operator, the traces of eP and $ePeP$ are real numbers and so are B_i and $B_i^\perp$.

The weight enumerators of Q are defined by

$$B(x, y) = \sum_{i=0}^n B_i x^{n-i} y^i, \quad B^\perp(x, y) = \sum_{i=0}^n B_i^\perp x^{n-i} y^i.$$

Theorem 5.2: *Let Q be an $((n, K))$ quantum code and $B(x, y)$, $B^\perp(x, y)$ be its weight enumerators. Then*

$$B(x, y) = \frac{1}{K} B^\perp \left(\frac{x + 3y}{2}, \frac{x - y}{2} \right).$$

Proof: The first proof was a mixture of physical and mathematical methods given in [39]. The following purely mathematical proof is given in [30]. Suppose that the matrices of e and P with respect to the basis $|j> = |j_0 \cdots j_{n-1}> (0 \le j \le 2^n - 1, j = j_0 + j_1 \cdot 2 + \cdots + j_{n-1} \cdot 2^{n-1})$ of $\mathbb{C}^{2^n}$ are

$$e = (e_{ij}), \quad P = (P_{ij}) \quad (0 \le i, j \le 2^n - 1).$$

Then $e|i> = \sum_j e_{ij} |j>$ and for $e = \sigma_0 \otimes \sigma_1 \otimes \cdots \otimes \sigma_{n-1}$ we have

$$e_{ij} = (\sigma_0)_{i_0 j_0} \cdots (\sigma_{n-1})_{i_{n-1} j_{n-1}}, \quad \sigma_i \in \{I_2, \sigma_x, \sigma_y, \sigma_z\}.$$

Therefore,

$$B(x, y) = \frac{1}{K^2} \sum_{t=0}^{n} x^{n-t} y^t \sum_{w_Q(e)=t} \sum_{i,j,k,l} P_{ji} e_{ij} P_{lk} e_{kl}$$

$$= \frac{1}{K^2} \sum_{i,j,k,l} P_{ji} P_{lk} \sum_{e} e_{ij} e_{kl} x^{n-w_Q(e)} y^{w_Q(e)}$$

$$= \frac{1}{K^2} \sum_{i,j,k,l} P_{ji} P_{lk} \prod_{t=0}^{n-1} b_{i_t j_t k_t l_t}(x, y),$$

where

$$b_{i_t j_t k_t l_t}(x, y) = x(I_2)_{i_t j_t}(I_2)_{k_t l_t}$$
$$+ y[(\sigma_x)_{i_t j_t}(\sigma_x)_{k_t l_t} + (\sigma_y)_{i_t j_t}(\sigma_y)_{k_t l_t} + (\sigma_z)_{i_t j_t}(\sigma_z)_{k_t l_t}].$$

Similarly we have

$$B^\perp(x, y) = \frac{1}{K} \sum_{i,j,k,l} P_{ji} P_{lk} \prod_{t=0}^{n-1} b^\perp_{i_t j_t k_t l_t}(x, y),$$

where

$$b^\perp_{i_t j_t k_t l_t}(x, y) = x(I_2)_{i_t l_t}(I_2)_{k_t j_t}$$
$$+ y[(\sigma_x)_{i_t l_t}(\sigma_x)_{k_t j_t} + (\sigma_y)_{i_t l_t}(\sigma_y)_{k_t j_t} + (\sigma_z)_{i_t l_t}(\sigma_z)_{k_t j_t}].$$

We need to prove

$$b_{ijkl}(x, y) = b^\perp_{ijkl}(\frac{x + 3y}{2}, \frac{x - y}{2})$$

for all 16 cases $i, j, k, l \in \{0, 1\}$. For example, $b_{0011}(x, y) = x - y$ and $b_{0011}^{\perp}(x, y) = 2y$. We leave the remaining cases to the reader. □

The name "weight enumerator" in Definition 5.1 can be justified by the following result.

Lemma 5.3: *If Q is an additive quantum code derived from an $(n, 2^{n-k})$ additive code C over $\mathbb{F}_4$, then*

$$B_i = \#\{c \in C : w_H(c) = i\}, \quad B_i^{\perp} = \#\{c \in C_{th}^{\perp} : w_H(c) = i\}.$$

Thus, Theorem 5.2 becomes the MacWilliams identity for additive codes over $\mathbb{F}_4$.

Proof: We identify C with $\bar{G}$ ($\subseteq \bar{E}$) by $\omega a + \bar{\omega} b \mapsto \bar{e} = (a|b) \in \bar{G}$ and $Q = Q(\chi)$ is the χ-eigenspace of G acting on $\mathbb{C}^{2^n}$ for some character $\chi \in \hat{G}$. We have $K = \dim_C Q = 2^k$. Then the projection operator $P : \mathbb{C}^{2^n} \to Q$ is

$$P = 2^{-(n-k)} \sum_{e \in G} \chi(e)e.$$

Then we know that for each $e \in E$,

$$T_r(eP) = 2^{-(n-k)} \sum_{e' \in G} \chi(e')T_r(ee')$$

$$= \begin{cases} \chi(e)T_r(P) = \chi(e)2^k, & \text{if } \bar{e} \in \bar{G} = C, \\ 0, & \text{otherwise.} \end{cases}$$

Therefore

$$B_i = \frac{1}{K^2} \sum_{\substack{e \in G \\ w_Q(e) = i}} 2^{2k} = \sum_{\substack{e \in G \\ w_Q(e) = i}} 1 = \sum_{\substack{c \in C \\ w_H(c) = i}} 1.$$

Similarly,

$$T_r(ePeP) = 2^{-2(n-k)} \sum_{a, b \in G} \chi(ab)T_r(eaeb)$$

$$= 2^{-2(n-k)} \sum_{a, b \in G} (-1)^{(\bar{e}, \bar{a})_s} \chi(ab)T_r(ab)$$

$$= 2^{-2(n-k)} \sum_{a \in G} (-1)^{(\bar{e}, \bar{a})_s} \sum_{b \in G} \chi(b)T_r(b)$$

$$= \begin{cases} 2^{-(n-k)} \sum_{b \in G} \chi(b)T_r(b) = T_r(P) = 2^k, & \text{if } \bar{a} \in (\bar{G})_s^{\perp}, \\ 0, & \text{otherwise.} \end{cases}$$

Therefore

$$B_i^{\perp} = \frac{2^k}{K} \sum_{\substack{\bar{a} \in (\bar{G})_s^{\perp} \\ w_Q(\bar{a}) = i}} 1 = \sum_{\substack{c \in (C)_{th}^{\perp} \\ w_H(c) = i}} 1.$$ $\square$

The following result shows that the weight enumerators of Q can determine $d(Q)$ as in the classical case even though Q is not additive.

Lemma 5.4: (1) $B_0 = B_0^{\perp} = 1$, $B_i^{\perp} \geq B_i \geq 0$ $(0 \leq i \leq n)$.

(2) Let t be the maximal integer such that $B_i = B_i^{\perp}$ $(0 \leq i \leq t)$. Then $d(Q) = t + 1$.

Proof: (1) $B_0 = \frac{1}{K^2} T_r^2(P) = 1$, $B_0^{\perp} = \frac{1}{K} T_r(P^2) = 1$.

From the definition we know that $B_i \geq 0$. Moreover, from the well-known inequality $T_r^2(A^{\dagger}B) \leq T_r(A^{\dagger}A)T_r(B^{\dagger}B)$ and $T_r(AB) = T_r(BA)$ we know that if $e^{\dagger} = e$,

$$T_r^2(eP) = T_r^2(PePP) \leq T_r(PePPeP) \cdot T_r(P^2) = T_r(ePeP) \cdot K, \quad (7)$$

from which $B_i^{\perp} \geq B_i$ can be derived.

(2) From the inequality (7) we know that

$$B_i^{\perp} = B_i \Leftrightarrow PeP = \lambda_e P \quad \text{for all } e \in E \text{ such that } w_Q(e) = i.$$

Then the conclusion (2) comes from the following lemma. $\square$

Lemma 5.5: A quantum code Q can correct $\leq l$ errors $\Leftrightarrow$ for each $e \in E_{2l}$, $PeP = \lambda_e P$ $(\lambda_e \in \mathbb{C})$.

Proof: ($\Leftarrow$) Suppose that $c, c' \in Q$ and $< c|c' > = 0$. We need to prove that $< c|ee'|c' > = 0$ for any $e, e' \in E_l$. Since $ee' \in E_{2l}$, we have $Pee'P = \lambda_{ee'} P$ $(\lambda_{ee'} \in \mathbb{C})$ by assumption. On the other hand, $Pc = c$ and $Pc' = c'$. Therefore

$$< c|ee'|c' > = < cP|ee'|Pc' > = < c|Pee'P|c' >$$
$$= < c|\lambda_{ee'} P|c' > = \lambda_{ee'} < c|c' > = 0.$$

($\Rightarrow$) Suppose that $c, c' \in Q$ such that $< c|c' > = 0$ and $< c|c > = < c'|c' > = 1$. Then $c \pm c' \in Q$ and $< c + c'|c - c' > = < c|c > - < c'|c' > = 0$. Each $e \in E_{2l}$ can be decomposed as $e = e_1 e_2$, where $e_1, e_2 \in E_l$. Since Q can correct $\leq l$ errors, we know that

$$0 = < c + c'|e|c - c' > = < c|e|c > - < c'|e|c' > . \quad (8)$$

Now we choose a basis $\{c_1, \ldots, c_K\}$ of Q such that $< c_i | c_j > = \overline{c_i} \cdot c_j^\top = \delta_{ij}$ ($1 \leq i, j \leq K$). From (8) we know that for each $e \in E_{2l}$, $< c_i | e | c_i > = \overline{c_i} e c_i^\top = \lambda_e \in \mathbb{C}$ is independent of i ($1 \leq i \leq K$). Thus, the matrix of P with respect to the basis $\{c_1, \ldots, c_K\}$ of Q is $P = \sum_{i=1}^{K} c_i^\top \overline{c_i}$. Then for each $e \in E_{2l}$,

$$
\begin{aligned}
PeP &= \left(\sum_i c_i^\top \overline{c_i} \right) e \left(\sum_j c_j^\top \overline{c_j} \right) = \sum_{i,j} c_i^\top (\overline{c_i} e c_j^\top) \overline{c_j} \\
&= \sum_{i,j} c_i^\top \overline{c_j} < c_i | e | c_j > = \lambda_e \sum_{i,j} c_i^\top \overline{c_j} \delta_{ij} = \lambda_e \sum_i c_i^\top \overline{c_i} = \lambda_e P. \quad \square
\end{aligned}
$$

Remark 5.6: From Theorem 5.2 we know that $K = \frac{1}{2^n} \sum_{i=0}^{n} B_i^\perp$.

5.2. *Quantum Singleton bound*

The Mac Williams identity can be used to derive bounds on the parameters of quantum codes by a traditional method in the classical case. Namely, we use the Krawtchouk polynomials.

Definition 5.7: For a fixed positive integer n, the *Krawtchouk polynomials* are defined by

$$
P_i(x) = \sum_{j=0}^{i} (-1)^j 3^{i-j} \binom{x}{j} \binom{n-x}{i-j} \quad (0 \leq i \leq n).
$$

By definition we know that $P_i(x) \in \mathbb{Q}(x)$ and $\deg P_i(x) = i$. The first three polynomials are

$$
P_0(x) = 1, \quad P_1(x) = 3n - 4x, \quad P_2(x) = \frac{1}{2}(16x^2 + 8x(1 - 3n) + 9n(n-1)).
$$

We list several properties of Krawtchouk polynomials and relations to weight enumerators of quantum codes in the following lemma.

Lemma 5.8: (1) (*Symmetric property*) $3^i \binom{n}{i} P_j(i) = 3^j \binom{n}{j} P_i(j)$ ($0 \leq i$, $j \leq n$).
   (2) (*Orthogonal relations*)

$$
\sum_{i=0}^{n} P_r(i) P_i(s) = 4^n \delta_{rs}, \quad \sum_{i=0}^{n} 3^i \binom{n}{i} P_r(i) P_s(i) = 4^n 3^r \binom{n}{r} \delta_{rs}.
$$

(3) *Any polynomial $f(x) \in \mathbb{R}[x]$ has a unique expression*

$$f(x) = \sum_{i=0}^{d} f_i P_i(x) \quad (f_i \in \mathbb{R}) \quad (d = \deg f),$$

and

$$f_i = \frac{1}{4^n} \sum_{j=0}^{n} f(j) P_j(i) = \frac{1}{4^n \cdot 3^i \binom{n}{i}} \sum_{j=0}^{n} 3^j \binom{n}{j} f(j) P_i(j).$$

(4) *Let $B(x,y) = \sum_{t=0}^{n} B_t x^{n-t} y^t$ and $B^{\perp}(x,y) = \sum_{t=0}^{n} B_t^{\perp} x^{n-t} y^t$ be the weight enumerators of an $((n,k,d))$ quantum code. Then*

$$B_i = \frac{1}{2^n K} \sum_{j=0}^{n} P_i(j) B_j^{\perp}.$$

Proof: (1) By Definition 5.7,

$$\sum_{i=0}^{n} P_i(j) z^j = \sum_{i=0}^{n} \sum_{\mu=0}^{i} (-1)^{\mu} 3^{i-\mu} \binom{n-j}{i-\mu} \binom{j}{\mu} z^i$$

$$= \sum_{\mu=0}^{n} (-1)^{\mu} \binom{j}{\mu} \sum_{i=\mu}^{n} 3^{i-\mu} \binom{n-j}{i-\mu} z^i$$

$$= \sum_{\mu=0}^{n} (-1)^{\mu} \binom{j}{\mu} z^{\mu} \sum_{i=0}^{n-j} 3^i \binom{n-j}{i} z^i$$

$$= (1-z)^j (1+3z)^{n-j} = (1+3z)^n \left(\frac{1-z}{1+3z} \right)^j,$$

and

$$\sum_{i,j=0}^{n} 3^j \binom{n}{j} P_i(j) z^i w^j = \sum_{j=0}^{n} \binom{n}{j} (3w)^j (1+3z)^n \left(\frac{1-z}{1+3z} \right)^j$$

$$= (1+3z)^n (1 + 3w \cdot \frac{1-z}{1+3z})^n = (1 + 3z + 3w - 3zw)^n.$$

The right-hand side of this equality is a symmetric polynomial over z and w. Thus

$$\sum_{i,j=0}^{n} 3^j \binom{n}{j} P_i(j) z^i w^j = \sum_{i,j=0}^{n} 3^j \binom{n}{j} P_i(j) z^j w^i = \sum_{i,j=0}^{n} 3^i \binom{n}{i} P_j(i) z^i w^j,$$

which implies that

$$3^j \binom{n}{j} P_i(j) = 3^i \binom{n}{i} P_j(i) \quad (0 \le i, j \le n).$$

(2) We have by the equality in (1) that

$$\sum_{r,s=0}^{n} \sum_{i=0}^{n} 3^i \binom{n}{i} P_r(i) P_s(i) z^r w^s$$

$$= \sum_{i=0}^{n} 3^i \binom{n}{i} (1-z)^i (1+3z)^{n-i} (1-w)^i (1+3w)^{n-i}$$

$$= (1+3z)^n (1+3w)^n \sum_{i=0}^{n} 3^i \binom{n}{i} \left(\frac{(1-z)(1-w)}{(1+3z)(1+3w)} \right)^i$$

$$= [(1+3z)(1+3w) + 3(1-z)(1-w)]^n$$

$$= (4+12zw)^n$$

$$= 4^n (1+3zw)^n.$$

Therefore

$$\sum_{i=0}^{n} 3^i \binom{n}{i} P_r(i) P_s(i) = \delta_{r,s} \cdot 4^n 3^r \binom{n}{r}.$$

By (1) we obtain the other orthogonal relation.

(3) Since $\deg P_i(x) = i$ $(i = 0, 1, 2, \dots)$, we have the unique expression. Then by the orthogonal relations,

$$\sum_{j=0}^{n} f(j) P_j(i) = \sum_{j=0}^{n} P_j(i) \sum_{k=0}^{n} f_k P_k(j) = \sum_{k=0}^{n} f_k \sum_{j=0}^{n} P_j(i) P_k(j) = 4^n f_i.$$

(4) From the MacWilliams identity,

$$\sum_{t=0}^{n} B_t x^{n-t} y^t = \frac{1}{K} \sum_{i=0}^{n} B_i^\perp \left(\frac{x+3y}{2} \right)^{n-i} \left(\frac{x-y}{2} \right)^i$$

$$= \frac{1}{2^n K} \sum_{i=0}^{n} B_i^\perp \sum_{\lambda, \mu \ge 0} \binom{n-i}{\lambda} \binom{i}{\mu} x^{n-i-\lambda} (3y)^\lambda x^{i-\mu} (-y)^\mu.$$

Thus, we have

$$B_t = \frac{1}{2^n K} \sum_{i=0}^{n} B_i^\perp \sum_{\substack{\lambda, \mu \geq 0 \\ \lambda + \mu = t}} \binom{n-i}{\lambda} \binom{i}{\mu} 3^\lambda (-1)^\mu$$

$$= \frac{1}{2^n K} \sum_{i=0}^{n} B_i^\perp P_t(i).$$

The other equality can be derived by the inverse transformation in (3). □

The next lemma shows that we can present a bound for K by choosing suitable $f(x)$ in $\mathbb{R}[x]$.

Lemma 5.9: *Let Q be an $((n, K, d))$ quantum code. If there exists a polynomial*

$$f(x) = \sum_{i=0}^{n} f_i P_i(x) \in \mathbb{R}[x] \quad (f_i \in \mathbb{R}),$$

such that
(i) $f_i \geq 0$ $(0 \leq i \leq n)$,
(ii) $f(j) > 0$ $(0 \leq j \leq d-1)$ and $f(j) \leq 0$ $(d \leq j \leq n)$,
then

$$K \leq \frac{1}{2^n} \max_{0 \leq j \leq d-1} \frac{f(j)}{f_j}.$$

Proof: From the assumptions and Lemma 5.8 we have the following estimation:

$$2^n K \sum_{i=0}^{d-1} f_i B_i \leq 2^n K \sum_{i=0}^{n} f_i B_i = \sum_{i=0}^{n} f_i \sum_{j=0}^{n} B_j^\perp P_i(j)$$

$$= \sum_{j=0}^{n} B_j^\perp \sum_{i=0}^{n} f_i P_i(j) = \sum_{j=0}^{n} B_j^\perp f(j)$$

$$\leq \sum_{j=0}^{d-1} B_j^\perp f(j) = \sum_{j=0}^{d-1} B_j f(j) \quad \text{(by Lemma 5.4 (2))}.$$

Therefore

$$2^n K \leq \frac{\sum_{j=0}^{d-1} B_j f(j)}{\sum_{j=0}^{d-1} f_j B_j} \leq \max_{0 \leq j \leq d-1} \frac{f(j)}{f_j}.$$

□

Now we present a bound for any quantum code which is an analogue of the Singleton bound for classical linear codes.

Theorem 5.10: (*Quantum Singleton bound*) *For any $((n, K, d))$ quantum code such that $d \le \frac{n}{2} + 1$, we have $K \le 2^{n-2d+2}$.*

Proof: Consider

$$f(x) = 4^{n-d+1} \prod_{j=d}^{n} \left(1 - \frac{x}{j}\right) = 4^{n-d+1} \frac{\binom{n-x}{n-d+1}}{\binom{n}{n-d+1}}.$$

The values $f(j)$ $(0 \le j \le n)$ satisfy the condition (ii) of Lemma 5.9. Next we compute the coefficients f_i in the expression $f(x) = \sum_{i=0}^{n} f_i P_i(x)$. For doing this, we consider

$$\sum_{\lambda=0}^{n} 3^{\lambda} \binom{n}{\lambda} f(\lambda) z^{\lambda} = \frac{4^{n-d+1}}{\binom{n}{n-d+1}} \sum_{\lambda=0}^{n} 3^{\lambda} \binom{n}{\lambda} \binom{n-\lambda}{n-d+1} z^{\lambda}$$

$$= 4^{n-d+1} \sum_{\lambda=0}^{n} \binom{d-1}{\lambda} (3z)^{\lambda} = 4^{n-d+1} (1 + 3z)^{d-1}.$$

On the other hand,

$$\sum_{\lambda=0}^{n} 3^{\lambda} \binom{n}{\lambda} f(\lambda) z^{\lambda} = \sum_{\lambda=0}^{n} 3^{\lambda} \binom{n}{\lambda} z^{\lambda} \sum_{i=0}^{n} f_i P_i(\lambda)$$

$$= \sum_{i,\lambda=0}^{n} 3^i \binom{n}{i} z^{\lambda} \sum_{i=0}^{n} f_i P_{\lambda}(i)$$

$$= \sum_{i=0}^{n} 3^i \binom{n}{i} f_i \sum_{\lambda=0}^{n} P_{\lambda}(i) z^{\lambda}$$

$$= \sum_{i=0}^{n} 3^i \binom{n}{i} f_i (1 + 3z)^n \left(\frac{1-z}{1+3z}\right)^i.$$

Thus

$$\sum_{i=0}^{n} 3^i \binom{n}{i} f_i \left(\frac{1-z}{1+3z}\right)^i = 4^{n-d+1}/(1 + 3z)^{n-d+1}.$$

Let $w = \frac{1-z}{1+3z}$, then $z = \frac{1-w}{1+3w}$ and

$$\sum_{i=0}^{n} 3^i \binom{n}{i} f_i w^i = \frac{4^{n-d+1}}{(\frac{4}{1+3w})^{n-d+1}} = (1+3w)^{n-d+1}$$

$$= \sum_{i=0}^{n-d+1} \binom{n-d+1}{i} 3^i w^i.$$

Therefore we obtain that $f_i = \binom{n-d+1}{i}/\binom{n}{i} = \binom{n-i}{d-1}/\binom{n}{d-1}$ $(0 \le i \le n)$ which satisfy the condition (i) of Lemma 5.9. Then we have

$$K \le \frac{1}{2^n} \max_{0 \le j \le d-1} \frac{f(j)}{f_j}$$

$$= \frac{1}{2^n} \max_{0 \le j \le d-1} \left\{ 4^{n-d+1} \frac{\binom{n-j}{n-d+1}}{\binom{n}{n-d+1}} \cdot \frac{\binom{n}{d-1}}{\binom{n-j}{d-1}} \right\}$$

$$= 2^{n-2d+2} \max_{0 \le j \le d-1} \frac{\binom{n-j}{n-d+1}}{\binom{n-j}{d-1}}.$$

Let $g(j) = \binom{n-j}{n-d+1}/\binom{n-j}{d-1}$. Then $\frac{g(j)}{g(j+1)} = \frac{n-j-d+1}{d-1-j} \ge 1$ by the assumption $d \le \frac{n}{2}+1$. Therefore $K \le 2^{n-2d+2}g(0) = 2^{n-2d+2}$. $\qquad \square$

6. Nonbinary Case

6.1. *Stabilizer nonbinary quantum codes*

The theory of binary $(p = 2)$ quantum codes, described in Sections 3–5, has been generalized to the nonbinary case $(p \ge 3)$ ([1, 21, 28, 31]). In this section we introduce the stabilizer nonbinary quantum codes. The concepts and results are parallel to the binary case without any essential difficulty, so we omit the proofs.

We fix an odd prime number p.

Definition 6.1: A *qubit* is a non-zero vector in $\mathbb{C}^p$,

$$0 \ne |v> = c_0|0> + c_1|1> + \cdots + c_{p-1}|p-1> \qquad (c_i \in \mathbb{C}),$$

where

$$\{|0>, |1>, \ldots, |p-1>\} = \{|a>: \ a \in \mathbb{F}_p\} \tag{9}$$

is a basis of $\mathbb{C}^p$.

An *n-qubit* is a non-zero vector in $(\mathbb{C}^p)^{\otimes n} = \mathbb{C}^{p^n}$,

$$0 \neq |v> = \sum_{a=(a_1,\ldots,a_n) \in \mathbb{F}_p^n} c_a |a> \quad (c_a \in \mathbb{C}),$$

where

$$\{|a> = |a_1 \cdots a_n> = |a_1> \otimes |a_2> \otimes \cdots \otimes |a_n>: \; a = (a_1,\ldots,a_n) \in \mathbb{F}_p^n\} \tag{10}$$

is a basis of $\mathbb{C}^{p^n}$.

A quantum error $\sigma_l \tau_s$ $(l, s \in \mathbb{F}_p)$ on a qubit is a unitary linear operator on $\mathbb{C}^p$ which acts on the basis (9) of $\mathbb{C}^p$ by

$$\sigma_l \tau_s |a> = \zeta_p^{sa} |a+l> \quad (a \in \mathbb{F}_p),$$

where $\zeta_p = e^{\frac{2\pi i}{p}}$.

A quantum error on an *n-qubit* is a unitary linear operator on $(\mathbb{C}^p)^{\otimes n} = \mathbb{C}^{p^n}$ written as

$$e = \zeta_p^t \omega_1 \otimes \omega_2 \otimes \cdots \otimes \omega_n \quad (\omega_i = \sigma_{l_i} \tau_{s_i}, t, l_i, s_i \in \mathbb{F}_p), \tag{11}$$

which acts on the basis (10) of $\mathbb{C}^{p^n}$ by

$$\begin{aligned}
e|a_1 \cdots a_n> &= \zeta_p^t \cdot (\omega_1 |a_1>) \otimes (\omega_2 |a_2>) \otimes \cdots \otimes (\omega_n |a_n>) \\
&= \zeta_p^{t+s_1 a_1 + \cdots + s_n a_n} (|a_1 + l_1> \otimes |a_2 + l_2> \otimes \cdots \otimes |a_n + l_n>) \\
&= \zeta_p^{t+(s,a)} |a+l> \quad (s = (s_1, \cdots, s_n), l = (l_1, \ldots, l_n) \in \mathbb{F}_p^n).
\end{aligned}$$

All errors e in (11) form a non-abelian error group E_n, $|E_n| = p^{2n+1}$. The error e can be expressed by

$$e = \zeta_p^\lambda X(a) Z(b) \quad (a = (a_1,\ldots,a_n), b = (b_1,\ldots,b_n) \in \mathbb{F}_p^n),$$

where

$$X(a) = \sigma_{a_1} \otimes \cdots \otimes \sigma_{a_n}, \quad Z(b) = \tau_{a_1} \otimes \cdots \otimes \tau_{a_n}.$$

Then for e and $e' = \zeta_p^{\lambda'} X(a') Z(b')$ we have

$$e'e = \zeta_p^{a' \cdot b - a \cdot b'} ee'. \tag{12}$$

From this we know that the center $C(E_n)$ of E_n is $\{\zeta_p^\lambda | \lambda \in \mathbb{F}_p\}$ and the quotient group $\widehat{E_n} = E_n/C(E_n)$ is abelian. The image of $e = \zeta_p^\lambda X(a) Z(b) \in E_n$ in $\overline{E_n}$ is denoted by

$$\bar{e} = (a|b) \quad (a, b \in \mathbb{F}_p^n).$$

Then $\overline{E_n}$ is naturally isomorphic to the additive group of $\mathbb{F}_p^{2n}$. We define the symplectic inner product $(\ ,\ )_s$ in $\mathbb{F}_p^{2n}$ by

$$((a|b), (a'|b'))_s = a \cdot b' - a' \cdot b = \sum_{i=1}^{n}(a_i b'_i - a'_i b_i) \in \mathbb{F}_p.$$

Then the formula (12) implies the following result.

Lemma 6.2: *A subgroup S of E_n is abelian if and only if $\overline{S}$ is a symplectic self-orthogonal subspace of $\mathbb{F}_p^{2n}$, namely $\overline{S} \subseteq (\overline{S})_s^{\perp}$ where*

$$(\overline{S})_s^{\perp} = \{v \in \mathbb{F}_p^{2n} | (v, c)_s = 0 \text{ for all } c \in \overline{S}\}.$$

For $e = \zeta_p^{\lambda} X(a) Z(b)$ $(a, b \in \mathbb{F}_p^n)$, we define the quantum weight of e and $\bar{e} = (a|b)$ by

$$w_Q(e) = w_Q(\bar{e}) = \sharp\{1 \leq i \leq n | (a_i, b_i) \neq (0, 0)\}.$$

Let

$$E_n(l) = \{e \in E_n | w_Q(e) \leq l\} \quad (0 \leq l \leq n).$$

Then $|E_n(l)| = \sum_{i=0}^{l} (p^2 - 1)^i \binom{n}{i}$.

Definition 6.3: *A subspace Q of $\mathbb{C}^{p^n}$ is called a quantum code with level p and length n. Let $K = \dim_{\mathbb{C}} Q$ and $k = \log_p K$. Then $0 \leq k \leq n$.*

A quantum code Q has *minimal distance* d (≥ 1) if $|v_1 >, |v_2 > \in Q$, $< v_1 | v_2 > = 0$ and $e \in E_n$, $w_Q(e) \leq d - 1$, imply $< v_1 | e | v_2 > = 0$ where $< v_1 | v_2 >$ denotes the hermitian inner product in $\mathbb{C}^{p^n}$.

A quantum code Q with parameters n, K (or k) and d can be expressed by $((n, K, d))_p$ or $[[n, k, d]]_p$.

A quantum code Q has ability to *correct* $\leq l$ *errors* if $|v_1 >, |v_2 > \in Q$, $< v_1 | v_2 > = 0$ and $e_1, e_2 \in E_n$, $w_Q(e_1) \leq l$, $w_Q(e_2) \leq l$, imply $< v_1 | e_1 e_2 | v_2 > = 0$.

Lemma 6.4: *An $[[n, k, d]]_p$ quantum code can correct $\leq \lceil \frac{d-1}{2} \rceil$ quantum errors.*

Lemma 6.5: *(Quantum Singleton bound) For any $[[n, k, d]]_p$ quantum code, $n \geq k + 2(d - 1)$.*

A quantum code Q with minimal distance d is called *pure* if the subspaces eQ ($e \in E_n$, $w_Q(e) \leq [\frac{d-1}{2}]$) are orthogonal to each other.

Lemma 6.6: *(Quantum Hamming bound) For any pure $((n, K, d))_p$ quantum code, $p^n \geq K(\sum_{i=0}^{l}(p^2 - 1)^i \binom{n}{i})$ where $l = [\frac{d-1}{2}]$.*

Theorems 4.6, 4.8 and 4.10 can be generalized to construct nonbinary quantum codes.

Theorem 6.7: *Suppose that C is a linear code in $\mathbb{F}_p^{2n}$, $k = dim_{\mathbb{F}_p}C$ and C is symplectic self-orthogonal ($C \subseteq C_s^{\perp}$). Then there exists a quantum code $[[n, n - k, d]]_p$, where*

$$d = \min\{w_Q(v)| \ v \in C_s^{\perp} \setminus C\}.$$

Theorem 6.8: *Suppose that there exists an $[n, k, d]_p$ linear code C in $\mathbb{F}_p^n$ and $C \supseteq C^{\perp}$, where $C^{\perp}$ is the dual code of C with respect to the ordinary inner product in $\mathbb{F}_p^n$: $(a, b) = \sum_{i=1}^{n} a_i b_i \in \mathbb{F}_p$. Then we have a quantum code $[[n, 2k - n, d]]_p$.*

Theorem 6.9: *If we have linear codes $C_1 = [n, k_1, d_1]_p$ and $C_2 = [n, k_2, d_2]_p$ in $\mathbb{F}_p^n$ such that $C_1^{\perp} \subseteq C_2$, then we have a quantum code $[[n, k_1 + k_2 - n, \min(d_1, d_2)]]_p$.*

At last we discuss how to construct quantum codes with level p from classical codes in $\mathbb{F}_{p^2}^n$. We define the *trace-hermitian inner product* $(\ , \ )_{th}$ in $\mathbb{F}_{p^2}^n$ as follows. For $a = (a_1, \ldots, a_n), b = (b_1, \ldots, b_n) \in \mathbb{F}_{p^2}^n$,

$$(a, b)_{th} = Tr((a^p, b)) = Tr\left(\sum_{i=1}^{n} a_i^p b_i\right) = \sum_{i=1}^{n}(a_i^p b_i + a_i b_i^p) \in \mathbb{F}_p.$$

For an additive subgroup C of $\mathbb{F}_{p^2}^n$, let

$$C_{th}^{\perp} = \{v \in \mathbb{F}_{p^2}^n | (v, c)_{th} = 0 \text{ for all } c \in C\}.$$

Theorem 6.10: *Let C be an additive subgroup of $\mathbb{F}_{p^2}^n$ and $C \subseteq C_{th}^{\perp}$, $|C| = p^{n-k}$. Then we have a quantum code $[[n, k, d]]_p$, where*

$$d = \min\{w_H(v)| \ v \in C_{th}^{\perp} \setminus C\}.$$

6.2. *Graphic method to construct nonbinary quantum codes*

In this section we introduce a new method to construct nonbinary quantum codes given by Schlingemann and Werner in [37]. It seems that many quantum codes can be constructed by this graphic (or matrix) method.

Let V be a finite set, $v = |V|$. Each symmetric matrix over $\mathbb{F}_p$

$$A = (a_{ij})_{i,j \in V} \quad (a_{ij} \in \mathbb{F}_p, \ a_{ij} = a_{ji}),$$

can be considered as a graph $G = G(A)$. The vertex set of G is V and each edge $\overline{ij}$ ($i, j \in V$) is weighted by $a_{ij} \ (= a_{ji}) \in \mathbb{F}_p$.

For two subsets X and Y of V, let A_{XY} be the submatrix of A

$$A_{XY} = (a_{ij})_{i \in X, j \in Y}.$$

We denote $\mathbb{F}_p^X = \mathbb{F}_p^{|X|}$ and $d^X = (d^x)_{x \in X}$ as a column vector in $\mathbb{F}_p^X$. For example, let $V = X \cup Y$ and $X \cap Y = \emptyset$, then a column vector in $\mathbb{F}_p^V = \mathbb{F}_p^v$ can be written as $d^V = \begin{pmatrix} d^X \\ d^Y \end{pmatrix}$ and

$$Ad^V = \begin{pmatrix} A_{XX} & A_{XY} \\ A_{YX} & A_{YY} \end{pmatrix} \begin{pmatrix} d^X \\ d^Y \end{pmatrix} = \begin{pmatrix} A_{XX}d^X + A_{XY}d^Y \\ A_{YX}d^X + A_{YY}d^Y \end{pmatrix}.$$

Theorem 6.11: ([37]) *Let p be an odd prime number, X and Y be disjoint sets, $|X| = k$, $|Y| = n \geq k + 2(d-1)$, $d \geq 2$. Let $A = (a_{ij})_{i,j \in X \cup Y}$ be a symmetric matrix over $\mathbb{F}_p$ with size $(n+k) \times (n+k)$. For each $E \subseteq Y$, $|E| = d - 1$, let $I = Y \setminus E$. Consider the following two statements for $d^X \in \mathbb{F}_p^X$ and $d^E \in \mathbb{F}_p^E$:*
  (1) $A_{IX}d^X + A_{IE}d^E = O^I$ (zero vector in $\mathbb{F}_p^I$);
  (2) $d^X = O^X$ and $A_{XE}d^E = O^X$.
If (1) implies (2) for any subset $E \subseteq Y, |E| = d - 1$, then there exists a quantum code $[[n, k, d]]_p$.

Proof: Here we present a proof with a more mathematical flavor.

Let $|X| = k$, $|Y| = n \geq k + 2(d-1)$, $d > 2$ and $X \cap Y = \emptyset$. Suppose that there exists a symmetric matrix $A = (a_{ij})_{i,j \in X \cup Y}$ over $\mathbb{F}_p$ satisfying the following condition:

For each $E \subseteq Y$, $|E| = d - 1$ and $I = Y \setminus E$, if $d^X \in \mathbb{F}_p^X$, $d^E \in \mathbb{F}_p^E$ and $A_{IX}d^X + A_{IE}d^E = O^I$, then

$$d^X = O^X \text{ and } A_{XE}d^E = O^X. \tag{13}$$

We define a $\mathbb{C}$–linear mapping

$$f : (\mathbb{C}^p)^{\otimes k} \to (\mathbb{C}^p)^{\otimes n}$$
$$|v> \mapsto f(|v>),$$

where for an element

$$|v> = \sum_{a=(a_1,\ldots,a_k)\in\mathbb{F}_p^k} c_a|a>$$

$$= \sum_{d^X\in\mathbb{F}_p^X} c(d^X)|d^X > \in (\mathbb{C}^p)^{\otimes k}, \quad (c_a, c(d^X)) \in \mathbb{C})$$

we define

$$f(|v>) = \sum_{d^Y\in\mathbb{F}_p^Y} f_v(d^Y)|d^Y > \in (\mathbb{C}^p)^{\otimes n},$$

where

$$f_v(d^Y) = \sum_{d^X\in\mathbb{F}_p^X} \zeta_p^{\frac{1}{2}((d^X)^t,(d^Y)^t)A\binom{d^X}{d^Y}} c(d^X) \in \mathbb{C}.$$

Let Q be the image of f, $Q = \{f(|v>) : |v> \in (\mathbb{C}^p)^{\otimes k}\}$. We show that Q is a quantum code $[[n, k, d]]_p$.

First we show that $\dim_{\mathbb{C}} Q = p^k$. Since $\dim_{\mathbb{C}}(\mathbb{C}^p)^{\otimes k} = p^k$ and f is a $\mathbb{C}$–linear mapping, we need to show that f is injective. Namely, if $|v> = \sum_{d^X\in\mathbb{F}_p^X} c(d^X)|d^X >$ and $f(|v>) = 0$, we need to prove that $c(d^X) = 0$ for all $d^X \in \mathbb{F}_p^X$.

Assume that $f(|v>) = 0$, which means that

$$\sum_{d^X} \zeta_p^{\frac{1}{2}((d^X)^t,(d^Y)^t)A\binom{d^X}{d^Y}} c(d^X) = 0 \quad \text{(for all } d^X \in \mathbb{F}_p^X). \tag{14}$$

Since

$$\frac{1}{2}((d^X)^t, (d^Y)^t)A\binom{d^X}{d^Y} = \frac{1}{2}(d^X)^t A_{XX} d^X$$
$$+ \frac{1}{2}(d^Y)^t A_{YY} d^Y + (d^Y)^t A_{YX} d^X,$$

the formula (14) becomes

$$\sum_{d^X} \alpha(d^X)\zeta_p^{(d^Y)^t A_{YX} d^X} = 0 \quad \text{(for all } d^Y \in \mathbb{F}_p^Y), \tag{15}$$

where

$$\alpha(d^X) = \zeta_p^{\frac{1}{2}(d^X)^t A_{XX} d^X} c(d^X). \qquad (16)$$

A_{YX} is a matrix over $\mathbb{F}_p$ with size $n \times k$ $(n > k)$. Now we prove that $\operatorname{rank}_{\mathbb{F}_p} A_{YX} = k$. If $\operatorname{rank}_{\mathbb{F}_p} A_{YX} < k$, then the k columns of A_{YX} are $\mathbb{F}_p$-linearly dependent. Thus, the k columns of A_{IX} are also $\mathbb{F}_p$-linearly dependent for each $I \subseteq Y$, $|T| = d - 1$. Therefore we have $d^X \neq O^X$ such that $A_{IX} d^X = O^I$. Let $d^E = O^E$. Then $A_{IX} d^X + A_{IE} d^E = O^I$. By the condition (13) we should have $d^X = O^X$ which contradicts our choosing $d^X \neq O^X$. Therefore $\operatorname{rank}_{\mathbb{F}_p} A_{YX} = k$. Then we have invertible matrices P and R over $\mathbb{F}_p$ with size $n \times n$ and $k \times k$, respectively, such that $A_{YX} = P \begin{pmatrix} I^k \\ 0 \end{pmatrix} R^{-1}$.

The formula (15) becomes

$$0 = \sum_{d^X} \alpha(d^X) \zeta_p^{(d^Y)^t P \begin{pmatrix} I^k \\ 0 \end{pmatrix} R^{-1} d^X} = \sum_{g^X \in \mathbb{F}_p^X} \alpha(Rg^X) \zeta_p^{(h^Y)^t \begin{pmatrix} I^k \\ 0 \end{pmatrix} g^X}$$

$$= \sum_{g^X \in \mathbb{F}_p^X} \alpha(Rg^X) \zeta_p^{h_1 g_1 + \cdots + h_k g_k} \quad (h_1, \ldots, h_k \in \mathbb{F}_p), \qquad (17)$$

where

$$g^X = (g_1, \ldots, g_k)^t = R^{-1} d^X, \quad (h^Y)^t = (h_1, \ldots, h_n) = (d^Y)^t P.$$

The formula (17) implies that for each $a^X = (a_1, \ldots, a_k) \in \mathbb{F}_p^X$,

$$0 = \sum_{(h_1, \ldots, h_k) \in \mathbb{F}_p^k} \zeta_p^{-(h_1 a_1 + \cdots + h_k a_k)} \sum_{g^X \in \mathbb{F}_p^X} \alpha(Rg^X) \zeta_p^{h_1 g_1 + \cdots + h_k g_k}$$

$$= \sum_{g^X \in \mathbb{F}_p^X} \alpha(Rg^X) \sum_{(h_1, \ldots, h_k) \in \mathbb{F}_p^k} \zeta_p^{h_1(g_1 - a_1) + \cdots + h_k(g_k - a_k)}$$

$$= p^k \alpha(Ra^X).$$

Since R is invertible, we know that $\alpha(d^X) = 0$ and $c(d^X) = 0$ (by (16)) for all $d^X \in \mathbb{F}_p^X$, which means that $|v> = 0$. Therefore we proved that $\dim_{\mathbb{C}} Q = p^k$.

Next we prove that Q has minimal distance d. Suppose that

$$|w> = f(|v>) = \sum_{d^Y} f_v(d^Y) |d^Y > \in Q,$$

$$|w'> = f(|v'>) = \sum_{d^Y} f_{v'}(d^Y) |d^Y > \in Q,$$

and $< w'|w> = 0$, where

$$f_v(d^Y) = \sum_{d^X} \zeta_p^{\frac{1}{2}((d^X)^t,(d^Y)^t)A\begin{pmatrix} d^X \\ d^Y \end{pmatrix}} c(d^X),$$

$$f_{v'}(d^Y) = \sum_{d^X} \zeta_p^{\frac{1}{2}((d^X)^t,(d^Y)^t)A\begin{pmatrix} d^X \\ d^Y \end{pmatrix}} c'(d^X).$$

We need to show that $< w'|e|w> = 0$ for each $e \in E_n$ $w_Q(e) \leq d-1$. For such e, with formula (11) in Section 6.1, we can find $E \subseteq Y$ $|E| = d-1$, $I = Y \setminus E$ such that

$$e|d^E, d^I> = \zeta_p^{(s^E, d^E)}|d^E + l^E, d^I> \quad (s^E, l^E \in \mathbb{F}_p^E).$$

Therefore

$$e|w> = \sum_{d^Y} f_v(d^Y)e|d^Y>$$

$$= \sum_{d^E, d^I} \sum_{d^Y} \zeta_p^{(s^E, d^E)+\frac{1}{2}((d^X)^t),(d^E)^t),(d^I)^t)A\begin{pmatrix} d^X \\ d^E \\ d^I \end{pmatrix}} c(d^X)|d^E + l^E, d^I>$$

$$= \sum_{d^E, d^I} \sum_{d^Y} \zeta_p^{(s^E, d^E)+\frac{1}{2}((d^X)^t,(d^E - l^E)^t,(d^I)^t)A\begin{pmatrix} d^X \\ d^E - l^E \\ d^I \end{pmatrix}} c(d^X)|d^E, d^I>,$$

and

$$< w'|e|w> = \sum_{d^E, d^I} \sum_{d^X,(d')^X} \zeta_p^m \overline{c'(d'^X)}c(d^X),$$

where

$$m = (s^E, d^E) + \frac{1}{2}((d^X)^t, (d^E - l^E)^t, (d^I)^t)A\begin{pmatrix} d^X \\ d^E - l^E \\ d^I \end{pmatrix}$$

$$- \frac{1}{2}((d'^X)^t, (d^E)^t, (d^I)^t)A\begin{pmatrix} d'^X \\ d^E \\ d^I \end{pmatrix}$$

$$= (s^E, d^E) + \frac{1}{2}(d^X)^t A_{XX} d^X - \frac{1}{2}(d'^X)^t A_{XX} d'^X$$
$$+ (d^X - d'^X)^t (A_{XE} d^E + A_{XI} d^I) - (d^X)^t A_{XE} l^E - (l^E)^t A_{EE} d^E$$
$$+ \frac{1}{2}(l^E)^t A_{EE} l^E - (l^E)^t A_{EI} d^I. \tag{18}$$

The terms in m related to d^I are

$$(d^X - d'^X)^t A_{XI} d^I - (l^E)^t A_{EI} d^I = (d^I)^t (A_{IX}(d^X - d'^X) - A_{IE} l^E).$$

Therefore, if $A_{IX}(d^X - d'^X) - A_{IE} l^E = a^I \neq O^I$, then

$$< w'|e|w> = \sum_{d^E, d^X, (d')^X} \cdots \sum_{d^I} \zeta_p^{(d^I, a^I)} = 0.$$

If $A_{IX}(d^X - d'^X) - A_{IE} l^E = O^I$, by condition (13) we have $d^X - d'^X = O^X$ and $A_{XE} l^E = O^X$, so that (18) becomes

$$m = (s^E, d^E) - (l^E)^t A_{EE} d^E + \frac{1}{2}(l^E)^t A_{EE} l^E - (l^E)^t A_{EI} d^I,$$

which is independent of $d^X (= d'^X)$. Therefore

$$< w'|e|w> = \alpha \cdot \sum_{d^X} \overline{c'(d'^X)} c(d^X) = \alpha < v'|v> \quad (\alpha \in \mathbb{C}). \tag{19}$$

On the other hand,

$$0 = < w'|w> = \sum_{d^Y} \sum_{d^X, d'^X} c(d^X) \overline{c'(d'^X)} \zeta_p^N, \tag{20}$$

where

$$N = \frac{1}{2}((d^X)^t, (d^Y)^t) A \begin{pmatrix} d^X \\ d^Y \end{pmatrix} - \frac{1}{2}((d'^X)^t, (d^Y)^t) A \begin{pmatrix} d'^X \\ d^Y \end{pmatrix}$$
$$= \frac{1}{2}(d^X)^t A_{XX} d^X - \frac{1}{2}(d'^X)^t A_{XX} d'^X + (d^Y)^t A_{YX}(d^X - d'^X).$$

From $\text{rank}_{\mathbb{F}_p} A_{YX} = k$ we can show as before that

$$\sum_{d^Y} \zeta_p^{(d^Y)^t A_{YX}(d^X - d'^X)} = \begin{cases} p^n, & \text{if } d^X = d'^X, \\ 0, & \text{otherwise.} \end{cases}$$

Then the formula (20) becomes

$$0 = p^n \sum_{d^X} c(d^X) \overline{c'(d^X)} = p^n < v'|v> .$$

By (19) we have $< w'|e|w > = 0$. This completes the proof of Theorem 6.11.   □

The condition (1) in Theorem 6.11 can be rewritten as

$$[A_{IX}\, A_{IE}]\begin{pmatrix} d^X \\ d^E \end{pmatrix} = O^I. \tag{21}$$

If $n = k + 2(d - 1)$, then $|X \cup E| = k + d - 1 = |I|$, so that $[A_{IX}\, A_{IE}]$ is a matrix of size $(k + d - 1) \times (k + d - 1)$. If this matrix is invertible over $\mathbb{F}_p$, then the condition (1) in Theorem 6.11 (which is the same as (21)) implies $d^X = O^X$ and $d^E = O^E$, so that the condition (2) of the theorem is satisfied. Thus, we have following result.

Theorem 6.12: *Let p be an odd prime number, X and Y be two disjoint sets, $|X| = k$, $|Y| = n = k + 2(d - 1)$, $d \geq 2$. Let $A = (a_{ij})_{i,j \in X \cup Y}$ be a symmetric matrix over $\mathbb{F}_p$. If the submatrix $[A_{IX}\, A_{IE}]$ is invertible over $\mathbb{F}_p$ for any subset E of Y, $|E| = d - 1$ and $I = Y \setminus E$, then there exists a quantum code $[[n, k, d]]_p$ (which reaches the Singleton bound).*

We may call the quantum codes constructed in Theorem 6.12 *graphic quantum codes*. One advantage of graphic quantum codes is shown in the following result.

Theorem 6.13: *If there exists a quantum code $[[n, k, d]]_p$ $(d \geq 2, n = k + 2(d - 1))$, then for each $i, 1 \leq i \leq d - 2$, there exists a (graphic) quantum code $[[n - i, k + i, d - i]]_p$.*

Proof: By assumption we know that there exists a symmetric matrix

$$A = (a_{ij})_{i,j \in X \cup Y} \quad (|X| = k, |Y| = n, X \cap Y = \emptyset)$$

over $\mathbb{F}_p$ such that for each $E \subseteq Y$, $|E| = d - 1, I = Y \setminus E$ the matrix $[A_{IX}\, A_{IE}]$ is invertible. Now we choose $T \subseteq Y$, $|T| = i$ and let $X' = X \cup T$, $Y' = Y \setminus T$. Then $|X'| = k + i$, $|Y'| = n - i$. For each $E' \subseteq Y'$, $|E'| = d - i - 1$ and $I' = Y' \setminus E'$, let $E = T \cup E' \subseteq Y$ and $I = I'$, then $|E| = d - 1, I = Y \setminus E$ and $X' \cup E' = X \cup E$, so that $[A_{I'X'}\, A_{I'E'}] = [A_{IX}\, A_{IE}]$ up to a permutation of columns. By assumption this matrix is invertible. Thus, there exists a graphic quantum code $[[n - i, k + i, d - i]]_p$.   □

Example 6.14: Rains [31] proved that there exists a quantum code $[[5, 1, 3]]_p$ for all $p \geq 3$. Schlingemann and Werner [37] present another proof by using Theorem 6.12. Let $X = \{x\}$, $Y = \{y_0, y_1, y_2, y_3, y_4\} = \{y_i \mid i \in \mathbb{Z}/5\mathbb{Z}\}$. Consider the following symmetric matrix

$$
A = \begin{array}{c}
 \\
x \\
y_0 \\
y_1 \\
y_2 \\
y_3 \\
y_4
\end{array}
\begin{array}{c}
\begin{array}{cccccc} x & y_0 & y_1 & y_2 & y_3 & y_4 \end{array} \\
\begin{bmatrix}
0 & 1 & 1 & 1 & 1 & 1 \\
1 & 0 & 1 & 0 & 0 & 1 \\
1 & 1 & 0 & 1 & 0 & 0 \\
1 & 0 & 1 & 0 & 1 & 0 \\
1 & 0 & 0 & 1 & 0 & 1 \\
1 & 1 & 0 & 0 & 1 & 0
\end{bmatrix}
\end{array}
$$

(graph G)

The graph G corresponding to A is shown on the right-hand side. For $E = \{y_i, y_j\}$ $(0 \le i \ne j \le 4)$ and $I = Y \setminus E$, we denote the determinant $A_{(ij)} = |A_{IX}\, A_{IE}|$. From the cyclic properties of the graph G we know that $A_{(ij)} = A_{(i+1, j+1)}$. Thus, we need to compute $A_{(01)}$ and $A_{(02)}$.

$$
A_{(01)} = \begin{vmatrix} 1 & 0 & 1 \\ 1 & 0 & 0 \\ 1 & 1 & 0 \end{vmatrix} = 1, \quad A_{(02)} = \begin{vmatrix} 1 & 1 & 1 \\ 1 & 0 & 1 \\ 1 & 1 & 0 \end{vmatrix} = 1.
$$

By Theorem 6.12, there exists a (graphic) quantum code $[[5, 1, 3]]_p$ for all $p \ge 3$. In fact, from Theorem 6.13 we know that there exists a $[[4, 2, 2]]_p$ for all $p \ge 3$.

With a little more number theory and combinatorics, we can show that there exist $[[6, 2, 3]]_p$, $[[5, 3, 2]]_p$ and $[[7, 3, 3]]_p$ for all $p \ge 3$ [13].

Theorem 6.15: *Let $d \ge 2$, $k \ge 0$, $n = k + 2(d-1)$. Then there exists a graphic quantum code $[[n, k, d]]_p$ for each prime number $p \ge \binom{n}{d-1} + 1$.*

Proof: Let $X \cap Y = \emptyset$, $|X| = k$, $|Y| = n$. Consider the set S of all zero-diagonal symmetric matrices over $\mathbb{F}_p$,

$$
S = \{A = (a_{ij})_{i,j \in X \cup Y} \mid a_{ij} \in \mathbb{F}_p, u_{ij} - a_{ji}, a_{ii} = 0\}.
$$

There are $\frac{1}{2}\binom{n}{2}$ independent entries a_{ij} in A, thus $|S| = p^{\frac{1}{2}\binom{n}{2}}$. For each $E \subseteq Y$, $|E| = d-1$ and $I = Y \setminus E$, the matrix $[A_{IX}\, A_{IE}]$ has $(k+d-1)^2$ independent entries since $I \cap (X \cup E) = \emptyset$. It is well known that the number of invertible matrices with size l over $\mathbb{F}_p$ is

$$|GL_l(\mathbb{F}_p)| = \prod_{i=0}^{l}(p^l - p^i).$$

Therefore

$$\#\{A \in S | \, |A_{IX}A_{IE}| \neq 0\} = |GL_{k+d-1}(\mathbb{F}_p)| \cdot p^{\frac{1}{2}\binom{n}{2} - (k+d-1)^2},$$

$$\#\{A \in S | \, |A_{IX}A_{IE}| = 0\} = p^{\frac{1}{2}\binom{n}{2}} - |GL_{k+d-1}(\mathbb{F}_p)| \cdot p^{\frac{1}{2}\binom{n}{2} - (k+d-1)^2}$$

$$= p^{\frac{1}{2}\binom{n}{2}} \left(1 - \prod_{i=1}^{k+d-1}\left(1 - \frac{1}{p^i}\right)\right).$$

There are $\binom{n}{d-1}$ subsets E of Y such that $|E| = d - 1$. Therefore, if

$$p^{\frac{1}{2}\binom{n}{2}} > \binom{n}{d-1} p^{\frac{1}{2}\binom{n}{2}}\left(1 - \prod_{i=1}^{k+d-1}\left(1 - \frac{1}{p^i}\right)\right), \qquad (22)$$

then there exists $A \in S$ such that $|A_{IX}A_{IE}| \neq 0$ for all $E \subseteq Y$, $|E| = d-1$ and $I = Y \setminus E$, which implies the existence of a graphic quantum code $[[n, k, d]]_p$ by Theorem 6.13. It is easy to see that the condition (22) is equivalent to $\prod_{i=1}^{k+d-1}(1 - \frac{1}{p^i}) > 1 - \binom{n}{d-1}^{-1}$. Moreover, if $p \geq \binom{n}{d-1} + 1$, then

$$\prod_{i=1}^{k+d-1}\left(1 - \frac{1}{p^i}\right) \geq 1 - \sum_{i=1}^{k+d-1}\frac{1}{p^i} > 1 - \sum_{i=1}^{\infty}\frac{1}{p^i}$$

$$= 1 - \frac{1}{p-1} \geq 1 - \binom{n}{d-1}^{-1}.$$

This completes the proof of Theorem 6.15.   □

References

1. A. Ashikhim and E. Knill, Non-binary quantum stabilizer codes, IEEE Trans. IT-47 (2001), 3065–3072.
2. A. Ashikhim and S. Litsyn, Upper bounds on the size of quantum codes, IEEE Trans. IT-45 (1999), 1206–1215.

3. A. Ashikhim, S. Litsyn, and M.A. Tsfasman, Asymptotically good quantum codes, quant-ph/0006061, 2000.

4. C.H. Bennett, D.P. Divincenzo, J.A. Smolin and W.K. Wootters, Mixed state entanglement and quantum error correction. Phys. Rev. A, 54 (1996), 3824–3851.

5. A.R. Calderbank, E.M. Rains, P.W. Shor and N.J.A. Sloane, Quantum error correction and orthogonal geometry, Phys. Rev. Lett., 78 (1997), 405–408.

6. A.R. Calderbank, E.M. Rains, P.W. Shor and N.J.A. Sloane, Quantum error correction via codes over $GF(4)$, IEEE Trans. IT-44 (1998), 1369–1387.

7. A.R. Calderbank, P.W. Shor, Good quantum error-correcting codes exist, Phys. Rev. A, 54 (1996), 1098–1105.

8. Hao Chen, Some good quantum error-correcting codes from algebraic-geometry codes, IEEE Trans. IT-47 (2001), 2059–2061.

9. Hao Chen and J.H. Lai, Construction of quantum error-correcting codes via algebraic-geometry codes, preprint, 1999.

10. H. Chen, S. Ling and C. Xing, Asymptotically good quantum codes exceeding the Ashikhim-Litsyn-Tsfasman bound, IEEE Trans. IT-47 (2001), 2055–2058.

11. H. Chen, S. Ling and C. Xing, Quantum codes from concatenated Reed-Solomon codes, preprint, 2001.

12. G. Cohen, S. Encheva and S. Litsyn, On binary constructions of quantum codes, IEEE Trans. IT-45 (1999), 2495–2498.

13. K. Feng, Quantum codes $[[6,2,3]]_p$ and $[[7,3,3]]_p$ $(p \geq 3)$ exist, to appear in IEEE Trans. IT (2002).

14. K. Feng and Zhi Ma, Quantum Gilbert-Varshamov bound, preprint, 2001.

15. D. Gottesman, Class of quantum error correcting codes saturating the quantum Hamming bound, Phys. Rev. A, 54 (1996), 1862–1868.

16. D. Gottesman, Stabilizator codes and quantum error correction, PhD thesis, CIT, 1997, quant-ph/9705052.

17. D. Gottesman, An Introduction to Quantum Error Correction, Proceedings of Symposia in Applied Mathematics, 2000.

18. M. Grassl and T. Beth, Quantum BCH codes, quant-ph/9910060, 1999.

19. M. Grassl, W. Geiselmann and T. Beth, Quantum Reed-Solomon codes, in Proceedings AAECC-13, LNCS 1719, Springer-Verlag, 1999, pp. 231–244.

20. J. Gruska, Quantum Computing, The McGraw-Hill Companies, London, 1999.

21. E.H. Knill, Non-binary unity error bases and quantum codes, quant-ph/9608048, May 1996.

22. E.H. Knill, Group representations, error bases and quantum codes, quant-ph/9608048, Aug. 1996.

23. E.H. Knill and R. Laflamme, Theory of quantum error-correcting codes, Phys. Rev. A, 55 (1997), 900–911.

24. R. Laflamme, C. Miquel, J. Paz and W.H. Zurek, Perfect quantum error correction code, Phys. Rev. Lett., 77 (1996), 198–201.

25. R. Lidl and H. Niederreiter, Introduction to Finite Fields and Their Applications, Cambridge Univ. Press, 1986.

26. J.H. van Lint, Introduction to Coding Theory, Springer-Verlag, New York, 1982.
27. F.J. MacWilliams and N.J.A. Sloane, The Theory of Error-Correcting Codes, North-Holland, Amsterdam, 1977.
28. R. Matsumoto and T. Uyematsu, Constructing quantum error-correcting codes for p^m-state systems from classical error-correcting codes, quant-phy/9911011, 1999.
29. A. Peres, Quantum Theory: Concepts and Methods, Kluwer Acad. Pub., 1993.
30. E.M. Rains, Quantum weight enumerators, IEEE Trans. IT-44 (1998), 1388–1394.
31. E.M. Rains, Non-binary quantum codes, IEEE Trans. IT-45 (1999), 1827–1832.
32. E.M. Rains, Quantum codes of minimal distance two, IEEE Trans. IT-45 (1999), 266–271.
33. E.M. Rains, Quantum shadow enumerators, IEEE Trans. IT-45 (1999), 2361–2366.
34. E.M. Rains, Polynomial invariants of quantum codes, IEEE Trans. IT-46 (2000), 54–59.
35. E.M. Rains, R.H. Hardin, P.W. Shor and N.J.A. Sloane, A nonadditive quantum code, Phys. Rev. Lett., 79 (1997), 953–954.
36. D. Schlingemann, Stabilizer codes can be realized as graph codes, arXir: quant-ph/0111080 v1, 2001.
37. D. Schlingemann and R.F. Werner, Quantum error-correcting codes associated with graphs, quant-ph/00012111, 2000. Phys. Rev. A, 65 (2001).
38. P.W. Shor, Scheme for reducing decoherence in quantum computer memory, Phys. Rev. A, 52 (1995), 2493–2496.
39. P.W. Shor and R. Laflamme, Quantum analog of the MacWilliams identities in class coding theory, Phys. Rev. Lett., 78 (1997), 1600–1602.
40. A.M. Steane, Multiple particle interference and quantum error correction, Proc. Roy. Soc. Lond. A, 452 (1996), 2551–2577.
41. A.M. Steane, Error correcting codes in quantum theory, Phys. Rev. Lett., 77 (1996), 793–797.
42. A.M. Steane, Simple quantum error correction codes, Phys. Rev. Lett., 77 (1996), 7.
43. A.M. Steane, Quantum Reed-Muller codes, IEEE Trans. IT-45 (1999), 1701–1704.
44. A.M. Steane, Enlargement of Calderbank-Shor-Steane quantum codes, IEEE Trans. IT-45 (1999), 2492–2495.
45. A. Thangaraj and S.W. Mclaughlin, Quantum codes from cyclic codes over $GF(4^m)$, quant-ph/008129, 2000.

PUBLIC KEY INFRASTRUCTURES

Dieter Gollmann

Microsoft Research, 7 J J Thomson Avenue,
Cambridge CB3 0FB, United Kingdom
E-mail: diego@microsoft.com

Introduction

Over the last few years, public key infrastructures (PKIs) have become a familiar sight at security exhibitions and received a lot of attention in the security literature, both in academic and commercial circles. To some, 1999 was the year of the PKI and the widespread deployment of PKIs was thought to be imminent. PKIs are being advertised as the patent solution to many Internet security problems, in particular as a means of creating trust for global e-commerce. This tutorial will not explain how to correctly install a PKI or try to evaluate the PKI *products* currently on offer – I cannot lay claim to any qualification for lecturing on these subjects – but will analyse what lies behind the acronym 'PKI'. We will examine the security problems PKIs claim to address and assess to which extent they can be solved at all by technical means. By the end of the tutorial, we hope the reader will have a better understanding of the security services one can expect from a PKI, and equally of the limitations of this particular technology.

The tutorial is structured into four parts. Part 1 lays the foundations, searching for a definition of the term PKI, giving a brief summary of the cryptographic background, and introducing the constituting elements of a PKI. Part 2 compares PKI standards like X.509, PKIX, PKCS, and SPKI. It also refers to legal matters relevant for trust in e-commerce. Part 3 examines policy issues like key generation and storage, naming, certificate validity, and revocation, and takes a quick look at the use of formal methods in the analysis of a PKI. Part 4 asks whether PKIs are necessary or sufficient to solve security problems in some typical applications areas.

143

I have no experience with PKI products, but have participated in relevant working groups and have collected information from those active in the field. If my attempts to keep the information in this tutorial accurate and recent have failed, I apologize to the reader. In particular, Internet Drafts and the URLs included for further reference are subject to frequent changes.

1. Foundations

1.1. *What is a PKI?*

To have a meaningful technical discussion on any subject, one first has to define the object of the conversation. In this spirit, we will search for a useful working definition of the term 'public key infrastructure'. We take our cue from the web pages of major PKI vendors and from some PKI users (government bodies piloting PKI projects).

> PKI products consist of software products designed to register, issue, and manage the public and private keys related to digital certificates throughout the life cycle of the certificate [Entrust].

This definition concentrates on the software support for public key cryptography. Compared to other proposals, it is to be commended for its precise focus. However, security is rarely ever achieved by technical means only, so one needs a wider infrastructure to reap the potential benefits of public key cryptography, as expressed by the next definition.

> A Public-Key Infrastructure (PKI) is a comprehensive structure consisting of protocols, services and standards that facilitate the use of public-key cryptography by allowing the secure distribution of public keys between communicating parties, and thereby supporting the exchange of sensitive information over an unsecured network such as the Internet [Secure Solutions Experts].

This is again a consistent definition, adding standards and services as components to the PKI. The next two quotes come from the same source.

> PKI provides the core framework for a wide variety of components, applications, policies and practices to combine and achieve the four principal security functions [confidentiality, integrity, authentication, non-repudiation] for commercial transactions.

> The technology to secure the Internet is known as Public Key Infrastructure (PKI) [Baltimore].

This quote makes a bold promise, and implies that public key cryptography is a prerequisite for securing the Internet. The same sentiment is reflected in the next pair of quotes.

> ... sensitive information and communications ... will require a range of security assurances. Fully and properly implemented, a PKI is a system of hardware, software, policies and people that can provide these assurances.

> The basis of PKI's security assurances is a sophisticated cryptographic technique known as public key cryptography. [United States General Accounting Office].

Indeed, any comprehensive security infrastructure has to consider the people involved in it and the procedures they are asked to follow. However, there is no logical reason why public key cryptography has to be used to protect sensitive information and communications.

> Federal agencies considering the use of public key technology will benefit by proceeding promptly to participate in building the Federal portion of the evolving worldwide PKI [NIST].

This final quote leads into yet another direction, envisaging a worldwide PKI as an infrastructure for issuing something akin to digital identity cards. The answers to the question 'What is a PKI?' are thus manifold. A PKI may be

- software for managing digital certificates,
- a system of hardware, software, policies and people providing security assurances,
- the technology for securing the Internet,
- a worldwide system of digital ID cards.

None of these is the 'correct' answer. Some may be more well-defined than others, but all have some merit. It is important to remember the differences, and whenever a PKI is encountered one has to establish which interpretation is intended at that particular moment before drawing any further conclusions.

1.2. *Public key cryptography*

It is assumed that the reader is already familiar with the basic concepts of public key cryptography. This summary is included as a brief reminder, and it creates an opportunity for pointing out some common misconceptions surrounding public key cryptography. The main technical reference for the cryptographic mechanisms mentioned is [14].

Figure 1 depicts the fundamental idea of symmetric key cryptography. Encryption protects documents on the way from A to B. A and B share a secret key. A procedure is required that allows A and B to obtain their shared key. For n parties to communicate, about n^2 keys are needed, $\frac{n(n-1)}{2}$ to be precise. Familiar symmetric key encryption algorithms are DES, IDEA, and now AES.

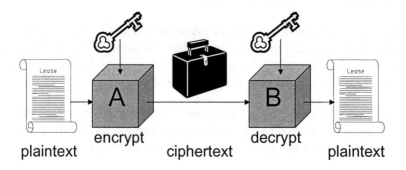

plaintext   encrypt   ciphertext   decrypt   plaintext

Fig. 1.   Schematic description of symmetric key cryptography.

Figure 2 depicts the fundamental idea of public key cryptography. Encryption protects documents on the way from A to B. B has a public encryption key and a private decryption key. A procedure is required that allows A to get an authentic copy of B's public key. For n parties to communicate, n key pairs are needed. The most familiar public key encryption algorithm is RSA. We assume the reader is familiar with its details.

Digital signatures protect the authenticity of documents. A has a public verification key and a private signature key (Figure 3). A procedure is required for B to get an authentic copy of A's public verification key. Successful verification of a signed document links the document to A's public key and implies that it had not been tampered with. Familiar digital signature algorithms are RSA and DSA.

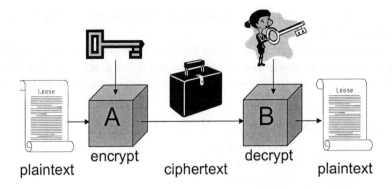

Fig. 2. Schematic description of public key cryptography.

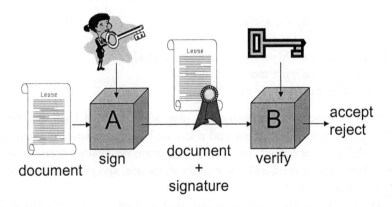

Fig. 3. Schematic description of digital key signatures.

1.3. *Misconceptions about digital signatures*

A number of misconceptions about digital signatures are prevalent in the cryptographic literature. The first is a technical point. Often the verification of a digital signature is explained as encryption with the public key, plus a comparison of the result with the document being verified. This description can be even found in X.509, which is probably a main source for the propagation of this fallacy.

The claim is not even true for RSA, which uses a public exponent e both for signature verification and for encryption. However, signature verification also requires some redundancy checks on the document. Otherwise, *existential forgeries* are possible where an attacker raises an arbitrary value

s to the public exponent e and presents the result $m = s^e \bmod n$ as a document with signature s.

The claim certainly does not hold for DSA. Signature generation in DSA takes as input a private key a, public values α, p, q, and the hash $h(m)$ of the document m. Guidance on the choice of parameters can be found in [14]. To sign m,

- select k at random, $0 < k < q$,
- compute $r = (\alpha^k \bmod p) \bmod q$,
- compute $s = k^{-1}(h(m) + ar) \bmod q$,
- the signature is the pair (r, s).

Signature verification takes as its input the signature (r, s), public values p, q, α, $y = \alpha^a \bmod p$, and the hash $h(m)$, and proceeds as follows:

- verify $0 < r < q$, $0 < s < q$,
- compute $w = s^{-1} \bmod q$,
- compute $u_1 = w \cdot h(m) \bmod q$, $u_2 = r \cdot w \bmod q$,
- compute $v = (\alpha^{u_1} y^{u_2} \bmod p) \bmod q$,
- accept if and only if $v = r$.

Beside this specific counterexample, observe in general that encryption takes as its input a document and a key and produces another document (the ciphertext) as output. In contrast, signature verification takes as its input a document, a verification key, and a signature, and produces a Boolean answer ('yes' or 'no').

The second misconception claims that a digital signature binds the signer (a person) to the document. As we have seen, digital signatures are mathematical evidence linking a document to a public verification key. The link between a public key and a person has to be established by procedural means, and could be recorded in a *certificate* (see Section 1.5). The holder of a private signature key has to protect the key from compromise, relying again on procedural means or on tamper-resistant devices. Finally, there has to be evidence that the signer actually wanted to sign the document the digital signature was computed for.

The third misconception claims that digital signatures are legally binding. Even when digital signatures are recognized by law, there is no guarantee that there is a court with jurisdiction. We return to this issue in Section 4.5.

The fourth misconception assumes that a certificate is needed to verify a digital signature. In fact, an authentic copy of the verification key is needed. The verification key may be stored in a certificate, but can also be stored in protected memory. Furthermore, to check a certificate, another verification key is needed. Ultimately, you need a root verification key whose authenticity is not guaranteed by a certificate.

1.4. *Electronic signatures*

To differentiate between digital signatures as a cryptographic mechanism and signatures as a concept at the application level, a convention is emerging that refers to digital signatures to indicate a cryptographic mechanism for associating documents with verification keys, and *electronic signature* to refer to any security service for associating documents with persons. Electronic signature services usually employ digital signatures as a building block, but could be implemented without them.

1.5. *Certificates*

Public key cryptosystems often rely on a public directory of user names and keys. Kohnfelder implemented the directory as a set of digitally signed data records containing a name and a public key, coining the term certificate for these records already in 1978 [13]. Certificates originally had a single function, binding between names and keys. Today, the term certificate is often used to denote a digitally signed statement [17]. Certificates are used to meet a wider range of objectives:

> Certificate: A signed instrument that empowers the Subject. It contains at least an Issuer and a Subject. It can contain validity conditions, authorization and delegation information [8].

A certificate is simply a signed document binding a *subject* to other information. Subjects can be people, keys, names, etc. Indeed, as the use of certificates changes, our view of what constitutes the subject in a given certificate may change too. We can distinguish between different types of certificates.

- *Identity (ID) certificate*: provides a binding (name, key); sometimes this is the default interpretation;
- *Attribute certificate*: provides a binding (authorization, name);
- *Authorization certificate*: provides a binding (authorization, key); authorization is a new aspect in our security considerations.

1.6. *Certification authorities*

Certificates are signed by an *Issuer. Certification Authority* (CA) is just another name for Issuer. Sometimes CA is used more narrowly for organizations issuing ID certificates [11]. Section 4 will touch on trust requirements for CAs. We write $\langle\langle X \rangle\rangle Y$ for a certificate for entity X issued by Y. *End entity* (EE) is a name for a PKI entity that does not issue certificates. *Cross certificates* are certificates issued by one CA for another CA to shorten certification paths (Figure 4).

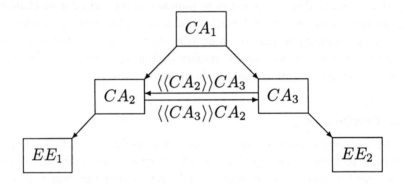

Fig. 4.   Cross certificates.

In general terms, a *Trusted Third Party* (TTP) is an entity providing a security service other parties rely on. A party maintaining a directory of public keys is a typical TTP. A party generating keys for others is a TTP. A CA that issues certificates relied on by others is a TTP. However, it is important to keep in mind that you do not have to be a TTP to issue certificates. Not every CA is a TTP.

1.7. *Registration authorities*

Before issuing an identity certificate to a person, the identity of that person has to be checked. This first step could be performed by a *Registration Authority* (RA). The CA would then only manage the certificate life cycle. For example, German post offices are RAs for Signtrust[a]. Note that RA and CA need not be separate entities. The tasks of an RA can be summarized as follows.

---

[a]http://www.signtrust.de

- Confirm the person's identity.
- Validate that the person is entitled to have the values requested in the certificate.
- Optionally, verify that the subject has possession of the private key associated with the public key requested for a certificate.

1.8. *Time stamping*

Applications may need independent evidence about the time a document has been signed. A *Time Stamp Authority* (TSA) is a TTP that provides a "proof-of-existence" for a particular datum at an instant in time. A TSA does not check the documents it certifies. An example of a time stamping protocol is TSP, the Time Stamp Protocol [2].

Summary

Security is a multi-faceted challenge.

- Do you want to send confidential messages?
- Do you want to request access to a service?
- Do you want to create evidence to be prepared for a potential dispute?
- Do you want protection from outsiders?
- Do you want protection from the party you are dealing with?

PKIs can help you address all these issues, but it is not necessarily a single PKI that solves all your problems, and some of these problems can be addressed quite efficiently without a PKI.

2. PKI Standards

This section gives an overview of some of the most important standards for public key infrastructures.

2.1. *X.509 – ISO|IEC/ITU-T 9594-8*

Today, X.509 version 3 is the most commonly used PKI standard. The original ITU-T Recommendation X.509 [12] was part of the X.500 Directory [6], which has since also been adopted as IS 9594-1. X.500 was intended as a global, distributed database of named entities such as people, computers, printers, etc., i.e. a global, on-line telephone book.

The information held by the Directory is typically used to facilitate communication between, with, or about objects such as application-entities, people, terminals and distribution lists.

X.509 certificates were intended to bind public keys (originally passwords) to X.500 path names (Distinguished Names) to note who has permission to modify X.500 directory nodes. X.500 was geared towards identity based access control:

Virtually all security services are dependent upon the identities of communicating parties being reliably known, i.e. authentication.

This view of the world predates applets and many new e-commerce scenarios, where a different kind of access control is more appropriate.

X.500 Distinguished Names identify object entries in the X.500 Directory. A distinguished name is defined as a sequence of *relative distinguished names* (attribute value assertions) representing an object and all its superior entities. For illustration, we mention some typical attributes:

- C ... country
- L ... locality
- O ... organisation
- OU ... organizational unit
- CN ... common name

X.509 uses the following terminology:

- User certificate (public key certificate, certificate): contains the public key of a user, together with some information, rendered unforgeable by encipherment with the secret key of the certification authority which issued it.
- Attribute certificate: contains a set of attributes of a user together with some other information, digitally signed under the private key of the CA.
- Certification authority: an authority trusted by one or more users to create and assign certificates.

Authorization attributes are linked to a cryptographic key via a common name in a user certificate and an attribute certificate (Figure 5).

Compared to previous versions, the X.509 v3 certificate format (Figure 6) includes extensions to increase flexibility. Extensions can be marked as *critical*. If a critical extension cannot be processed by an implementation, the certificate must be rejected. Non-critical extensions may

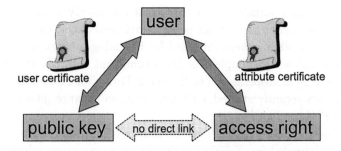

Fig. 5.   Access control in X.509.

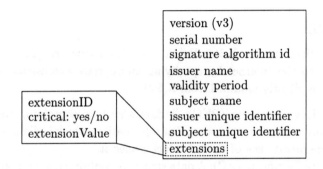

Fig. 6.   X.509 v3 certificate format.

be ignored. Critical extensions can be used to standardize policy with respect to the use of certificates.

X.509 v3 — Evaluation

X.509 is not a good source to learn about public key cryptography. Many explanations are RSA-centric and concepts like *data recipient entity authentication* appear without justification and proper definitions. X.509 is geared towards identity certificates. For access control, access rights are assigned to names, not to keys. Only certificate revocation through CRLs is specified. CAs are 'trusted' and have to be 'authorized'. CAs are distinguished from end entities. X.509 no longer assumes that CAs are organized in a global hierarchy.

X.509 has been criticised for using ASN.1 notation, which was deemed to be too resource consuming. Today, XML is often the format of choice for defining data structures, so this argument hardly applies anymore. X.509 has been criticised for not being flexible enough, so extensions have been added. X.509 has been criticised for becoming too flexible by including extensions, so that implementations compliant with the standard are not necessarily interoperable. The European Forum for Electronic Business (EEMA) has recently started a pki-Challenge to advance interoperability between PKI products[b].

Distinguish between the X.509 certificate format and the application X.509 was intended for. X.509 v3 is simply a data structure, which has the potential to be used for more than one purpose. The data structure does not tell you how you have to use it.

2.2. *PKIX*

PKIX is the Internet X.509 Public Key Infrastructure [11]. It adapts X.509 v3 to the Internet by specifying appropriate extensions. The terminology is slightly modified from X.509.

- Public Key Certificate (PKC): A data structure containing the public key of an end-entity and some other information, which is digitally signed with the private key of the CA which issued it.
- Attribute Certificate (AC): A data structure containing a set of attributes for an end-entity and some other information, which is digitally signed with the private key of the Attribute Authority which issued it.
- Public Key Infrastructure (PKI): The set of hardware, software, people, policies and procedures needed to create, manage, store, distribute, and revoke PKCs based on public-key cryptography.
- Privilege Management Infrastructure (PMI): A collection of ACs, with their issuing Attribute Authorities, subjects, relying parties, and repositories.

PKC extensions include:

- Authority Key Identifier: for issuers with multiple signing keys.
- Subject Key Identifier: typically a hash of the key.
- Key Usage: encipher, sign, non-repudiation, ...
- Private Key Usage Period: if different from certificate validity.
- Certificate Policies: policy under which certificate was issued and purpose for which it may be used.

---

[b]http://www.eema.org/pki-challenge/index.asp

- Subject Alternative Name: additional identities bound to certificate subject.

PKIX authorization follows the X.509 model. Access control requires an AC and a PKC. In X.509, 'authorization conveys privilege from one entity that holds such privilege, to another entity. Users of public key-based systems must be confident that, any time they rely on a public key, the subject that they are communicating with owns the associated private key.' Certificates can provide this assurance if 'owns' means 'is entitled to use'. Certificates cannot provide this assurance if 'owns' means 'is in possession of'.

PKIX — Evaluation

PKIX is a 'name centric' PKI. It adapts X.509 v3 for the Internet by standardizing extensions and specifying profiles for certificates, CRLs (certificate revocation lists), and CRL extensions. There is also a profile for the use of X.509 attribute certificates in Internet protocols and there are many protocols for certificate management. For PKIX related documents, see http://www.ietf.org/ids.by.wg/pkix.html. Among those, the PKIX roadmap [3] contains a useful discussion of PKI theory.

2.3. *SPKI*

The Simple Public Key Infrastructure (SPKI) [8] was developed to address some of the perceived shortcomings of X.509. In particular, there were concerns about the global name spaces once assumed in X.509, and it was also felt that for authorization it would be cleaner to assign authorization attributes directly to cryptographic keys rather than taking the detour via a user names. SPKI specifies

> a standard form for digital certificates whose main purpose is authorization rather than authentication.

Trust management takes the same approach, as expressed in this quote from [9]:

> In answering the question "is the key used to sign this request authorized to take this action?", a trust management system should not have to answer the question "whose key is it?".

In this world, *authorization certificates* link keys to authorization attributes. An identity certificate linking a name to a key would still have a role when

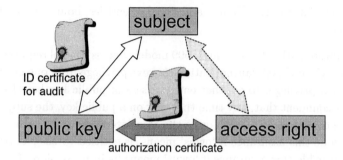

Fig. 7. Access control in SPKI.

auditing access decisions, but it is no longer necessary for access control (Figure 7).

The following definitions are given in [8]:

- Certificate: a signed instrument that empowers the subject; contains at least issuer and subject
- Issuer: signer of the certificate and source of empowerment
- Subject: the thing empowered by a certificate or ACL entry

The SPKI policy is expressed in its *tuple reduction algorithms*. Tuples are abstract notation for certificates or ACL entries. For example, a SDSI 5-tuple has the following fields:

- Issuer: public key (or "Self")
- Subject: public key, name identifying a public key, hash of an object, ...
- Delegation: TRUE or FALSE
- Authorization: access rights
- Validity dates: not-before date and not-after date

A tuple reduction algorithm evaluates 'certificate chains'. No algorithm for finding relevant tuples is specified, the algorithm processes the tuples presented to it. Authorizations and validity periods can only be reduced. Here is an example of a reduction rule:

Input: <Issuer1,Subject1,D1,Auth1,Val1>
       <Issuer2,Subject2,D2,Auth2,Val2>

IF Subject1 = Issuer2 AND D1 = TRUE
   THEN output <Issuer1, Subject2, D2, AIntersect(Auth1,Auth2),
                           VIntersect(Val1,Val2)>

SPKI — Evaluation

The SPKI Certificate Theory [8] is recommended reading on names, access control, etc. SPKI is oriented towards access control and away from global CA hierarchies. There is a separation of concerns. ID certificates are used for accountability. Attribute and authorisation certificates are used for access control. Certificates facilitate distributed storage of Access Control Lists. SPKI standardizes (prescribes?) some policy decisions. For example, only permissions held by the delegator can be delegated. Such a scheme does not support separation of duties very well.

2.4. *PKCS*

PKCS is set of industry standards published by RSA Inc.[c]. For example, PKCS#10 gives a syntax for certification request. A request contains a distinguished name, a public key, and optionally a set of attributes, collectively signed by the entity requesting certification. Certification requests are sent to a CA, which transforms the request into an X.509 public-key certificate.

2.5. *Legal matters*

Many countries have enacted legislation aimed at electronic commerce that is also relevant for PKIs and for the role digital signatures can play in securing electronic transactions. We only give a brief and very selective overview.

The European Union has adopted the Directive 1999/93/EC of 13 December 1999 on a Community framework for electronic signatures. Its aim is to facilitate the use of electronic signatures and contribute to their legal recognition. The Directive is not a law itself but has to be implemented in national law by EU states. For our purpose, the annexes of the Directive have to be mentioned.

- Annex I: requirements for qualified certificates
- Annex II: requirements for certification-service-providers issuing qualified certificates
- Annex III: requirements for secure signature devices

In the US relevant legislation exists both at federal and at state level:

---

[c]http://www.rsasecurity.com/rsalabs/pkcs/index.html

- E-SIGN: Electronic Signatures in Global and National Commerce Act [7] (federal law)
- UCITA: Uniform Computer Information Transactions Act: uniform commercial code for software licenses and other computer information transactions [15] (state law passed in Virginia and Maryland)
- UETA: Uniform Electronic Transactions Act [16] (state law passed in a number of states)

2.6. *Qualified certificates*

In the parlance of RFC 3039 [18], a *qualified certificate* is a PKC issued to a natural person. The primary purpose is identifying a person with high level of assurance in public non-repudiation services. A qualified certificate contains an unmistakable identity based on a real name or a pseudonym of the subject and exclusively indicates non-repudiation as key usage. The CA has to include a statement about the certificate purpose in the certificate policy. Two ways of constructing a qualified certificate are accepted:

- by a policy in the certificate policies extension
- by a private Qualified Certificates Statements extension

Once again, implementations compliant with RFC 3039 may not be interoperable.

Qualified Certificates in the sense of the EU Directive vouch for *advanced electronic signatures*. Advanced electronic signatures created on secure-signature-creation devices should 'satisfy the legal requirements of a signature in relation to data in electronic form in the same manner as a handwritten signature satisfies those requirements in relation to paper-based data' [ETSI TS 101 862]. This document specifies a qualified certificate profile to help in implementing the EU Directive. The following requirements are stated.

- The issuer field must contain country name.
- There must be an indication that the certificate is issued as a qualified certificate.
- There must be a limit to the value of transactions.
- There must be a limit to the retention period of registration information.
- The certificate contains the name of the signatory or a pseudonym, which shall be identified as such.

RFC 3039 defines a certificate format, but cannot imply compliance with the Directive.

2.7. *Summary*

Standards for certificate formats like X.509 v3 are widely supported. Should there be standards for more than certificate formats? How much should be assumed about the usage of certificates? Different applications have different requirements. The more aspects are considered, the more extensive the documents become (see PKIX). Different (partial) implementations of the same standard are then less likely to be interoperable.

3. Points of Policy

The previous section covered certificate formats. This section turns to the *certificate policies* and operational rules that may apply in a PKI. Background for this discussion is the certificate life cycle shown in Figure 8.

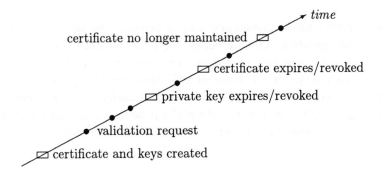

Fig. 8.   Certificate life cycle.

3.1. *Key generation and storage*

Key generation is an important security aspect. Strong cryptography fails if keys are predictable. Weak pseudo random generators are a frequent cause of failure.

- Keys could be generated by the users (certificate subjects), but do they have the competence to do so?

- Keys could be generated by the CA or some other TTP, but can the CA be trusted with the users' private keys?

As so often, there is no correct answer and it will depend on the actual situation which solution is preferable. When users generate their own keys, *proof of possession* (POP) of the private key may be required when a certificate is being issued. POP is recommended when the certificate is to be used for non-repudiation.

Key storage is equally an important security aspect. If a private signing key is compromised, the link to the certificate subject is broken. If a bogus verification key is inserted, users are misled when verifying certificates. The integrity of verification keys can be protected through:

- cryptographic means: e.g. digital signatures, but root verification keys are not protected by cryptographic means;
- physical and operating system security;
- out-of-band checks: fingerprints of keys published in some other medium, e.g. in a book.

The confidentiality of private keys can be protected by means of physical security or operating system security. Keys can be stored on tokens like smart cards, but you have to consider the consequences of the token being lost or stolen. Keys can be stored on a PC protected by a password. Cryptographic protection is then only as strong as security on the PC, and to a large degree limited by the strength of the password. Long keys protected under a short password could be as insecure as the password.

3.2. *Naming*

Subjects have to be given names. For the interoperability of PKI products, there needs to be agreement on syntax. Examples from PKIX are

- X.500 distinguished name
- RFC822 e-mail address
- DNS domain name
- URI: uniform resource identifier

For interoperability of systems, there needs to be agreement on semantics. SPKI makes a strong case that globally unique names are relatively useless in open systems like the Internet. Names have only local meaning.

3.3. *Validity*

Certificates have expiry dates and validity periods. It is a common misconception to think that a certificate cannot be used after it has expired. Deciding what should be done with expired certificates is a policy decision. As an example from the physical world, consider possible entry policies for holders of an EU passport:

- The passport has to be valid n months beyond entry.
- The passport has to be valid until exit.
- The passport has expired less than a year ago (entry into EU countries).

When evaluating a certificate chain we have to consider that certificates expire or may be revoked. The *shell model* takes a conservative approach. All certificates have to be valid at the time of evaluation. In the *chain model* the issuer's certificate has to be valid at the time it issues certificates.

Figure 9 illustrates the shell model. Certificate $\langle\langle EE\rangle\rangle CA_3$ is valid at time t_1 as all three certificates are valid, but is invalid at time t_2 as certificate $\langle\langle CA_2\rangle\rangle CA_1$ has expired. With the shell model, CAs should only issue certificates that expire before their own certificate. If a top level certificate expires or is revoked, all certificates signed by the corresponding private key have to be re-issued under a new key. The shell model is appropriate for certificates defining hierarchical address spaces. It is also the policy implemented in SPKI.

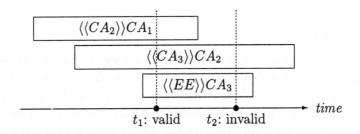

Fig. 9.   Certificate validity in the shell model.

Figure 10 illustrates the chain model. Certificate $\langle\langle EE\rangle\rangle CA_3$ is valid at times t_1 and t_2 as $\langle\langle CA_3\rangle\rangle CA_2$ was valid when $\langle\langle EE\rangle\rangle CA_3$ was issued, and $\langle\langle CA_2\rangle\rangle CA_1$ was valid when $\langle\langle CA_3\rangle\rangle CA_2$ was issued. The chain model requires a time-stamping service (some means of reliably establishing when a certificate was issued). If a top level certificate expires or is revoked, cer-

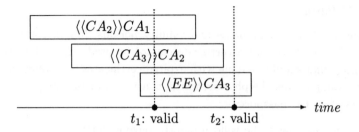

Fig. 10.   Certificate validity in the chain model.

tificates signed by the corresponding private key remain valid. For example, consider an organisation issuing membership certificates signed by a manager. When the manager leaves and his certificate is revoked, it should not be necessary to re-issue all membership certificates.

Policies beyond the shell and chain model have been proposed, and we can generalize our observations for end entity certificates to other types of signed documents.

Never forget that a certificate cannot tell the end user what the end user's policy is. A certificate can tell the end user what the CA's policy is and it may limit the CA's liability. Be aware that policy decisions have consequences. In the shell model, certificates have to be re-issued when a top level certificate has been revoked. In the chain model certificates should be time-stamped.

3.4. *Revocation*

Certificates may have to be revoked if a corresponding private key is compromised or if a fact the certificate vouches for no longer is valid. *Certification Revocation Lists* (CRLs) are a solution to this problem and are used in X.509. CRLs are distributed in regular intervals or on demand. A *Delta-CRL* records only the changes since the issue of the last CRL. CRLs make sense if on-line checks are not possible or too expensive.

When on-line checks are feasible, CRLs can be queried on-line. Indeed, when on-line checks are feasible, certificate status can be queried on-line. Examples are the PKIX Online Certificate Status Protocol — OCSP [RFC 2560] and so called *positive lists* of valid certificates in the German signature infrastructure. Today, on-line checks are often the preferred solution.

When a CA issues certificates for its own use, e.g. for access control, only a local CRL is required.

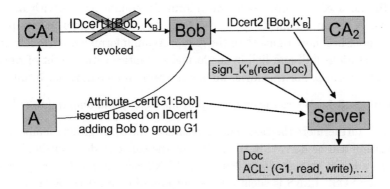

Fig. 11.   Revocation – a scenario.

Figure 11 poses a question about revocation and attribute certificates. Bob had acquired an attribute certificate on the basis of an identity certificate IDcert1[Bob,K_B] that is now no longer valid. Bob uses this attribute certificate and another identity certificate, IDcert2[Bob,K'_B], when requesting access to the resource *Doc*. Should the request be granted? Should the attribute certificate be revoked?

The answers depend on your policy. Is the attribute certificate linked to Bob's name or to an identity certificate? Does the server differentiate between CA_1 and CA_2? Would your policy change if IDcert1 and IDcert2 had been issued by the same CA? You cannot look at Figure 11 to learn your policy. You can look at it to learn about issues your policy may have to address.

3.5. *Formal methods and PKIs*

Formal methods can help to analyse the effects of complex policy rules. We will use SDSI [17] as an example. SDSI is a PKI for access control, using groups (and access control lists) to express security policies. We write a SDSI certificate as $k : n \to p$, denoting that principal p is added to group n in the local name space of k, or that n is a name for principal p in the local name space of k. (We will sometimes omit the identifier of the local name space in our examples.) We write k's n for the name n in the local name space of k. SDSI specifies an algorithm for deciding whether a principal is a member of a given group. We will give brief overviews of two formal analyses of SDSI. In the first case, a logic of access control was tailored to SDSI [1]. The second case is a 'precise' logic (LLNS) [10] for SDSI.

In the access control logic, we informally equate names with access rights. A group certificate $n \to p$ gives p the access rights of group n. A member has more rights than the group. Again informally, we equate the interpretations $[[n]]$ and $[[p]]$ in the formal semantics with the set of rights given to n and p. For this logic, we have a theorem matching our intuition.

Theorem 3.1: *We can derive $n \to p$ in the logic if and only if $[[n]] \subset [[p]]$.*

It is furthermore the case that all results of the SDSI group membership algorithm can be derived in the logic, but the converse does not hold.

To address this apparent deficiency, LLNS was intended as a 'precise' logic for SDSI. LLNS is silent on the purpose of certificates. A group certificate $n \to p$ is simply read as 'p is member of n'. For our purpose, we equate groups with email lists. The group certificate $n \to p$ adds p to the list n. The group contains more entries than a member. Informally, we equate the interpretations $[[n]]$ and $[[p]]$ in the formal semantics with the set of addresses in n and p. Also for this logic, we have a theorem matching our intuition.

Theorem 3.2: *We can derive $n \to p$ in the logic if and only if $[[n]] \supset [[p]]$.*

It is the case that LLNS captures SDSI group membership exactly. We are now in the interesting situation of having two logics that make contradicting observations about the world. This raises a few questions.

- Is LLNS the more reasonable formalisation of SDSI as it models SDSI precisely?
- Is LLNS a logic for access control?
- Are the SDSI rules compatible with the SDSI goals?

To see which logic is 'right', we tell a short story.

> MIT is an authority for *Ron*, a cryptographer who changes keys weekly to reduce key exposure. In week 1, MIT issues MIT: *Ron* $\to k_1$ and *Ron* issues k_1: *Butler* $\to$ *Lampson*. In week 2, MIT issues MIT: *Ron* $\to k_2$ and *Ron* issues k_2: *Lampson* $\to k_3$. Re-issuing *Ron*'s first certificate with k_2 would defeat the purpose of changing keys. Later, *Ron* receives a request signed by key k_3 for a resource *Butler* is allowed to access. Should the request be granted?

The conclusion MIT's *Ron's Butler* $\to k_3$ that would grant access follows in the access control logic because both k_1 and k_2 'speak for' *Ron*, but not from SDSI group membership, and it cannot be derived in LLNS. In

SDSI, a local name can be bound to more than one key. The wording in [17] suggests that the above conclusion would be acceptable.

The conclusion we must reach is that SDSI group membership does not define a complete access control scheme, and that one has to take care when interpreting the results of formal analysis.

3.6. *Summary*

The rules your PKI implements are given by your policy, not by a law of nature. Most PKI standards make assumptions about the world and standardize policy to some extent. Complex relationships between PKI entities may imply complex evaluation and revocation rules. Formal methods can help in the analysis of complex systems, but you have to make sure that the formal analysis captures your intentions.

4. PKI as a Solution

Having dealt with the technical aspects internal to a PKI, we finally turn to applications that seem to be promising areas for PKI deployment. We will check whether PKIs are necessary or sufficient to deal with the security problems encountered.

4.1. *Secure email*

Messages arrive from arbitrary sources. There is usually some indication of the name of the sender, but you may not know that person or organisation. Authenticating the sender requires an authentic copy of the verification key, e.g. a certificate. Certificates bind a name and/or email address to an encryption and/or verification key. Three classes of certificates are common today:

- Class-1: online enrolment, no identity checks; some CAs therefore do not include a name in the certificate;
- Class-2: CA checks identity and address against third party information; checks that the name is 'real' but not that it is really the applicant;
- Class-3: applicant has to appear in person at CA.

Email is signed to authenticate the sender. Signed email is authenticated on receipt, either when it arrives in the mailbox or when it is first opened. Signed email is rarely a contract that has to be archived or re-authenticated at a later date or be presented as evidence in a dispute. (Signed) email may be evidence held against you.

Messages are mailed to known addresses. End-to-end encryption of email requires an authentic copy of the recipient's encryption key. Alternatively, email may be encrypted between the sender (receiver) and a mail server.

Consider now the special case of secure company email. Staff has external access to the company mail server. To secure access to the mail server, each user gets a private key/public key pair and a certificate for their public keys, which is presented on access to the server. A PKI is deployed to support this application. The company could issue the certificates, so no TTP is needed. The public keys could be stored in a protected list at the mail server, so no certificates are needed. Users could be issued with secret keys (one key per user), so no public key cryptography is needed. The company can secure access to its mail server without deploying a PKI.

The secret key solution and the public key solutions both use one key per user. For the stated security goal, compromise of secret keys and private keys are equivalent. The PKI solution has no advantage compared to a conventional secret key solution. This situation may change when the requirements change, e.g.

- staff should get access to company resources isolated from a central server (revocation will have to be addressed),
- support for collaboration between companies,
- end-to-end secure email,
- support for non-repudiation.

4.2. *Non-repudiation*

Handwritten signatures on paper documents signify the intent of the signer and thus bind the signer to the document. In the electronic world, non-repudiation protocols should create evidence that binds a (legal) person to a document.

As discussed in Section 1.3, digital signatures bind a document to a public verification key. A certificate can bind the public verification key to a name. Operational procedures at the CA check the correspondence between the name and a person. Further operational procedures check that the person holds the private key matching the public key in the certificate. Computer security protects the private key from compromise and misuse. With this level of indirection, it becomes doubtful whether digital signatures can really prove intent [19].

Proper non-repudiation requires a complex infrastructure. The concept of a PKI is relevant when seen as 'a system of hardware, software, policies and people'. CAs are part of this system. This is generally appreciated and policies for CAs are receiving considerable attention. End users are also part of this system. Can they fulfil their security obligations? Can they plead ignorance or incompetence if they want to renege on a commitment? The answers will again depend on the type of end user envisaged. Companies participating in a PKI may well be expected to meet higher standards of professionalism than home users. On the technical level, end user devices are part of the system. Securing end user devices is a difficult task and may prove impossible when the devices become too complex.

It is important to keep in mind that signing email is inherently different from signing a contract. Email is signed to authenticate the source, not to indicate intent. Automatic signing of email for authentication is reasonable, automatic contract signing is not. Contracts are signed to indicate intent (non-repudiation). In this case, the certificate and the key have to be valid at time contract is signed. Hence, a time stamping service is required. A contract will be verified at some time in the future. Thus, contracts have to be archived. Certificates (verification keys) have to be archived. Software for displaying the contract has to be archived.

4.3. *Access control*

X.509-style PKIs support identity-based access control. Identity-based access control is a legacy from previous eras when *closed systems* were dominant. In the Internet age, we are dealing frequently with *open systems* requiring different models of access control. Moreover, identity-based access control raises privacy and data protection issues. Still, organisations are used to identity-based access control, maybe augmented by Role Based Access Control (RBAC), so there is a demand for this type of systems.

For access control on the Internet, the world is still searching for appropriate security policies that are not based on user identities. We have the Java sandbox model and code signing. We have a choice between authorization certificates assigning rights to keys and attribute certificates assigning rights to names. We may consider 'anonymous' access to services paid for in e-cash. Trust management systems such as KeyNote [5] deal with 'delegation' of access decisions to third parties.

In access control, certificates form a loop starting from the party that makes the final access decision (Figure 12). Identity certificates can be used for auditing purposes to augment access control. CAs are part of your local PKI and should adhere to your security policy. For access control within an organisation, you can be your own CA, and would not even need certificates.

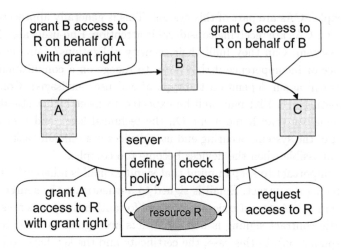

Fig. 12. Certificate loop in access control.

Applications like secure email provide communications security. The theoretical investigations in Section 3.5 could imply that at a fundamental level communications security and access control are following incompatible rules, and be a warning sign of potential pitfalls when using a single PKI to support both access control and communications security. In particular, it is potentially much more difficult to federate the access control systems of several organisations than to federate their secure email systems.

4.4. *Digital identity cards*

Specific certificates such as the qualified certificates of the EU Directive are intended to vouch for the identity of a person conducting an electronic transaction, thus taking the role of identity cards on the Internet. Such 'digital identity cards' can be useful for e-government, e.g. to protect access to personal data (health information) or to control access to services and benefits. By their very nature, government services are mainly offered locally to residents of a country.

Digital identity cards do not seem to be directly useful for e-commerce. It is often unnecessary to know the precise identity of an entity, relying on identities (of persons) may create privacy problems, and the necessary infrastructure may be too expensive to support.

When CAs act as TTPs issuing identity (qualified) certificates, the parties relying on these certificates may want to have some guarantees that the

CA performs its job properly. A regulatory framework may therefore state the liability of a CA or even provide for the accreditation of CAs. Accreditation is not required in general by the EU Directive. In specific cases a CA may be governed by rules pertinent to the application. This point is raised, for example, in a discussion paper recently published by the UK Financial Services Authority [4].

4.5. *Legal assurance*

Assume that electronic (digital) signatures are accepted as evidence in any court of law. Assume properly regulated CAs issue qualified certificates. Assume there is a dispute. Which court has jurisdiction? Depending on the circumstances, there may be no court, there may be exactly one court (it helps when all parties reside in the same country), or there may be several courts.

In our legal systems location is an important factor, but e-commerce and the Internet are notorious for 'ignoring' national boundaries. Where does an electronic transaction take place? To adjudicate on electronic transactions, the problem is less the recognition of electronic signatures, but determining a court with jurisdiction.

Consider the current situation in the United States regarding the electronic delivery of goods. UCITA gives the *licensor* (provider) the right to determine where to litigate a dispute. However, Iowa has passed UETA with an explicit provision that for consumers in Iowa, the case is heard in Iowa. Legal uncertainty is not caused by doubts about the validity of digital signatures, but by conflicting laws.

This example illustrates a general dilemma faced by those having to regulate e-commerce. Is it more important to protect the interests of consumers or the interest of (small) companies?

4.6. *Trust*

Lack of trust supposedly inhibits e-commerce, and PKIs are being marketed as the means of creating trust. Unfortunately, trust is a word with too many meanings. Trust can refer to a quality of a personal relationship, or to a lack of fear from the unknown, or to expectation of behaviour. In a commercial setting, trust can be related to the fact that there is recourse to arbitration or to the legal system. In access control (trust management), trust stands for the right to access a service or resource. In many security discussions these aspects get hopelessly mixed up.

Can a PKI create trust? Would you trust someone because he has a Verisign[d] certificate (or an ID card)? A certificate from a trusted CA can certainly reduce uncertainty. Trust in a CA may mean that it is expected to follow proper procedures. Trust in a CA may mean that there is legal recourse in case of damages. If the CA has authority over certificate subjects, uncertainty may be reduced even more.

Trust management is a fashionable term for certain types of access control systems, and a PKI can be part of such an access control system. Access rights may be based on trust, but may also be assigned for other reasons, e.g. on the need to do a job. It is not wise to pretend that all authorizations are expressions of trust.

In general, PKIs will *support* existing trust relationships, whatever they may be, more than *creating* trust relationships on their own.

4.7. The PKI paradox

PKIs promise to put trust into *global* e-commerce so that you can securely interact with parties you have never met before from all places in the world. Paradoxically, PKIs work best *locally*. Concerning trust between people, you know most about the people and institutions closest to you. Concerning legal issues, the situation is clearest when all parties, CA included, are under the same jurisdiction. For access control, you can be your own CA.

Commerce uses chains of local 'trust' relations, it does not need a global dispenser of trust. For example, Identrus[e] is setting up a PKI for B2B that is built upon local relationships (Figure 13).

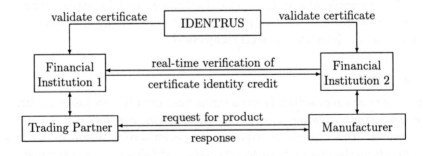

Fig. 13.   The Identrus model.

---

[d]http://www.verisign.com
[e]http://www.identrus.com

Summary

There are several options for deploying a PKI. There are *local PKIs*, e.g. when using public key cryptography for access control. In this case, all that is required is a protected local list of user names and verification keys. Neither certificates nor a third party are needed.

The credit card system is an example for a *private PKI*. Customer and merchant have a pre-established relationship with the card issuer, who is the CA issuing certificates. The CA is party to the transactions, not an independent TTP.

Qualified certificates are intended for use in a *public PKI*. Certificates are issued by CAs to certificate holders. The certificate verifiers have no pre-established relationships with the CA.

The PKI sector is facing a number of challenges. At the technical level, the interoperability of PKI products has to be addressed. In this area, the EEMA pki-Challenge has already been mentioned. At the application level, e-government, secure email, single log-on for access control, and non-repudiation for e-commerce are the main areas of promise for PKIs. In commercial settings, the business case for a PKI has to be examined. Does your application need a PKI? There are numerous cases of technically successful PKI pilots that did not lead to a full PKI deployment. At Information Security Solutions Europe 2001 a rumour went round that PKIs are dead.

As concluding advice, analyse your security problems first, then look for a solution that fits your situation. Do not buy a 'solution' and then try to fit your organisation around it. PKIs have many applications. The application tells you how to use a PKI, if a PKI is useful at all. The PKI should not tell you what you ought to do.

Without a killer application, the PKI will become the application killer.

Acronyms

AA	Attribute Authority [11]
AC	Attribute Certificate [11]
CA	Certification Authority
CRL	Certificate Revocation List
EE	End Entity
OCSP	PKIX Online Certificate Status Protocol
PKC	Public Key Certificate [11]
PKCS	Public-Key Cryptography Standards (RSA Security Inc.)

PKI	Public Key Infrastructure
PKIX	Internet X.509 Public Key Infrastructure [11]
PMI	Privilege Management Infrastructure [11]
RA	Registration Authority
SDSI	Simple Distributed Security Infrastructure [17]
SPKI	Simple Public Key Infrastructure [8]
TSA	Time Stamp Authority
TSP	Time Stamp Protocol
TTP	Trusted Third Party

Certificate Types

- Attribute certificate (X.509, PKIX)
- Authorization certificate (SPKI)
- Identity certificate (frequently mentioned, e.g. in SPKI, probably equivalent to user certificate)
- Public key certificate (PKIX)
- Qualified certificate (EU Directive, RFC 3039)
- User certificate (X.509) (also public key certificate, certificate)

References

1. Martín Abadi, On SDSI's linked local name spaces. *Journal of Computer Security*, 6:3–21, 1998.
2. C. Adams, P. Cain, D. Pinkas, and R. Zuccherato. Internet X.509 Public Key Infrastructure Time Stamp Protocols, October 2000. Internet Draft draft-ietf-pkix-time-stamp-11.txt.
3. Alfred W. Aresenault and Sean Turner, Internet X.509 Public Key Infrastructure – Roadmap, November 2000. Internet Draft draft-ietf-pkix-roadmap-06.txt.
4. Financial Services Authority. The FSA's approach to the regulation of e-commerce, June 2001. http://www.fsa.gov.uk/pubs/discussion/dp6.pdf.
5. Matt Blaze, Joan Feigenbaum, John Ioannidis, and Angelos D. Keromytis, The KeyNote Trust-Management System Version 2, September 1999. RFC 2704.
6. CCITT. *The Directory – Overview of Concepts, Models and Services*, 1988. CCITT Rec X.500.
7. US Congress. *Electronic Signatures in Global and National Commerce Act*, 2000. Pub. L No. 106-229, 114 Stat. 464 (2000) (codified as 15 U.S.C. 7001-7006, 7021, 7031) (enacted S. 761).
8. Carl M. Ellison, Bill Frantz, Butler Lampson, Ron Rivest, Brian M. Thomas, and Tatu Ylonen, *SPKI Certificate Theory*, November 1998. Internet Draft.

9. Joan Feigenbaum, Overview of the AT&T Labs trust-management project. In *Security Protocols, LNCS 1550*, pages 45–50. Springer Verlag, 1998.
10. J. Y. Halpern and R. van der Meyden, A logic for SDSI linked local name spaces. In *Proceedings of the 12th IEEE Computer Security Foundations Workshop*, pages 111–122, 1999.
11. Russell Housley, Warwick Ford, Tim Polk, and David Solo, Internet X.509 Public Key Infrastructure – Certificate and CRL Profile, January 1999. RFC 2459.
12. International Organization for Standardization. *Information technology — Open Systems Interconnection — The Directory-Authentication Framework*. Genève, Switzerland, June 1997. ISO/IEC 9594-8—ITU-T Rec X.509 (1997 E).
13. Loren M. Kohnfelder, Towards a practical public-key cryptosystem, May 1978. MIT S.B. Thesis.
14. A.J. Menezes, P.C. van Oorschot, and S.A. Vanstone. *Handbook of Applied Cryptography*. CRC Press, Boca Raton, FL, 1997.
15. National Conference of Commissioners of Uniform State Laws. *Uniform Computer Information Transactions Act*, 1999. See also www.ucitaonline.com.
16. National Conference of Commissioners of Uniform State Laws. *Uniform Electronic Transactions Act*, 1999. See also www.uetaonline.com.
17. Ron Rivest and Butler Lampson, SDSI – a Simple Distributed Security Infrastructure. Technical report, 1996.
    http://theory.lcs.mit.edu/~cis/sdsi.html.
18. Stefan Santesson, Tim Polk, Petra Barzin, and Magnus Nystrom. Internet X.509 Public Key Infrastructure — Qualified Certificates Profile, January 2001. RFC 3039.
19. Jane K. Winn, The emperor's new clothes: The shocking truth about digital signatures and Internet commerce, 2001. Revised Draft, faculty.smu.edu/jwinn/shocking-truth.htm.

COMPUTATIONAL METHODS IN PUBLIC KEY
CRYPTOLOGY

Arjen K. Lenstra

Citibank, N.A. and Technische Universiteit Eindhoven
1 North Gate Road, Mendham, NJ 07945-3104, U.S.A.
E-mail: arjen.lenstra@citigroup.com

These notes informally review the most common methods from computational number theory that have applications in public key cryptology.

Contents

1. Introduction   177
2. Public Key Cryptography   178
   2.1. Problems that are widely believed to be hard   178
   2.2. RSA   180
   2.3. Diffie-Hellman protocol   181
   2.4. The ElGamal family of protocols   181
   2.5. Other supposedly hard problems   182
3. Basic Computational Methods   184
   3.1. Integer arithmetic   184
        3.1.1. Remark on moduli of a special form   185
   3.2. Montgomery arithmetic   186
        3.2.1. Montgomery representation   186
        3.2.2. Montgomery addition and subtraction   186
        3.2.3. Montgomery multiplication   187
        3.2.4. Montgomery squaring   189
        3.2.5. Conversion to Montgomery representation   189
        3.2.6. Subtraction-less Montgomery multiplication   189
        3.2.7. Computing the auxiliary inverse for Montgomery arith-
               metic   190
        3.2.8. Montgomery inversion   190
   3.3. Euclidean algorithms   190
        3.3.1. Euclidean algorithm   191
        3.3.2. Extended Euclidean algorithm   191

3.3.3. Binary Euclidean algorithm   192
3.3.4. Binary extended Euclidean algorithm   193
3.3.5. Lehmer's method   193
3.3.6. Chinese remaindering   194
3.3.7. RSA with Chinese remaindering   195
3.3.8. Exponentiation based inversion   196
3.4. Exponentiation   196
3.4.1. Square and multiply exponentiation   196
3.4.2. Window exponentiation   197
3.4.3. Sliding window exponentiation   198
3.4.4. Multi-exponentiation   199
3.4.5. Miscellaneous exponentiation tricks   199
3.5. Prime generation   201
3.5.1. The prime number theorem   201
3.5.2. Probabilistic compositeness testing   201
3.5.3. Probabilistic compositeness test using Montgomery arith-
        metic   202
3.5.4. Generating industrial primes   203
3.5.5. Remark on provable primality   204
3.5.6. Prime generation with trial division   204
3.5.7. Prime generation with a wheel   204
3.5.8. Prime generation with sieving   205
3.6. Finite field arithmetic   205
3.6.1. Basic representation   206
3.6.2. Finding an irreducible polynomial   207
3.6.3. Optimal normal basis   207
3.6.4. Inversion using normal bases   208
3.6.5. Finding a generator   208
3.7. Elliptic curve arithmetic   209
3.7.1. Elliptic curves and elliptic curve groups   209
3.7.2. Remark on additive versus multiplicative notation   210
3.7.3. Elliptic curve group representation   211
3.7.4. Elliptic curves modulo a composite   211
3.7.5. Elliptic curve modulo a composite taken modulo a prime 212
3.7.6. Generating elliptic curves for cryptographic applications 212
4. Factoring and Computing Discrete Logarithms   213
4.1. Exponential-time methods   214
4.1.1. Exhaustive search   214
4.1.2. Pollard's $p-1$ method   214
4.1.3. The Silver-Pohlig-Hellman method   215
4.1.4. Shanks' baby-step-giant-step   216
4.1.5. Pollard's rho and lambda methods   216
4.2. Subexponential-time methods   218
4.2.1. The L function   218
4.2.2. Smoothness   219
4.2.3. Elliptic curve method   220

4.2.4.	The Morrison-Brillhart approach	222
4.2.5.	Quadratic sieve	224
4.2.6.	Historical note	227
4.2.7.	Number field sieve	228
4.2.8.	Index calculus method	229
Acknowledgment		231
References		232

1. Introduction

Cryptology consists of cryptography and cryptanalysis. Cryptography refers to the design and application of information protection methods. Cryptanalysis is the evaluation of the strength of cryptographic methods.

Roughly speaking, there are two types of cryptology: symmetric key cryptology and public key cryptology. In the former a single key is shared (and kept secret) by the communicating parties. It is used for encryption by the sender, and for decryption by the recipient. Examples are the Data Encryption Standard (DES) and the Advanced Encryption Standard (AES). In public key cryptology each party has a key that consists of two parts, a public and a secret (or private) part. The public part can be used to encrypt information, the corresponding secret part to decrypt. Alternatively, the secret key is used to sign a document, the corresponding public key to verify the resulting signature. Furthermore, a widely shared public key can be used to establish a common secret among two parties, for instance a key for a symmetric system. Examples of public key cryptosystems are the Diffie-Hellman key agreement protocol and the ElGamal and RSA encryption and signature schemes (Section 2).

The effectiveness – or security – of symmetric key systems relies on the secrecy of the shared symmetric key and the alleged infeasibility to decrypt without access to the key. Similarly, public key systems rely on the secrecy of the secret key and the infeasibility to decrypt (or sign) without access to the secret key. For public key systems this implies that it should be infeasible to derive the secret key from its corresponding public key. On the other hand, it must be possible to derive the public key from the secret key, or else public key systems could not be realized. Thus, public key systems involve some type of function that is effectively one-way. Given the secret key it must be possible to compute the corresponding public key reasonably efficiently. But deriving the secret from the public key must be impossible, practically speaking.

All currently popular public key systems rely on problems from computational number theory that are widely believed to be hard. As a consequence, computational methods involving integers play an important role in public key cryptology. They are used both to obtain efficient implementations (cryptography) and to provide guidance in key size selection (cryptanalysis). These notes review some of the most important methods from computational number theory that have applications in public key cryptology. Section 2 briefly outlines some of the most popular public key cryptosystems. The most important basic arithmetic methods required for efficient implementations of those systems are described in Section 3. Cryptanalytic methods relevant for the methods from Section 2 are sketched in Section 4.

2. Public Key Cryptography

In this section the basic versions of some of the most popular public key encryption and signature schemes are informally sketched. In Sections 2.2 and 2.4 the same public keys are used for the encryption and signature protocols. This is done for expository purposes only. In general it cannot be recommended to use a single key both for encryption and signature purposes. In practical applications many other peculiarities must be dealt with. They are not taken into account in the descriptions below, see [80].

2.1. *Problems that are widely believed to be hard*

So far only two supposedly hard problems have found widespread applications in public key cryptography:

(1) Integer factorization: given a positive composite integer n, find a nontrivial factor of n.
(2) Discrete logarithm: given a generator g of an appropriately chosen group and $h \in \langle g \rangle$, find an integer t such that

$$\underbrace{g \times g \times \ldots \times g}_{t} = g^t = h.$$

Here it is assumed that the group law is written multiplicatively and indicated by the symbol $\times$. The integer t is unique modulo the order of g, referred to as the *discrete logarithm* of h with respect to g, and denoted by $\log_g(h)$. In the literature it is also referred to as the *index* of h with respect to g.

The integer factorization problem is certainly not always hard, not even if n is large. Indeed, most randomly selected integers (say of some fixed large size) have a relatively small non-trivial factor that can quickly be found.

The point is, however, that a hard instance of the integer factoring problem can easily be generated: just pick two sufficiently large primes (Section 3.5) and compute their product. Given just the product there does not seem to be an efficient way to retrieve the primes. In Section 4 the question is addressed what is meant by 'sufficiently large'.

Similarly, the discrete logarithm problem is not always hard. Computing discrete logarithms in $\langle g \rangle$ is straightforward if $g = 1$ generates the additive group of integers $\mathbf{Z}/m\mathbf{Z}$, for any positive integer m. A less trivial example is when g generates the multiplicative group $\mathbf{F}_{p^\ell}^*$ of a finite field $\mathbf{F}_{p^\ell}$. In this case, the hardness of the discrete logarithm problem in $\langle g \rangle$ depends on the size of the largest prime factor of $p^\ell - 1$ (Section 4.1), the size of p^ℓ itself, and the characteristic p of $\mathbf{F}_{p^\ell}$ (Section 4.2). Alternatively, g may just generate a subgroup of $\mathbf{F}_{p^\ell}^*$ (see [105]). In that case the security depends on the size of the largest prime factor of the order of g (which divides $p^\ell - 1$), the characteristic p, and the size of p^d for the smallest $d \leq \ell$ (and dividing ℓ) such that the order of g divides $p^d - 1$ (see [64]). It follows that g should be chosen so that $d = \ell$. It is easy to construct finite fields with multiplicative groups (or subgroups thereof) in which the discrete logarithm problem is believed to be intractable. For appropriately chosen subgroups compression methods based on traces such as LUC [113] and XTR [73] can be used to represent the subgroup elements.

Another popular group where the discrete logarithm problem may be sufficiently hard is the group of points of a properly chosen elliptic curve over a finite field [59, 81]. This has the advantage that the methods from Section 4.2 do not seem to apply. Therefore it is generally believed that the finite field can be chosen much smaller than in the earlier example where $g \in \mathbf{F}_{p^\ell}^*$. In particular for high security applications this should lead to more manageable public key cryptosystems than the ones based on factoring, multiplicative groups, or anything else that is known to be susceptible to the methods from Section 4.2. Appropriate elliptic curves are much harder to find than hard to factor integers or good multiplicative groups. This is a serious obstacle to widespread deployment of elliptic curves in public key cryptography.

A proof that integer factorization is indeed a hard problem has never been published. The alleged difficulty of integer factorization is just a belief and, for the moment at least, nothing more than that. Computing discrete logarithms, on the other hand, can be proved to be hard. The proof, however, applies only to an abstract setting without practical relevance [88, 108]. Thus, also the alleged difficulty of the discrete logarithm problems referred to above is just a belief.

On a quantum computer, factoring and computing discrete logarithms can be done in polynomial time [107]. Believing that factoring and computing discrete logarithms are hard problems implies belief in the impossibility of quantum computing [40].

2.2. *RSA*

The RSA cryptosystem [102] is named after its inventors Rivest, Shamir, and Adleman. It relies for its security on the difficulty of the integer factoring problem. Each user generates two distinct large primes p and q (Section 3.5), integers e and d such that $ed \equiv 1 \bmod (p-1)(q-1)$ (Section 3.3), and computes $n = pq$. The pair of integers (n, e) is made public. The corresponding p, q, and d are kept secret. This construction satisfies the requirement mentioned in Section 1 that the public key can efficiently be derived from the secret key. The primes p and q can be derived from (n, e) and d. After generation of (n, e) and d, they are in principle no longer needed (but see 3.3.7).

Encryption. A message $m \in \mathbf{Z}/n\mathbf{Z}$ intended for the owner of public key (n, e) is encrypted as $E = m^e \in \mathbf{Z}/n\mathbf{Z}$. The resulting E can be decrypted by computing $D = E^d \in \mathbf{Z}/n\mathbf{Z}$ (see also 3.3.7). It follows from $ed \equiv 1 \bmod (p - 1)(q - 1)$, Fermat's little theorem, and the Chinese remainder theorem that $D = m$.

Signature generation. A message $m \in \mathbf{Z}/n\mathbf{Z}$ can be signed as $S = m^d \in \mathbf{Z}/n\mathbf{Z}$. The signature S on m can be verified by checking that $m = S^e \in \mathbf{Z}/n\mathbf{Z}$.

In practice the public exponent e is usually chosen to be small. This is done to make the public operations (encryption and signature verification) fast. Care should be taken with the use of small public exponents, as shown in [31, 32, 30, 49]. The secret exponent d corresponding to a small e is in general of the same order of magnitude as n. There are applications for which a small d (and thus large e) would be attractive. However, small private exponents have been shown to make RSA susceptible to attacks [12, 119].

A computational error made in the RSA private operation (decryption and signature generation) may reveal the secret key [11]. These operations should therefore always be checked for correctness. This can be done by applying the corresponding public operation to the result and checking that the outcome is as expected. So-called fault attacks are rendered mostly ineffective by this simple precaution. So-called timing attacks [60] can be made less effective by 'blinding' the private operation. Blinding factors b and b_e

are selected by randomly generating $b \in \mathbf{Z}/n\mathbf{Z}$ and computing $b_e = b^{-e}$. Direct computation of x^d for some $x \in \mathbf{Z}/n\mathbf{Z}$ is replaced by the computations of the blinded value $b_e x$, the private operation $y = (b_e x)^d$, and the final outcome $by = x^d$.

2.3. Diffie-Hellman protocol

A key agreement protocol is carried out by two communicating parties to create a shared key. This must be done in such a way that an eavesdropper does not gain any information about the key being generated. The Diffie-Hellman protocol [38] is a key agreement protocol. It is not an encryption or signature scheme.

Let g be a publicly known generator of an appropriately chosen group of known order. To create a shared key, parties A and B proceed as follows:

(1) A picks an integer $a \in \{2, 3, \dots, \text{order}(g) - 1\}$ at random, computes g^a, and sends g^a to B.
(2) B receives g^a, picks an integer $b \in \{2, 3, \dots, \text{order}(g) - 1\}$ at random, computes g^b and g^{ab}, and sends g^b to A.
(3) A receives g^b and computes g^{ba}.
(4) The shared key is $g^{ab} = g^{ba}$.

The Diffie-Hellman problem is the problem of deriving g^{ab} from g, g^a, and g^b. It can be solved if the discrete logarithm problem in $\langle g \rangle$ can be solved: an eavesdropper can find a from the transmitted value g^a and the publicly known g, and compute g^{ab} based on a and g^b. Conversely, it has not been proved in generality that the Diffie-Hellman problem is as hard as solving the discrete logarithm problem. If g generates a subgroup of the multiplicative groups of a finite field the problems are not known to be equivalent. Equivalence has been proved for the group of points of an elliptic curve over a finite field [13].

A related problem is the Decision Diffie-Hellman problem of efficiently deciding if $g^c = g^{ab}$, given g, g^a, g^b, and g^c. For multiplicative groups of finite fields this problem is believed to be hard. It is known to be easy for at least some elliptic curve groups [55].

2.4. The ElGamal family of protocols

Let g be a publicly known generator of an appropriately chosen group of known order. Let $h = g^t$ for a publicly known h and secret integer t. ElGamal encryption and signature protocols [41] based on a public key consisting of (g, h) come in many different variations. Basic variants are as follows.

Encryption. A message $m \in \langle g \rangle$ intended for the owner of public key (g, h) is encrypted as $(g^k, m \times h^k)$ for a randomly selected $k \in \{2, 3, \ldots, \mathrm{order}(g) - 1\}$. It follows that the party with access to $t = \log_g(h)$ can compute m as

$$m = \frac{m \times h^k}{(g^k)^t}.$$

Signature generation (based on DSA, the Digital Signature Algorithm). Let $\mathrm{order}(g) = q$ for a publicly known prime number q and $t = t \bmod q \in \mathbf{Z}/q\mathbf{Z}$. Let f be a publicly known function from $\langle g \rangle$ to $\mathbf{Z}/q\mathbf{Z}$. To sign a message $m \in \mathbf{Z}/q\mathbf{Z}$, compute $r = f(g^k) \in \mathbf{Z}/q\mathbf{Z}$ for a randomly selected $k \in \{2, 3, \ldots, q - 1\}$ with $f(g^k) \neq 0$ and

$$s = \frac{m + tr}{k} \in \mathbf{Z}/q\mathbf{Z}.$$

The signature consists of $(r, s) \in (\mathbf{Z}/q\mathbf{Z})^2$, unless $s = 0$ in which case another k is selected. Signature $(r, s) \in (\mathbf{Z}/q\mathbf{Z})^2$ is accepted if both r and s are non-zero and $r = f(v)$ where

$$v = g^e \times h^d,$$

with $w = s^{-1}$, $e = mw$, and $d = rw$, all in $\mathbf{Z}/q\mathbf{Z}$. This follows from

$$\frac{m + tr}{s} = k \text{ in } \mathbf{Z}/q\mathbf{Z}$$

and

$$g^e \times h^d = g^{mw} \times g^{trw} = g^{\frac{m+tr}{s}} = g^k.$$

The encryption requirement that $m \in \langle g \rangle$ can be relaxed by replacing $m \times h^k$ by the encryption of m using some symmetric key system with key h^k (mapped to the proper key space). In the signature scheme, the usage of m in the definition of s can be replaced by the hash of m (mapped to $\mathbf{Z}/q\mathbf{Z}$), so that m does not have to be in $\mathbf{Z}/q\mathbf{Z}$.

2.5. *Other supposedly hard problems*

Most practical public key system are based on one of only two supposedly hard problems, integer factoring and discrete logarithms. As shown in Section 4 many integer factoring methods allow variants that solve the discrete logarithm problem. The complexity of the two problems may therefore be related. From a practical point of view this is an undesirable situation with a potential for disastrous consequences. If factoring integers and computing discrete logarithms turn out to be easier than anticipated, most existing security solutions can no longer be used.

For that reason there is great interest in hard problems that are suitable for public key cryptography and that are independent of integer factoring and discrete logarithms. Candidates are not easy to find. Once they have been found it is not easy to convince the user community that they are indeed sufficiently hard. Suitable problems for which the hardness can be guaranteed have not been proposed, so that security is a matter of perception only: the longer a system survives, i.e., no serious weaknesses are detected, the more users are willing to believe its security. It will be hard for any newcomer to acquire the level of trust currently enjoyed by the integer factoring and discrete logarithm problems. Some of the latest proposals that are currently under scrutiny are the following.

Lattice based systems. A lattice is a discrete additive subgroup of a fixed dimensional real vector space. Over the past two decades, lattices have been used to great effect in cryptanalysis, most notably to attack so-called knapsack based cryptosystems. More recently, their cryptographic potential is under investigation. The hard problems one tries to exploit are the following:

  – The shortest vector problem: find a shortest non-zero element of the lattice, with respect to an appropriately chosen norm.
  – The closest vector problem: given an element of the vector space, find a lattice element closest to it, under an appropriately chosen norm.

See [89] for an overview of lattice based cryptology and references to relevant literature.

NTRU and NSS. Let $N > 0$, $q > 0$, and $d < N/2$ be three appropriately chosen integer security parameters. Given a polynomial $h \in (\mathbf{Z}/q\mathbf{Z})[X]$ of degree $< N$, it is supposedly hard to find polynomials $f, g \in (\mathbf{Z}/q\mathbf{Z})[X]$ of degrees $< N$, such that

$$hf = g \bmod (X^N - 1).$$

Here both f and g have d coefficients equal to 1, another d coefficients equal to $(-1 \bmod q)$, and $N - 2d$ zero coefficients. Due to the way h is constructed, it is known that such f and g exist. The encryption scheme NTRU [50] and the signature scheme NSS [51] are meant to rely on this problem of recovering f and g from h. This problem can be seen as a lattice shortest vector problem [34]. It was not designed as such. NSS as proposed in [51] is known to have several independent weaknesses [43, 114].

Braid groups. Let Σ_n be the group of n-permutations, for an integer $n > 0$. Let $\sigma_i \in \Sigma_n$ with $i \in \{1, 2, \ldots, n-1\}$ be the permutation that swaps the i-th and $(i+1)$-st element. The n-braid group B_n is the subgroup of Σ_n generated by σ_1, σ_2, ..., σ_{n-1}. In [57] public key systems are described that rely for their security on the difficulty of the following conjugacy problem in B_n: given $x, y \in B_n$ such that $y = bxb^{-1}$ for some unknown $b \in B_m$ with $m \leq n$, find $a \in B_m$ such that $y = axa^{-1}$. See also [6].

3. Basic Computational Methods

This section reviews some of the computational tools for arithmetic on integers that are required to implement the methods from Sections 2.2, 2.3, and 2.4. For a more complete treatment, including runtime analyses, refer to [9, 56, 76, 80].

3.1. *Integer arithmetic*

The primes used in realistic implementations of number theoretic cryptographic protocols (Section 2) may be many hundreds, and even thousands, of bits long. Thus, the elements of the various groups involved in those protocols cannot be represented by ordinary 32-bit or 64-bit values as available on current computers. Neither can they directly be operated upon using commonly available instructions, such as addition or multiplication modulo 2^{32} on 32-bit integer operands (as common in a programming language such as C).

Implementations of many public key cryptosystems therefore have to rely on a set of multiprecision integer operations that allow arithmetic operations on integers of arbitrary or sufficiently large fixed sizes. It is assumed that routines are available implementing addition, subtraction, multiplication, and remainder with division on integer operands of any size. The design of these basic routines is not the subject of these notes, because most design decisions will, to a large extent, depend on the hardware one intends to use. A variety of different packages is available on the Internet. A basic software only package, written in C, is freely available from the present author [63]. See [109] for a more advanced, but also free, C++ package.

Throughout this section it is assumed that non-negative integers are represented on a computer (or other hardware platform) by their *radix B* representation. Here B is usually a power of 2 that is determined by the wordsize of the computer one is using. On a 32-bit processor $B = 2^{32}$ is a possible, but not always the best, choice. Thus, an integer $m \geq 0$ is represented as

$$m = \sum_{i=0}^{s-1} m_i B^i \text{ for some } s > 0 \text{ and } m_i \in \{0, 1, \ldots, B-1\}.$$

The m_i are referred to as the *blocks* of m. The least and most significant blocks are m_0 and m_{s-1}, respectively. The representation is called *normalized* if $m = m_0 = 0$ and $s = 1$ or if $m \neq 0$ and $m_{s-1} \neq 0$. If a representation is normalized, s is called its *block-length*. A normalized representation is unique.

For integers $m > 0$ and n, the quotient $q = [n/m]$ and the remainder $r = n \bmod m$ are the unique integers with $n = qm + r$ and $r \in \{0, 1, \ldots, m-1\}$. The ring of integers $\mathbf{Z}/m\mathbf{Z}$ is represented by and identified with the set $\{0, 1, \ldots, m-1\}$ of least non-negative residues modulo m. For any integer n, its remainder modulo m is regarded as an element of $\mathbf{Z}/m\mathbf{Z}$ and denoted by $n \bmod m$. It follows that for $u, v \in \mathbf{Z}/m\mathbf{Z}$ the sum $u + v \in \mathbf{Z}/m\mathbf{Z}$ can be computed at the cost of an integer addition and an integer subtraction:

$$\text{compute } w = u + v \in \mathbf{Z} \text{ and } x = w - m,$$

then the sum equals $x \in \mathbf{Z}/m\mathbf{Z}$ if $x \geq 0$ or $w \in \mathbf{Z}/m\mathbf{Z}$ if $x < 0$. The intermediate result $w = u + v$ is at most $2m - 2$, i.e., still of the same order of magnitude as m. Similarly, $u - v \in \mathbf{Z}/m\mathbf{Z}$ can be computed at the cost of an integer subtraction and an integer addition. The computation of the product $uv = (uv) \bmod m \in \mathbf{Z}/m\mathbf{Z}$ is more involved. It requires the computation of a product of non-negative integers $< m$, resulting in an intermediate result of twice the order of magnitude of m, followed by the computation of a remainder of the intermediate result modulo m. Because for large m the latter computation is often costly (and cumbersome to implement in hardware), various faster methods have been developed to perform calculations in $\mathbf{Z}/m\mathbf{Z}$. The most popular of these methods is so-called Montgomery arithmetic [82], as described in Section 3.2 below.

3.1.1. *Remark on moduli of a special form*

Computing the remainder modulo m of the intermediate result uv can be done quickly (and faster than using Montgomery arithmetic) if m is chosen such that it has a special form. For instance, if the most significant block m_{s-1} is chosen to be equal to 1, and $m_i = 0$ for $s/2 < i < s - 1$, then the reduction modulo m can be done at about half the cost of the computation of the product uv. Such moduli can be used in most applications mentioned in Section 2, including RSA [65]. Their usage has so far seen only limited applications in cryptography.

3.2. *Montgomery arithmetic*

Let $m > 2$ be an integer coprime to B and let $m = \sum_{i=0}^{s-1} m_i B^i$ be its normalized radix B representation. It follows that for common choices of B the integer m must be odd. In cryptographic applications m is usually a large prime or a product of large primes, so requiring m to be odd is not a serious restriction.

3.2.1. *Montgomery representation*

Let R be the smallest power of B that is larger than m, so $R = B^s$. This R is referred to as the *Montgomery radix*. Its value depends on the value of the ordinary radix B. The Montgomery representation $\tilde{x}$ of $x \in \mathbf{Z}/m\mathbf{Z}$ is defined as

$$\tilde{x} = (xR \bmod m) \in \mathbf{Z}/m\mathbf{Z}.$$

Montgomery arithmetic in $\mathbf{Z}/m\mathbf{Z}$ is arithmetic with the Montgomery representations of elements of $\mathbf{Z}/m\mathbf{Z}$ (but see 3.2.6).

3.2.2. *Montgomery addition and subtraction*

Let $u, v \in \mathbf{Z}/m\mathbf{Z}$ be represented by their Montgomery representations $\tilde{u}$ and $\tilde{v}$. The Montgomery sum of $\tilde{u}$ and $\tilde{v}$ is the Montgomery representation of the sum $u + v$, i.e., the element $\tilde{z} \in \mathbf{Z}/m\mathbf{Z}$ for which the corresponding z satisfies $z = u + v \in \mathbf{Z}/m\mathbf{Z}$. From

$$\tilde{z} = zR \bmod m = (u + v \bmod m)R \bmod m$$
$$\equiv (uR \bmod m + vR \bmod m) \bmod m$$
$$= (\tilde{u} + \tilde{v}) \bmod m$$

it follows that Montgomery addition is the same as ordinary addition in $\mathbf{Z}/m\mathbf{Z}$:

$$\tilde{z} = (\tilde{u} + \tilde{v}) \bmod m.$$

It therefore requires just an integer addition and an integer subtraction. Similarly, Montgomery subtraction is the same as ordinary subtraction in $\mathbf{Z}/m\mathbf{Z}$.

3.2.3. *Montgomery multiplication*

The Montgomery product of $\tilde{u}$ and $\tilde{v}$ is the element $\tilde{z} \in \mathbf{Z}/m\mathbf{Z}$ for which $z = (uv) \bmod m$:

$$
\begin{aligned}
\tilde{z} = zR \bmod m &= ((uv) \bmod m)R \bmod m \\
&= (uR \bmod m)v \bmod m \\
&= (uR \bmod m)(vR \bmod m)R^{-1} \bmod m \\
&= \tilde{u}\tilde{v}R^{-1} \bmod m.
\end{aligned}
$$

Thus, the Montgomery product of $\tilde{u}$ and $\tilde{v}$ is their ordinary integer product divided by the Montgomery radix R modulo m. This can be computed as follows.

Let $w \in \mathbf{Z}$ be the integer product of $\tilde{u}$ and $\tilde{v}$ regarded as elements of $\mathbf{Z}$ (as opposed to $\mathbf{Z}/m\mathbf{Z}$) and let

$$
w = \sum_{i=0}^{2s-1} w_i B^i \le (m-1)^2
$$

be w's not necessarily normalized radix B representation. This w must be divided by R modulo m. Division by R modulo m is equivalent to s-fold division by B modulo m, since $R = B^s$. If $w_0 = 0$, division by B can be carried out by shifting out the least significant block $w_0 = 0$ of w:

$$
w/B = \sum_{i=0}^{2s-2} w_{i+1} B^i.
$$

If $w_i \ne 0$, then a multiple t of m is found such that $(w + t) \bmod B = 0$. The resulting $w + t$ can be divided by B by shifting out its least significant block. But because $w \equiv w + t \bmod m$, dividing $w + t$ by B is equivalent to dividing w by B modulo m. The same process is then applied to the resulting number (w/B or $(w + t)/B$) until the division by B has been carried out s times.

It remains to explain how an appropriate multiple t of m is found with $(w + t) \bmod B = 0$, in the case that $w_0 \ne 0$. Because m and B are coprime, so are $m_0 = m \bmod B$ and B. It follows that there exists an integer $m_0^{-1} \in$

$\{0, 1, \ldots, B - 1\}$ such that $m_0 m_0^{-1} \equiv 1 \bmod B$ (see 3.2.7). Consider the integer

$$t = ((B - w_0)m_0^{-1} \bmod B)m = \sum_{i=0}^{s} t_i B^i.$$

From $m_0 m_0^{-1} \equiv 1 \bmod B$ it follows that

$$t_0 = t \bmod B = ((B - w_0)m_0^{-1} \bmod B)m_0 \bmod B = B - w_0.$$

The integer t is a multiple of m and

$$\begin{aligned}
(w + t) \bmod B &= (w \bmod B + t \bmod B) \bmod B \\
&= (w_0 + t_0) \bmod B \\
&= (w_0 + B - w_0) \bmod B \\
&= 0.
\end{aligned}$$

Thus, $w + t$ is divisible by B.

Although the result of the s-fold division by B modulo m is equivalent to wR^{-1} modulo m, it is not necessarily an element of $\mathbf{Z}/m\mathbf{Z}$. In the course of the above process at most $\sum_{i=0}^{s-1}(B-1)mB^i = m(R-1)$ is added to the original w, after which R is divided out. Therefore, the result is bounded by

$$\frac{w + m(R - 1)}{R} \leq \frac{(m - 1)^2 + m(R - 1)}{R} < 2m.$$

It follows that the result can be normalized to $\mathbf{Z}/m\mathbf{Z}$ using at most a single subtraction by m.

The division of w by R modulo m requires code that is very similar to the code for ordinary multiplication of the integers $\tilde{u}$ and $\tilde{v}$. Also, the time required for the division of w by R modulo m is approximately the same as the time required for the computation of w itself (i.e., the multiplication). Thus, the total runtime of Montgomery multiplication modulo m is approximately equal to the time required for two ordinary multiplications of integers of size comparable to m.

The multiplication and division by R modulo m can be merged in such a way that no intermediate result is larger than B^{s+1}. This makes Montgomery multiplication ideally suited for fast and relatively simple hardware implementations.

3.2.4. *Montgomery squaring*

The Montgomery square of $\tilde{u}$ is the Montgomery product of $\tilde{u}$ and $\tilde{u}$. The square $w = \tilde{u}^2$ (where $\tilde{u}$ is regarded as an element of $\mathbf{Z}$) can be computed in slightly more than half the time of an ordinary product of two integers of about the same size. However, the reduction of the resulting w by R modulo m cannot take advantage of the fact that w is a square. It takes the same time as in the general case. As a result it is reasonable to assume that the time required for a Montgomery squaring is 80% of the time required for a Montgomery product, with the same modulus [27].

3.2.5. *Conversion to Montgomery representation*

Given $u \in \mathbf{Z}/m\mathbf{Z}$, its Montgomery representation $\tilde{u}$ can be computed as the Montgomery product of u and $R^2 \bmod m$. This follows from the fact that the Montgomery product is the ordinary integer product divided by the Montgomery radix R modulo m:

$$(uR^2 \bmod m)R^{-1} \bmod m = uR \bmod m = \tilde{u}.$$

Similarly, given $\tilde{u}$, the corresponding u can be computed as the Montgomery product of $\tilde{u}$ and 1 (one):

$$\tilde{u}R^{-1} \bmod m = uRR^{-1} \bmod m = u.$$

It follows that conversion to and from the Montgomery representation each require a Montgomery multiplication. It is assumed that the square $R^2 \bmod m$ of the Montgomery radix is computed once and for all per modulus m. In typical applications of Montgomery arithmetic conversions are carried out only before and after a lengthy computation (Section 3.4). The overhead caused by the conversions is therefore minimal. In some applications conversions can be avoided altogether, see 3.5.3.

3.2.6. *Subtraction-less Montgomery multiplication*

If R is chosen such that $4m < R$ (as opposed to $m < R$), then the subtraction that may be needed at the end of each Montgomery multiplication can be mostly avoided. If Montgomery multiplication with modulus m as in 3.2.3 is applied to two non-negative integers $< 2m$ (as opposed to $< m$), then w as in 3.2.3 is $\leq (2m - 1)^2$. The result of the division of w by R modulo m is therefore bounded by

$$\frac{w + m(R-1)}{R} \leq \frac{(2m-1)^2 + m(R-1)}{R} < 2m$$

because $4m < R$. Thus, if $4m < R$ and the Montgomery product with modulus m is computed of two Montgomery representations non-uniquely represented as elements of $\mathbf{Z}/2m\mathbf{Z}$, then the outcome is again an element of $\mathbf{Z}/2m\mathbf{Z}$ if the subtraction is omitted. This implies that, in lengthy operations involving Montgomery products, the subtractions are not needed if all operations are carried out in $\mathbf{Z}/2m\mathbf{Z}$ instead of $\mathbf{Z}/m\mathbf{Z}$. In this case, all intermediate results are non-uniquely represented as elements of $\mathbf{Z}/2m\mathbf{Z}$ instead of $\mathbf{Z}/m\mathbf{Z}$. At the end of the computation each relevant result can then be normalized to $\mathbf{Z}/m\mathbf{Z}$ at the cost of a single subtraction.

3.2.7. *Computing the auxiliary inverse for Montgomery arithmetic*

The computation of m_0^{-1} mod B may be carried out once and for all per modulus m, so it does not have to be done very efficiently. As shown in the C-program fragment in Figure 1, this computation does not require the more involved methods from Section 3.3.

```
unsigned long m0, m0shift, mask, product, result;
<m0 = 'the least-significant block of the Montgomery modulus m'>
if (!(m0&1)) exit(0);
for (m0shift=(m0<<1), mask = 2, product = m0, result = 1; mask;
     m0shift <<= 1, mask <<= 1)
  if (product & mask) {
    /* invariant: product == m0 * result */
    product += m0shift;
    result += mask;
  }
```

Fig. 1.   Computation of m_0^{-1} mod B, for $B = 2^k$ and k the bit-length of an unsigned long.

3.2.8. *Montgomery inversion*

Given the Montgomery representation $\tilde{u}$ of u with respect to modulus m, the Montgomery representation of u^{-1} (modulo m, obviously) is computed by inverting $\tilde{u}$ modulo m (Section 3.3) and by Montgomery multiplying the result by R^3 mod m (a value that may be computed once and for all per modulus m).

3.3. *Euclidean algorithms*

The greatest common divisor $\gcd(m, n)$ of two non-negative integers m and n (not both zero) is defined as the largest positive integer that divides both

m and n. If $m > 0$ and $n = 0$ then $\gcd(m, n) = m$. If $m, n > 0$, then $\gcd(m, n)$ can be computed by finding the prime factorization of m and n and by determining the product of all common factors. A method that is in general much faster is the following algorithm due to Euclid, from about 300 BC.

3.3.1. *Euclidean algorithm*

The Euclidean algorithm is based on the observation that d divides m and n if and only if d divides $m - kn$ and n for arbitrary $k \in \mathbf{Z}$. It follows that if $n > 0$ then

$$\gcd(m, n) = \gcd(n, m \bmod n).$$

Subsequently, if $m \bmod n = 0$, then $\gcd(m, n) = \gcd(n, 0) = n$. However, if $m \bmod n \neq 0$, then

$$\gcd(m, n) = \gcd(n, m \bmod n) = \gcd(m \bmod n, n \bmod (m \bmod n)).$$

Thus, $\gcd(m, n)$ can be computed by replacing (m, n) by $(n, m \bmod n)$ until $n = 0$, at which point m is the greatest common divisor of the original m and n.

The integer n strictly decreases per iteration. This guarantees termination. The most convenient way to prove fast termination is to use least absolute (as opposed to least non-negative) residues, and to use absolute values. Then $\lceil \log_2(n) \rceil$ strictly decreases per iteration, so that $O(\log(n))$ iterations suffice. If one incorporates the effect of the decreasing size of the operands, then the overall runtime can be seen to be $O(\log(m) \log(n))$.

3.3.2. *Extended Euclidean algorithm*

If the quotients $\lfloor m/n \rfloor$ are explicitly computed in the Euclidean algorithm, then it can be extended so that it not only computes $d = \gcd(m, n)$, but an integer e such that $ne = d \bmod m$ as well. This is particularly useful when m and n are coprime. In that case $d = 1$ and e is the multiplicative inverse of n modulo m. The ability to quickly compute multiplicative inverses is important in many public key cryptosystems (Section 2 and 3.3.7).

If $ne = d \bmod m$, then $(ne - d)/m$ is an integer, say f, and $ne - fm = \gcd(m, n)$. In the literature this is the identity that is typically solved by the extended Euclidean algorithm.

To solve $ne = d \bmod m$, proceed as follows. Let $s = m$ and $t = n$, then the following two initial equivalences modulo m hold trivially:

$$n \cdot 0 \equiv s \bmod m,$$

$$n \cdot 1 \equiv t \bmod m.$$

This is the initial instance of the following pair of equivalences modulo m:

$$n \cdot u \equiv s \bmod m,$$

$$n \cdot v \equiv t \bmod m.$$

Given such a pair of equivalences, a new pair with 'smaller' right hand sides can be found by applying the ordinary Euclidean algorithm to s and t. If $t \neq 0$ and $s = qt + r$ with $0 \leq r < t$, then $n \cdot (u - vq) \equiv r \bmod m$, so that the new pair is given by

$$n \cdot v \equiv t \bmod m,$$

$$n \cdot (u - vq) \equiv r \bmod m.$$

Thus, per iteration the 4-tuple (s, t, u, v) is replaced by $(t, s \bmod t, v, u - v[s/t])$. This terminates with $t = 0$ and s equal to $d = \gcd(m, n)$, at which point u equals the desired e with $ne \equiv d \bmod m$.

3.3.3. *Binary Euclidean algorithm*

The efficiency of the Euclidean algorithms from 3.3.1 and 3.3.2 depends on the efficiency of the division operations. On many platforms divisions on small operands are relatively slow compared to additions, shifts (i.e., division or multiplication by a power of 2), and multiplications on small operands. The binary variants, here and in 3.3.4, avoid divisions. They are often faster when the operands are small. For large operands the difference in efficiency between the binary and ordinary versions is much smaller – the ordinary version may even turn out to be faster – because the quotients are on average often quite small. This results in divisions that are often, and for larger block-lengths, fast compared to shifts over several blocks.

The binary variant of the Euclidean algorithm works as follows. Assume that $m > n$ are both positive and odd: common factors 2 are removed and later included in the gcd, any remaining factors 2 can be removed. Because m and n are odd, $m - n > 0$ is even and the odd part $w > 0$ of $m - n$ is at most equal to $(m - n)/2$. Furthermore, $\gcd(m, n) = \gcd(n, w)$, the pair n, w is strictly 'smaller' than the pair m, n, and n and w are again both odd. As soon as $n = w$, then n is the desired gcd. Otherwise replace (m, n) by (n, w) if $n > w$ and by (w, n) if $w > n$.

3.3.4. *Binary extended Euclidean algorithm*

This combines the ideas from 3.3.2 and 3.3.3. Let $m > n$ be positive and odd, as in 3.3.3, and let

$$n \cdot u \equiv s \bmod m,$$
$$n \cdot v \equiv t \bmod m$$

be a pair of equivalences modulo m as in 3.3.2. Initially $u = 0$, $v = 1$, $s = m$, and $t = n$. Let $s - t = 2^k w$ for some odd w and $k > 0$, then

$$n \cdot ((u - v)/2^k \bmod m) \equiv w \bmod m.$$

Because m is odd, $(u - v)/2^k \bmod m$ can be computed quickly by applying the following k times:

$$x/2 \bmod m = \begin{cases} x/2 & \text{if } x \text{ is even} \\ (x + m)/2 & \text{otherwise.} \end{cases}$$

The resulting iteration is particularly fast for $m, n < B$ (i.e., for m and n that can, in C, be represented by a single **unsigned long** each). For larger m and n it may be faster or slower than the method from 3.3.2, depending on the characteristics of the computer one is using. For large m and n, however, the method from 3.3.5 below is often much faster than either the binary or ordinary method.

3.3.5. *Lehmer's method*

Let again

$$n \cdot u \equiv s \bmod m,$$
$$n \cdot v \equiv t \bmod m$$

be a pair of equivalences modulo m, with initial values $u = 0$, $v = 1$, $s = m$, and $t = n$. The sequence of operations carried out by the extended Euclidean algorithm described in 3.3.2 depends on the quotients $[s/t]$. Each operation requires an update of s, t, u and v that is relatively costly (since it involves large integers). D.H. Lehmer observed that the process can be made faster by postponing the costly updates. This is done by bounding the sequence of quotients in a cheap way both from below and from above. Only if the two bounds are different the expensive update steps are performed.

Assume that s and t have the same block-length. Let $\bar{s}$ and $\bar{t}$ be the most significant blocks of s and t, respectively (Section 3.1). Application of extended Euclidean algorithm 3.3.2 to $\bar{s}$ and $\bar{t} + 1$ leads to a sequence of quotients, and application to $\bar{s} + 1$ and $\bar{t}$ leads to another sequence of quotients. From

$$\frac{\bar{s}}{\bar{t}+1} < \frac{s}{t} < \frac{\bar{s}+1}{\bar{t}}$$

it follows that as long as the two resulting quotient sequences are the same, the same quotients are obtained if the extended Euclidean algorithm would be applied to s and t. After k iterations the initial pair $(\bar{s}, \bar{t} + 1)$ is transformed into

$$(a_k\bar{s} + b_k(\bar{t} + 1), c_k\bar{s} + d_k(\bar{t} + 1)),$$

for $a_k, b_k, c_k, d_k \in \mathbf{Z}$ that depend on the sequence of quotients. If the first k quotients are the same, then the initial pair $(\bar{s} + 1, \bar{t})$ is transformed into

$$(a_k(\bar{s} + 1) + b_k\bar{t}, c_k(\bar{s} + 1) + d_k\bar{t}),$$

the pair (s, t) would be transformed into

$$(a_k s + b_k t, c_k s + d_k t),$$

and therefore (u, v) into

$$(a_k u + b_k v, c_k u + d_k v).$$

Thus, in Lehmer's method the extended Euclidean algorithm is simultaneously applied to two small initial pairs (without keeping track of the corresponding u's and v's) that depend, as described above, on the 'true' large pair (s, t), while keeping track of the small linear transformation (i.e., the a_k, b_k, c_k, and d_k) on one of the small pairs. As soon as the quotients become different the last linear transformation for which the quotients were the same is applied to (s, t) and the corresponding (u, v). If necessary, the process is repeated (by first determining two new small initial pairs depending on the updated smaller pair (s, t)). The only 'expensive' steps are the occasional updates of the 'true' (s, t) and (u, v). All other steps involve only small numbers. The number of small steps before a large update is proportional to the number of bits in the radix B (Section 3.1).

For large m and n Lehmer's method is substantially faster than either the ordinary or binary versions of the extended Euclidean algorithm. A disadvantage is that it requires a much larger amount of code.

3.3.6. Chinese remaindering

Chinese remaindering is not a Euclidean algorithm, but an application of the ability to compute modular inverses that was not explicitly described in

Section 2. Let p and q be two positive coprime integers, and let $r(p) \in \mathbf{Z}/p\mathbf{Z}$ and $r(q) \in \mathbf{Z}/q\mathbf{Z}$. According to the Chinese remainder theorem there exists a unique $r \in \mathbf{Z}/pq\mathbf{Z}$ such that $r \bmod p = r(p)$ and $r \bmod q = r(q)$. This r can be computed as follows:

$$r = r(p) + p\left((r(q) - r(p))(p^{-1} \bmod q) \bmod q\right),$$

where $r(p)$ and $r(q)$ are regarded as elements of $\mathbf{Z}$ and the result is regarded as an element of $\mathbf{Z}/pq\mathbf{Z}$. The integer $p^{-1} \bmod q$ can be computed using one of the extended Euclidean algorithms described above, or using the method from 3.3.8 if q is prime.

If $\gcd(p, q) = d > 1$, a very similar approach can be used, under the necessary and sufficient condition that $r(p) \equiv r(q) \bmod d$, to find r such that $r \bmod p = r(p)$ and $r \bmod q = r(q)$. Apply the above to $p' = p/d$, $r(p') = 0$, $q' = q/d$, and $r(q') = (r(q) - r(p))/d$ to find r' that is 0 modulo p' and $(r(q) - r(p))/d$ modulo q'. Then $r = r(p) + dr'$, since $dr' \equiv 0 \bmod p$ and $dr' \equiv r(q) - r(p) \bmod q$. The resulting r is unique modulo pq/d.

3.3.7. *RSA with Chinese remaindering*

As an application of Chinese remaindering, consider the private RSA operation, described in Section 2.2:

given integers n, d and an element $w \in \mathbf{Z}/n\mathbf{Z}$, compute $r = w^d \in \mathbf{Z}/n\mathbf{Z}$.

This computation takes about $c \log_2(d) \log_2(n)^2$ seconds (Section 3.4), for some constant c depending on the computer one is using. Since d is usually of the same order of magnitude as n, this becomes $c \log_2(n)^3$ seconds. With Chinese remaindering it can be done about 4 times faster, assuming $n = pq$ for primes p, q of about the same order of magnitude:

(1) Compute $d(p) = d \bmod (p-1)$ and $d(q) = d \bmod (q-1)$. These values depend on n and d only, so they can be computed once and for all.
(2) Compute $w(p) = w \bmod p \in \mathbf{Z}/p\mathbf{Z}$ and $w(q) = w \bmod q \in \mathbf{Z}/q\mathbf{Z}$.
(3) Compute $r(p) = w(p)^{d(p)} \in \mathbf{Z}/p\mathbf{Z}$ and $r(q) = w(q)^{d(q)} \in \mathbf{Z}/q\mathbf{Z}$.
(4) Use Chinese remaindering to compute $r \in \mathbf{Z}/pq\mathbf{Z} = \mathbf{Z}/n\mathbf{Z}$ such that $r \bmod p = r(p)$ and $r \bmod q = r(q)$. This r equals $w^d \in \mathbf{Z}/n\mathbf{Z}$.

The exponentiations in Step 3 take about

$$c \log_2(d(p)) \log_2(p)^2 + c \log_2(d(q)) \log_2(q)^2 \approx 2c \log_2(p)^3$$

seconds. The other steps are negligible. Since $\log_2(p) \approx \frac{1}{2} \log_2(n)$ the total runtime becomes $\frac{c}{4} \log_2(n)^3$ seconds.

If n is the product of t primes, then a total speed-up of a factor t^2 is obtained. For RSA it may therefore be attractive to use moduli that consist of more than two prime factors. This is known as *RSA multiprime*. Care should be taken to make sure that the resulting moduli are not susceptible to an attack using the elliptic curve factoring method (4.2.3).

3.3.8. *Exponentiation based inversion*

If m is prime, then multiplicative inverses modulo m can be computed using exponentiations modulo m (Section 3.4), based on Fermat's little theorem. If $x \in \mathbf{Z}/m\mathbf{Z}$ then $x^m \equiv x \bmod m$ due to m's primality, so that, if $x \neq 0$,

$$x^{m-2} \equiv x^{-1} \bmod m.$$

This way of computing inverses takes $O(\log_2(m)^3)$ operations as opposed to just $O(\log_2(m)^2)$ for the methods based on the Euclidean algorithm. Nevertheless, it may be an option in restricted environments where exponentiation must be available and the space for other code is limited. When combined with Chinese remaindering it can also be used to compute inverses modulo composite m with known factorization.

3.4. *Exponentiation*

As shown in Section 2 exponentiation is of central importance for many public key cryptosystems. In this subsection the most important exponentiation methods are sketched. See [46] for a complete treatment.

Let g be an element of some group G that is written multiplicatively (with the notation as in Section 2.1). Suppose that $g^e \in G$ must be computed, for some positive integer e. Let $e = \sum_{i=0}^{L-1} e_i 2^i$ with $e_i \in \{0,1\}$, $L \geq 1$, and $e_{L-1} = 1$.

3.4.1. *Square and multiply exponentiation*

The most common and simplest exponentiation method is the square and multiply method. It comes in two variants depending on the order in which the bits e_i of e are processed: left-to-right if the e_i are processed for $i = L-1$, $L-2$, ... in succession, right-to-left if they are processed the other way around. The left-to-right variant is the simplest of the two. It works as follows.

(1) Initialize r as $g \in G$.
(2) For $i = L - 2, L - 3, \dots, 0$ in succession, do the following:

(a) Replace r by $r \times r$.

(b) If $e_i = 1$, then replace r by $g \times r \in G$.

(3) The resulting r equals $g^e \in G$.

In some applications it may be hard to access the e_i from $i = L - 2$ downwards. In that case, it may be more convenient to use the right-to-left variant:

(1) Initialize s as $g \in G$ and r as the unit element in G.

(2) For $i = 0, 1, \ldots, L - 1$ in succession, do the following:

(a) If $e_i = 1$, then replace r by $s \times r \in G$.

(b) If $i < L - 1$, then replace s by $s \times s$.

(3) The resulting r equals $g^e \in G$.

If G is the group of the integers (under addition, i.e., g^e actually equals the ordinary integer product eg), then the right-to-left method corresponds to the ordinary schoolbook multiplication method applied to the binary representations of g and e.

For randomly selected L-bit exponents either method performs $L - 1$ squarings in G and on average about $\frac{L}{2}$ multiplications in G. If $G = \mathbf{Z}/m\mathbf{Z}$, the group law denoted by $\times$ is multiplication modulo m, and Montgomery arithmetic is used, then one may expect that the average runtime of square and multiply exponentiation is about the same as the time required for $1.3 \log_2(e)$ multiplications in $\mathbf{Z}/m\mathbf{Z}$ (see 3.2.4). In many practical circumstances this can be improved using so-called windows exponentiation methods.

3.4.2. *Window exponentiation*

The window exponentiation method is similar to left-to-right exponentiation, except that the bits of e are processed w bits at a time, for some small *window size* $w \in \mathbf{Z}_{>0}$.

(1) Let

$$e = \sum_{i=0}^{\bar{L}-1} \bar{e}_i 2^{wi}, \text{ for } \bar{e}_i \in \{0, 1, \ldots, 2^w - 1\} \text{ and } \bar{e}_{\bar{L}-1} \neq 0,$$

if $\bar{e}_i \neq 0$, let $\bar{e}_i = \bar{d}_i 2^{\ell_i}$ with $\bar{d}_i < 2^w$ odd and $0 \leq \ell_i < w$, and if $\bar{e}_i = 0$ let $\ell_i = w$.

(2) For all odd positive $d < 2^w$ compute $g(d) = g^d$.

(3) Initialize r as $g(\bar{d}_{\bar{L}-1})$.

(4) For $j = 1, 2, \dots, \ell_{\bar{L}-1}$ in succession replace r by $r \times r$.

(5) For $i = \bar{L} - 2, \bar{L} - 3, \dots, 0$ in succession, do the following:

    (a) If $\bar{e}_i \neq 0$, then

         • For $j = 1, 2, \dots, w - \ell_i$ in succession replace r by $r \times r$.

         • Replace r by $g(\bar{d}_i) \times r$.

    (b) For $j = 1, 2, \dots, \ell_i$ in succession replace r by $r \times r$.

(6) The resulting r equals $g^e \in G$.

For L-bit exponents, the number of squarings in G is $L-1$, as in square and multiply exponentiation. The number of multiplications in G is $2^{w-1} - 1$ for Step 2 and at most $\frac{L}{w}$ for Step 5, for a total of about $2^{w-1} + \frac{L}{w}$. It follows that the optimal window size w is proportional to $\log\log(e)$. For $w = 1$ window exponentiation is identical to left-to-right square and multiply exponentiation.

3.4.3. *Sliding window exponentiation*

In window exponentiation 3.4.2 the exponent e is split into consecutive 'windows' of w consecutive bits, irrespective of the bit pattern encountered. A slight improvement can be obtained by breaking e up into odd windows of at most w consecutive bits, where the windows are not necessarily consecutive and may be separated by zero bits. There are various ways to achieve this. A convenient way is a greedy approach which can easily be seen to be optimal. It determines the 'next' odd window of at most w consecutive bits in the following way:

- Scan the exponent (from left to right) for the first unprocessed one bit (replacing the prospective result r by $r \times r$ for each zero bit encountered).
- As soon as a one bit is found, determine the largest odd window of at most w consecutive bits that starts with the one bit just found. Let d be the odd integer thus determined, of length $k \leq w$.
- Replace r by $g(d) \times r^{2^k}$.

This method and any other variant that determines the w-bit windows in a flexible way is referred to as sliding window exponentiation. See [24] for a complete analysis of this and related methods.

3.4.4. *Multi-exponentiation*

Multi-exponentiation refers to the computation of $g^e \times h^d$ for some other element $h \in G$ and $d = \sum_{i=0}^{L-1} d_i 2^i \in \mathbf{Z}_{>0}$ with $d_i \in \{0,1\}$ (Section 2.4). Obviously, this can be done at the cost of two separate exponentiations, followed by a multiplication in G. It is also possible to combine the two exponentiations. This results in the following multi-exponentiation method.

(1) Compute $f = g \times h \in G$.
(2) Initialize r as the unit element in G.
(3) For $i = L - 1, L - 2, \ldots, 0$ in succession, do the following:
    (a) Replace r by $r \times r$.
    (b) If $e_i = 1$, then
        • If $d_i = 1$ replace r by $f \times r \in G$.
        • Else if $d_i = 0$ replace r by $g \times r \in G$.
    (c) Else if $e_i = 0$, then
        • If $d_i = 1$ replace r by $h \times r \in G$.
(4) The resulting r equals $g^e \times h^d \in G$.

There are L squarings in G and, on average for random L-bit e and d, about $\frac{3L}{4}$ multiplications in G. If $G = \mathbf{Z}/m\mathbf{Z}$ and Montgomery arithmetic is used, then one may expect that the average runtime of the above multi-exponentiation is about the same as the time required for $1.55 \log_2(e)$ multiplications in $\mathbf{Z}/m\mathbf{Z}$ (see 3.2.4). Thus, multi-exponentiation is only about 20% slower than single square and multiply exponentiation. Improvements can be obtained by using (sliding) windows. All resulting methods can be generalized to compute the product of more than two powers.

3.4.5. *Miscellaneous exponentiation tricks*

Because exponentiation methods are of such great importance for public key cryptography, the cryptographic literature contains a wealth of tricks and special-purpose methods. Some of the most important are sketched here.

Exponentiation with precomputation. Suppose that g^e must be computed for the same g and many different e's (of bit-length at most L). In this case it may be worthwhile to precompute and store the L values g^{2^i} for $i = 0, 1, \ldots, L-1$. Each application of the right-to-left square and multiply exponentiation then requires only $(\sum_{i=0}^{L-1} e_i) - 1 \approx (\log_2(e))/2$ multiplications in G. However, storing L precomputed values may be prohibitive. A single precomputed value has, relatively speaking, a much higher pay-off. Given $h = g^{2^{L/2}}$ and writing $e = \bar{e} + 2^{L/2}\bar{\bar{e}}$, the value $g^e = g^{\bar{e}} \times h^{\bar{\bar{e}}}$ can be computed using multi-exponentiation 3.4.4. This takes $\frac{L}{2}$ squarings and on average about $\frac{3L}{8}$ multiplications in G. For $G = \mathbf{Z}/m\mathbf{Z}$ and using Montgomery arithmetic this becomes about $0.78\log_2(e)$ multiplications in G. Because h can be computed at the cost of $\frac{L}{2}$ squarings in G, this method can also be used to reduce the runtime for ordinary square and multiply exponentiation in general (i.e., not for fixed g): it is reduced from L squarings and about $\frac{L}{2}$ multiplications to $\frac{L}{2} + \frac{L}{2} = L$ squarings and about $\frac{3L}{8}$ multiplications in G. See [19] for more involved and efficient precomputation methods.

Exponentiation with signed exponent representation. Let $e = \sum_{i=0}^{L-1} e_i 2^i$ be the binary representation of an L-bit exponent. Any block of k consecutive one bits with maximal $k > 1$, i.e., $e_i = e_{i+1} = \ldots = e_{i+k-1} = 1$ and $e_{i+k} = 0$, can be transformed into $e_i = -1$, $e_{i+1} = \ldots = e_{i+k-1} = 0$, and $e_{i+k} = 1$, without affecting the value of e. Applying this transformation from right to left results in a signed-bit representation of e of length $\leq L + 1$ where no two consecutive signed-bits are non-zero (and where a signed-bit is in $\{-1, 0, 1\}$). Such a representation is unique, for random e the number of non-zero signed-bits (the *weight*) is $\log_2(e)/3$ on average, and it is of minimal weight among the signed-bit representations of e.

It follows that if g^{-1} is available, g^e can be computed using a variation of square and multiply exponentiation at the cost of L squarings and on average $\log_2(e)/3$ multiplications in G. This may be worthwhile for groups G where the computation of g^{-1} given $g \in G$ can be performed at virtually no cost. A common example is the group of points of an elliptic curve over a finite field. For a combination with sliding windows, see [24].

Exponentiation with Frobenius. Let $\mathbf{F}_{p^\ell}$ be a finite field of characteristic p. Often elements of $\mathbf{F}_{p^\ell}$ can be represented in such a way that the Frobenius map that maps $x \in \mathbf{F}_{p^\ell}$ to $x^p \in \mathbf{F}_{p^\ell}$ is essentially for free (3.6.3). If that is the case, $G = \mathbf{F}_{p^\ell}^*$ or a subgroup thereof, and $e > p$, the computation of g^e can take advantage of the free Frobenius map. Let $e = \sum_{i=0}^{L'-1} f_i p^i$ be the radix p representation of e, then g^{f_i} for $0 \leq i < L'$ can be computed at the cost of $\log_2(p)$ squarings in G and, on average, about $\frac{L'}{2}\log_2(p) \approx \log_2(e)/2$ multiplications in G (using L'-fold right-to-left square and multiply exponentiation). The value g^e then follows with an additional $L' - 1$ multiplications in G and $L' - 1$ applications of the Frobenius map.

If G is the group of points over $\mathbf{F}_{p^\ell}$ of an elliptic curve defined over $\mathbf{F}_p$, then the Frobenius endomorphism on G may be computable for free. As shown in [58] this may be used to speed-up exponentiation (or rather scalar multiplication, as it should be referred to (Remark 3.7.2)).

3.5. *Prime generation*

3.5.1. *The prime number theorem*

The ability to quickly generate primes of almost any size is of great importance in cryptography. That this is possible is in the first place due to the prime number theorem:

Let $\pi(x)$ be the prime counting function, the number of primes $\leq x$. The function $\pi(x)$ behaves as $\frac{x}{\log(x)}$ for $x \to \infty$.

The prime number theorem says that a randomly selected L-bit number has reasonably large probability, namely more than $\frac{1}{L}$, to be prime, for any L of cryptographic interest. It then remains to find out if the candidate is prime or not. That can be done quickly, in all practical circumstances, using a variation of Fermat's little theorem, as shown in 3.5.2 below. Thus, searching for primes can be done quickly.

The prime number theorem also holds in arithmetic progressions:

Let $\pi(x, m, a)$ for integers m, a with $\gcd(m, a) = 1$ be the number of primes $\leq x$ that are congruent to a modulo m. The function $\pi(m, x, a)$ behaves as $\frac{x}{\phi(m) \log(x)}$ for $x \to \infty$, where $\phi(m)$ is the Euler phi function, the number of positive integers $\leq m$ which are relatively prime to m.

It follows that primes in certain residue classes, such as (3 mod 4) or 1 modulo some other prime, can be found just as quickly by limiting the search to the desired residue class. It also follows that the search can be made more efficient by restricting it to numbers free of small divisors in certain fixed residue classes.

3.5.2. *Probabilistic compositeness testing*

Fermat's little theorem says that

$$a^p = a \text{ for } a \in \mathbf{Z}/p\mathbf{Z} \text{ and prime } p.$$

It follows that if n and a are two integers for which $\gcd(n, a) = 1$ and

$$a^{n-1} \not\equiv 1 \bmod n,$$

then n cannot be a prime number. Fermat's little theorem can thus be
used to prove the compositeness of a composite n without revealing a non-
trivial factor of n. Based on the methods from Section 3.4 this can be done
efficiently, if a suitable a can be found efficiently. The latter is in general,
however, not the case: there are infinitely many composite numbers n, so-
called Carmichael numbers, for which $a^{n-1} \equiv 1 \bmod n$ for all a coprime
to n. This was shown in [5].

Suitable a's always exist, and can easily be found, for Selfridge's variation of Fermat's little theorem (commonly referred to as the Miller-Rabin
test):

> If p is an odd prime, $p - 1 = 2^t \cdot u$ for integers t, u with u odd, and
> a is an integer not divisible by p, then
>
> either $a^u \equiv 1 \bmod p$ or $a^{2^i u} \equiv -1 \bmod p$ for some i with $0 \le i < t$.

Let n be an odd composite number and $n = 2^t \cdot u$ for integers t, u with u
odd. An integer $a \in \{2, 3, \dots, n-1\}$ is called a witness to the compositeness
of n if

$$a^u \not\equiv 1 \bmod n \text{ and } a^{2^i u} \not\equiv -1 \bmod n \text{ for } i \text{ with } 0 \le i < t.$$

For odd composite n and a randomly selected $a \in \{2, 3, \dots, n - 1\}$ there
is a probability of at least 75% that a is witness to the compositeness
of n, as shown in [101]. It follows that in practice the compositeness of any
composite n can efficiently be proved:

Probabilistic compositeness test

> Let n be an odd positive number, and let k be a positive integer. Ran-
> domly and independently select at most k values $a \in \{2, 3, \dots, n - 1\}$
> until an a is found that is a witness to n's compositeness, or until all
> a's have been tried and no witness has been found. In the former case
> a proof of n's compositeness has been found and n is declared to be
> composite. In the latter case declare 'failure'.

If n is an odd composite, then the chance that a randomly selected a is
not a witness is less than $\frac{1}{4}$. Therefore, the chance that an odd composite
n escapes to be declared composite by the probabilistic compositeness test
with inputs n and k is less than $\left(\frac{1}{4}\right)^k$.

3.5.3. *Probabilistic compositeness test using Montgomery arithmetic*

If n is subjected to a probabilistic compositeness test, the computations of
a^u and $a^{2^i u}$ can be carried out in Montgomery arithmetic, even though a is

generated as an ordinary integer in $\{2, 3, \ldots, n-1\}$ and **not** converted to its Montgomery representation. All one has to do is change the '$\not\equiv 1$' and '$\not\equiv -1$' tests to '$\not\equiv R$' and '$\not\equiv -R$', respectively, where R is the Montgomery radix corresponding to n, as in 3.2.1.

3.5.4. *Generating industrial primes*

Combined with the prime number theorem (3.5.1), the probabilistic compositeness test from 3.5.2 makes it possible to generate L-bit primes quickly. Randomly select an odd L-bit number, and subject it to the probabilistic compositeness test with some appropriately chosen value k. If the number is declared to be composite, one tries again by selecting another L-bit number at random. Otherwise, if the probabilistic compositeness test declares 'failure', the selected number is called a *probable prime*, because if it were composite that would have been proved with overwhelming probability (namely, larger than $1 - (\frac{1}{4})^k$).

A probable prime – or, following Carl Pomerance, an *industrial prime* – is a number for which many attempts to prove its compositeness failed, not a number that was proved to be prime. In all practical circumstances, however, it is a prime number and if k is chosen properly it can be used without reservations in cryptographic applications. If odd L-bit random numbers are selected uniformly and independently, then the resulting (probable) primes will be uniformly distributed over the L-bit primes. See [80] for a further discussion on the probability that a probable prime is composite, and the choice of k.

Based on the prime number theorem, one may expect that on average approximately $\frac{L}{2} \log(2)$ composite n's are selected before a (probable) prime is found. It may also be expected that, for the composites, the first attempted witness is indeed a witness (because for random odd composite n the probability that the first choice is a witness is much higher than 75%). It follows that the total number of exponentiations modulo n (required for the computation of $a^u \bmod n$) is about $\frac{L}{2} \log(2) + k$: one for each 'wrong' guess, and k for the final 'correct' choice. In the remainder of this subsection it is shown how the $\frac{L}{2} \log(2)$ term can be reduced, at the cost of less costly computations or by giving up the uniformity of the distribution of the resulting primes.

Similar methods and estimates apply to the generation of primes with specific properties, such as primes that are 3 mod 4 or 1 modulo some other (large) prime.

3.5.5. *Remark on provable primality*

Probable primes can be proved to be prime – if they are indeed prime – using a general purpose primality test. Compared to the probabilistic compositeness test from 3.5.2, these tests are rather involved. They are hardly relevant for cryptology, and are therefore beyond the scope of these notes. See [9, 67] or [3, 15, 25, 26, 44, 86] for more details. As shown in [4], primes can be recognized in polynomial time.

It is also possible to generate primes uniformly in such a way that very simple primality proofs (based on Pocklington's theorem and generalizations thereof [9, 15, 67]) can be applied to them. See [77] for details.

3.5.6. *Prime generation with trial division*

Most random odd numbers have a small factor. Therefore, most 'wrong' guesses in 3.5.4 can be cast out much faster by finding their smallest prime factor than by attempting to find a witness. This can, for instance, conveniently be done by subjecting each candidate to trial division with the primes up to some bound B (rejecting a candidate as soon as a divisor is found). The ones for which no small factor is found are subjected to the probabilistic compositeness test. The choice for B depends on the desired length L and the relatively speeds of trial division and modular exponentiation. It is best determined experimentally. This approach leads to a substantial speed-up.

3.5.7. *Prime generation with a wheel*

Let P be the product of a small number of (small) primes. Trial division with the primes dividing P can be avoided by selecting L-bit random numbers of the form $aP + b$ for appropriately sized random integers a and b with b coprime to P. Selection of b can be done by picking random b with $0 < b < P$ until $\gcd(b, P) = 1$. For small P the computation of $\gcd(b, P)$ is very fast, so that this may on average turn out to be faster than trial division with the primes in P. Alternatively, if P is small enough, b can be selected from the precomputed and stored set $C_P = \{c : 0 < c < P, \gcd(c, P) = 1\}$. Given a candidate value $aP + b$ that is coprime to P, it can either right away be subjected to the probabilistic compositeness test, or it may be faster to perform trial division with larger primes first.

If for each choice of a a value b with $0 < b < P$ is randomly selected (from C_P or not), the resulting primes are again uniformly distributed. If for each choice of a all values in C_P are tried in order, then this method is referred to as a *wheel*. Primes resulting from a wheel-based search are no longer uniformly distributed. For RSA based cryptography this may be considered to be a (negligibly small) security risk, since the prime factors are the secrets and should in principle be indistinguishable from random primes. For discrete logarithm based cryptography usage of such primes does in general not pose an additional security risk.

3.5.8. *Prime generation with sieving*

An even faster way to perform the trial divisions is by using a *sieve*. It also results in even greater non-uniformity, but the security impact (even for RSA) is still negligible. Let $(s(i))_{i=0}^{[cL]-1}$ be an array of $[cL]$ bits, for a small constant c. Initially $s(i) = 0$ for all i. Let n be a random L-bit integer and let B' be the sieving bound. For all primes $p \le B'$, set the bit $s(j)$ to one if $n + j$ is divisible by p. This is done by replacing $s(kp - (n \bmod p))$ by one for all integers k such that $0 < kp - (n \bmod p) < [cL]$. For each p this requires the computation of $n \bmod p$ (once) and about $\frac{cL}{p}$ additions and bit assignments. This process is referred to as *sieving* and the array s is called a *sieve*. After sieving with all $p \le B'$, the locations j for which $s(j)$ has not been touched, i.e., $s(j) = 0$, correspond to integers $n + j$ that are not divisible by the primes $\le B'$. Those integers $n + j$ are subjected to the probabilistic compositeness test, until a probable prime is found. Based on 3.5.1 it may be expected that this will happen if c is large enough (say 2 or 3).

In general sieving works much faster than individual trial division such as in 3.5.6. As a result the optimal sieving bound B' is usually substantially larger than the optimal trial division bound B. Given L, the best B' is best determined experimentally.

3.6. *Finite field arithmetic*

Fast arithmetic in finite fields plays an important role in discrete logarithm based cryptography, both using a traditional multiplicative group of a finite field, and using the group of an elliptic curve over a finite field. For prime fields refer to Sections 3.1, 3.2, and 3.3 for addition, multiplication, and inversion in $\mathbf{Z}/m\mathbf{Z}$ with m prime (so that $\mathbf{Z}/m\mathbf{Z}$ is a finite field). In principle prime fields suffice for most cryptographic applications. Extension fields, however, have certain practical advantages over prime fields because they may allow faster arithmetic without compromising security. This subsection briefly reviews some of the most important notions and methods for arithmetic in finite extensions of prime fields. Let $\mathbf{F}_{p^\ell}$ denote a finite field of p^ℓ elements, where p is prime and $\ell > 1$. So, $\mathbf{F}_p$ is the prime field underlying $\mathbf{F}_{p^\ell}$.

3.6.1. Basic representation

Let $f(X) \in \mathbf{F}_p[X]$ be an irreducible polynomial of degree ℓ and let α be such that $f(\alpha) = 0$. Then $\{\alpha^0, \alpha^1, \alpha^2, \ldots, \alpha^{\ell-1}\}$ forms a basis for $\mathbf{F}_{p^\ell}$ over $\mathbf{F}_p$, i.e., the set

$$S = \{\sum_{i=0}^{\ell-1} a_i \alpha^i : a_i \in \mathbf{F}_p\}$$

is isomorphic to $\mathbf{F}_{p^\ell}$. This follows from the fact that $\#S = p^\ell$ (otherwise f would not be irreducible) and because an addition and a multiplication can be defined on S that make it into a finite field. For $a = \sum_{i=0}^{\ell-1} a_i \alpha^i, b = \sum_{i=0}^{\ell-1} b_i \alpha^i \in S$ the sum $a + b$ is defined as

$$a + b = \sum_{i=0}^{\ell-1} (a_i + b_i) \alpha^i \in S$$

and $-a = \sum_{i=0}^{\ell-1} -a_i \alpha^i \in S$. The product ab is defined as the remainder modulo $f(\alpha)$ of the polynomial $\left(\sum_{i=0}^{\ell-1} a_i \alpha^i\right) \times \left(\sum_{i=0}^{\ell-1} b_i \alpha^i\right)$ of degree $2\ell - 2$ in α. Thus $ab \in S$. Because f is irreducible, $\gcd(a, f) = 1$ for any $a \in S \setminus \{0\}$. The inverse $a^{-1} \in S$ can therefore be found by generalizing the extended Euclidean algorithm (3.3.2) from $\mathbf{Z}$ to $\mathbf{F}_p[X]$. Note that $S = (\mathbf{F}_p[X])/(f(X))$.

To make the reduction modulo $f(\alpha)$ easier to compute it is customary to assume that f is monic, i.e., that its leading coefficient equals one. Also, it is advantageous for the reduction to select f in such a way that it has many zero coefficients, for the same reason that moduli of a special form are advantageous (3.1.1). A 'Montgomery-like' approach to avoid the reduction modulo f, though easy to realize, is not needed because it would not be faster than reduction modulo a well chosen f.

If $g(X) \in \mathbf{F}_p[X]$ is another irreducible polynomial of degree ℓ with, say, $g(\beta) = 0$, then elements of $\mathbf{F}_{p^\ell}$ can be represented as $\sum_{i=0}^{\ell-1} b_i \beta^i$ with $b_i \in \mathbf{F}_p$. An effective isomorphism between the two representations of $\mathbf{F}_{p^\ell}$ can be found by finding a root of $g(X)$ in $(\mathbf{F}_p[X])/(f(X))$. This can be done efficiently using a method to factor polynomials over finite fields [56].

To represent the elements of a finite field $\mathbf{F}_{p^{\ell k}}$ the above construction can be used with an irreducible polynomial in $\mathbf{F}_p[X]$ of degree ℓk. Alternatively $\mathbf{F}_{p^{\ell k}}$ can be constructed as a k-th degree extension of $\mathbf{F}_{p^\ell}$ using an irreducible polynomial in $\mathbf{F}_{p^\ell}[X]$ of degree k. It depends on the circumstances and sizes of p, ℓ, and k which of these two methods is preferable in practice.

3.6.2. *Finding an irreducible polynomial*

A random monic polynomial of degree ℓ in $\mathbf{F}_p[X]$ is irreducible with probability about $\frac{1}{\ell}$. This result is comparable to the prime number theorem (3.5.1 and the definition of norm in 4.2.2). Irreducibility of polynomials in $\mathbf{F}_p[X]$ can be tested in a variety of ways [56]. For instance, $f \in \mathbf{F}_p[X]$ of degree ℓ is irreducible if and only if

$$\gcd(f, X^{p^i} - X) = 1 \text{ for } i = 1, 2, \dots, [\ell/2].$$

This follows from the facts that $X^{p^i} - X \in \mathbf{F}_p[X]$ is the product of all monic irreducible polynomials in $\mathbf{F}_p[X]$ of degrees dividing i, and that if f is irreducible it has a factor of degree at most $[\ell/2]$. This irreducibility condition can be tested efficiently using generalizations of the methods described in Section 3.3. Irreducible polynomials can thus be found quickly. In practice, however, it is often desirable to impose some additional restrictions on f. Some of these are sketched in 3.6.3 below.

3.6.3. *Optimal normal basis*

If the elements of $\mathbf{F}_{p^\ell}$ can be represented as $\sum_{i=0}^{\ell-1} a_i \alpha^{p^i}$, with $a_i \in \mathbf{F}_p$ and $f(\alpha) = 0$ for some irreducible $f(X) \in \mathbf{F}_p[X]$ of degree ℓ, then $\{\alpha, \alpha^p, \dots, \alpha^{p^{\ell-1}}\}$ is called a normal basis for $\mathbf{F}_{p^\ell}$ over $\mathbf{F}_p$. A normal basis representation has the advantage that p-th powering consists of a single circular shift of the coefficients:

$$\left(\sum_{i=0}^{\ell-1} a_i \alpha^{p^i}\right)^p = \sum_{i=0}^{\ell-1} a_i^p \alpha^{p^{i+1}} = \sum_{i=0}^{\ell-1} a_{((i-1) \bmod \ell)} \alpha^{p^i},$$

since $a_i^p = a_i$ for $a_i \in \mathbf{F}_p$ and $\alpha^{p^\ell} = \alpha$. As shown in 3.4.5 this may make exponentiation in $\mathbf{F}_{p^\ell}$ cheaper.

As an example, let $\ell + 1$ be prime, and let p be odd and a primitive root modulo $\ell + 1$, i.e., $p \bmod (\ell + 1)$ generates $(\mathbf{Z}/(\ell + 1)\mathbf{Z})^*$. The polynomial

$$f(X) = \frac{X^{\ell+1} - 1}{X - 1} = X^\ell + X^{\ell-1} + \dots + X + 1$$

is irreducible over $\mathbf{F}_p$. If $f(\alpha) = 0$, then $f(\alpha^{p^i}) = 0$ for $i = 0, 1, \dots, \ell - 1$, and $\alpha^i = \alpha^{i \bmod (\ell+1)}$ because $\alpha^{\ell+1} = 1$. With the fact that p is a primitive root modulo $\ell + 1$ it follows that $\{\alpha^i : 1 \le i \le \ell\}$ is the same as $\{\alpha^{p^i} : 0 \le i < \ell\}$. Because the latter is a normal basis, p-th powering using the basis $\{\alpha^i : 1 \le i \le \ell\}$ is just a permutation of the coefficients. Furthermore, with the basis $\{\alpha^i : 1 \le i \le \ell\}$ the reduction stage of the multiplication in $\mathbf{F}_{p^\ell}$ is easy. It takes just $\ell - 2$ additions and ℓ subtractions in $\mathbf{F}_p$, since $\alpha^i = \alpha^{i \bmod (\ell+1)}$ and $\alpha^0 = -\alpha^\ell - \alpha^{\ell-1} - \dots - \alpha$. This is an example of an *optimal* normal basis. The above construction fully characterizes optimal normal bases for odd characteristics. For a definition of an optimal basis and characteristic 2 existence results and constructions, see for instance [79].

3.6.4. *Inversion using normal bases*

As shown in [52], normal bases can be used for the computation of inverses in $\mathbf{F}_{p^\ell}^*$. Let p be odd and $x \in \mathbf{F}_{p^\ell}^*$ then

$$x^{-1} = (x^r)^{-1} x^{r-1}$$

for any integer r. The choice $r = \frac{p^\ell - 1}{p - 1}$ makes the computation of x^{-1} particularly fast: first compute x^{r-1} (as described below), compute $x^r = x x^{r-1}$, note that $x^r \in \mathbf{F}_p$ because its $(p-1)$-st power equals 1, compute $(x^r)^{-1} \in \mathbf{F}_p$ using a relatively fast inversion in $\mathbf{F}_p$, and finally compute $x^{-1} = (x^r)^{-1} x^{r-1}$ using ℓ multiplications in $\mathbf{F}_p$. Since $r - 1 = p^{\ell-1} + p^{\ell-2} + \dots + p^2 + p$, computation of x^{r-1} takes about $\log_2(m-1)$ multiplications in $\mathbf{F}_{p^\ell}$ and (free) p^j-th powerings for various j, by observing that

$$\left(x^{p+p^2+\dots+p^j}\right)\left(x^{p+p^2+\dots+p^j}\right)^{p^j} = x^{p+p^2+\dots+p^{2j}}.$$

Refer to [52] for further details on the computation of x^{r-1} and a similar construction for characteristic 2.

3.6.5. *Finding a generator*

A primitive normal basis for $\mathbf{F}_{p^\ell}$ over $\mathbf{F}_p$ is a normal basis as in 3.6.3 where α is a generator (or *primitive element*) of $\mathbf{F}_{p^\ell}^*$. In this case the irreducible polynomial $f(X)$ with $f(\alpha) = 0$ is called primitive. Primitive normal bases always exist. In general there is, however, no fast method to find a generator of $\mathbf{F}_{p^\ell}^*$. If the factorization of $p^\ell - 1$ is known, then the best method is to pick random elements $x \in \mathbf{F}_q^*$ until $x^{(p^\ell-1)/q} \ne 1$ for all distinct prime factors q of $p^\ell - 1$.

In cryptographic applications a generator of the full multiplicative group is hardly ever needed. Instead, a generator of an order q subgroup of $\mathbf{F}_{p^\ell}^*$ suffices, for some prime q dividing $p^\ell - 1$. Such a generator can be found by picking random elements $x \in \mathbf{F}_q^*$ until $x^{(p^\ell-1)/q} \neq 1$, in which case $x^{(p^\ell-1)/q}$ is the desired generator. The prime q must be chosen as a factor of the ℓ-th cyclotomic polynomial. For $q > \ell$ this guarantees that q does not divide $p^d - 1$ for any $d \leq \ell$ (and dividing ℓ) so that $\langle g \rangle$ cannot be embedded in a proper subfield of $\mathbf{F}_{p^\ell}$ (see [64]). The ℓ-th cyclotomic polynomial $\phi_\ell(X)$ is defined as follows:

$$X^\ell - 1 = \prod_{1 \leq d \leq \ell, d \text{ dividing } \ell} \phi_d(X).$$

3.7. Elliptic curve arithmetic

Elliptic curves were introduced in cryptanalysis with the invention of the elliptic curve factoring method (4.2.3) in 1984 [74]. This triggered applications of elliptic curves in cryptography and primality testing.

Early cryptographic applications concentrated on elliptic curves over fields of characteristic 2. This restriction was inspired by the relatively high computational demands of elliptic curve cryptography, and the possibility to realize competitive hardware implementations in characteristic 2. Nowadays they still enjoy a wide popularity. However, larger characteristic elliptic curves are becoming increasingly common in cryptography. These notes focus on the very basics of elliptic curves over fields of characteristic > 3. For a more general and complete treatment refer to [10, 110].

3.7.1. Elliptic curves and elliptic curve groups

Let $p > 3$ be prime. Any pair $a, b \in \mathbf{F}_{p^\ell}$ such that $4a^3 + 27b^2 \neq 0$ defines an *elliptic curve* $E_{a,b}$ over $\mathbf{F}_{p^\ell}$. Let $E = E_{a,b}$ be an elliptic curve over $\mathbf{F}_{p^\ell}$. The *set of points* $E(\mathbf{F}_{p^\ell})$ over $\mathbf{F}_{p^\ell}$ of E is informally defined as the pairs $x, y \in \mathbf{F}_{p^\ell}$ satisfying the *Weierstrass equation*

$$y^2 = x^3 + ax + b$$

along with the *point at infinity* $\mathcal{O}$. More precisely, let a projective point $(x : y : z)$ over $\mathbf{F}_{p^\ell}$ be an equivalence class of triples $(x, y, z) \in (\mathbf{F}_{p^\ell})^3$ with $(x, y, z) \neq (0, 0, 0)$. Two triples (x, y, z) and (x', y', z') are equivalent if $cx = x'$, $cy = y'$, and $cz = z'$ for $c \in \mathbf{F}_{p^\ell}^*$. Then

$$E(\mathbf{F}_{p^\ell}) = \{(x : y : z) : y^2 z = x^3 + axz^2 + bz^3\}.$$

The unique point with $z = 0$ is the point at infinity and denoted $\mathcal{O} = (0 : 1 : 0)$. The other points correspond to solution $\frac{x}{z}, \frac{y}{z} \in \mathbf{F}_{p^\ell}$ to the Weierstrass equation. They may or may not be normalized to have $z = 1$.

The set $E(\mathbf{F}_{p^\ell})$ is an abelian group, traditionally written additively. The group law is defined as follows. The point at infinity is the zero element, i.e., $P + \mathcal{O} = \mathcal{O} + P = P$ for any $P \in E(\mathbf{F}_{p^\ell})$. Let $P, Q \in E(\mathbf{F}_{p^\ell}) \setminus \{\mathcal{O}\}$ with normalized representations $P = (x_1 : y_1 : 1)$ and $Q = (x_2 : y_2 : 1)$. If $x_1 = x_2$ and $y_1 = -y_2$ then $P + Q = \mathcal{O}$, i.e., the opposite $-(x : y : z)$ is given by $(x : -y : z)$. Otherwise, let

$$\lambda = \begin{cases} \frac{y_1 - y_2}{x_1 - x_2} & \text{if } x_1 \neq x_2 \\[2ex] \frac{3x_1^2 + a}{2y_1} & \text{if } x_1 = x_2 \end{cases}$$

and $x = \lambda^2 - x_1 - x_2$, then $P + Q = (x : \lambda(x_1 - x) - y_1 : 1)$. This *elliptic curve addition* allows an easy geometric interpretation. If $P \neq Q$, then $-(P + Q)$ is the third point satisfying the Weierstrass equation on the line joining P and Q. If $P = Q$, then $-(P + Q)$ is the second point satisfying the equation on the tangent to the curve in the point $P = Q$. The existence of this point is a consequence of the condition that a, b defines an elliptic curve over $\mathbf{F}_{p^\ell}$, i.e., $4a^3 + 27b^2 \neq 0$.

Two elliptic curves $a, b \in \mathbf{F}_{p^\ell}$ and $a', b' \in \mathbf{F}_{p^\ell}$ are *isomorphic* if $a' = u^4 a$ and $b' = u^6 b$ for some $u \in \mathbf{F}_{p^\ell}^*$. The corresponding isomorphism between $E_{a,b}(\mathbf{F}_{p^\ell})$ and $E_{a',b'}(\mathbf{F}_{p^\ell})$ sends $(x : y : z)$ to $(u^2 x : u^3 y : z)$.

Computing in the group of points of an elliptic curve over a finite field involves computations in the underlying finite field. These computations can be carried out as set forth in Section 3.6. For extension fields bases satisfying special properties (such as discussed in 3.6.3) may turn out to be useful. In particular optimal normal bases in characteristic 2 (not treated here for elliptic curves) are advantageous, because squaring in the finite fields becomes a free operation.

3.7.2. *Remark on additive versus multiplicative notation*

In Section 3.4 algorithms are described to compute g^e for g in a group G, where g^e stands for $g \times g \times \ldots \times g$ and $\times$ denotes the group law. In terms of 3.7.1, the group G and the element g correspond to $E(\mathbf{F}_{p^\ell})$ and some $P \in E(\mathbf{F}_{p^\ell})$, respectively, and $\times$ indicates the elliptic curve addition. The latter is indicated by $+$ in 3.7.1 for the simple reason that the group law in $E(\mathbf{F}_{p^\ell})$ is traditionally written additively. Given the additive notation in $E(\mathbf{F}_{p^\ell})$ it is inappropriate to denote $P + P + \ldots + P$ as P^e and to refer to this operation as exponentiation. The usual notation eP is used instead. It is referred to as 'scalar multiplication' or 'scalar product'. In order to compute eP, the methods from Section 3.4 to compute g^e can be applied. Thus, 'square and multiply exponentiation' can be used to compute eP. Given the additive context it may, however, be more appropriate to call it 'double and add scalar multiplication'.

3.7.3. *Elliptic curve group representation*

There are several ways to represent the elements of $E(\mathbf{F}_{p^\ell}) \setminus \{\mathcal{O}\}$ and to perform the elliptic curve addition. For the group law as described in 3.7.1 the elements can be represented using *affine coordinates*, i.e., as $(x, y) \in (\mathbf{F}_{p^\ell})^2$, where (x, y) indicates the point $(x : y : 1)$. If affine coordinates are used the group law requires an inversion in $\mathbf{F}_{p^\ell}$. This may be too costly. If *projective coordinates* are used, i.e., elements of $E(\mathbf{F}_{p^\ell}) \setminus \{\mathcal{O}\}$ are represented as $(x : y : z)$ and not necessarily normalized, then the inversion in $\mathbf{F}_{p^\ell}$ can be avoided. However, this comes at the cost of increasing the number of squarings and multiplications in $\mathbf{F}_{p^\ell}$ per application of the group law.

Another representation that is convenient in some applications is based on the *Montgomery model* of elliptic curves. In the Montgomery model the coordinates of the group elements satisfy an equation that is slightly different from the usual Weierstrass equation. An isomorphic curve in the Weierstrass model can, however, easily be found, and vice versa. As a consequence, group elements can be represented using just two coordinates, as in the affine case. Furthermore, an inversion is not required for the group law, as in the projective case. The group law is more efficient than the group law for the projective case, assuming that the difference $P - Q$ is available whenever the sum $P + Q$ of P and Q must be computed. This condition makes it impossible to use the ordinary square and multiply exponentiation (Remark 3.7.2) to compute scalar products. Refer to [14, 83] for detailed descriptions of the Montgomery model and a suitable algorithm to compute a scalar multiplication in this case.

Refer to [27] for a comparison of various elliptic curve point representations in cryptographic applications.

3.7.4. *Elliptic curves modulo a composite*

The field $\mathbf{F}_{p^\ell}$ in 3.7.1 and 3.7.3 can be replaced by the ring $\mathbf{Z}/n\mathbf{Z}$ for a composite n (coprime to 6). An elliptic curve over $\mathbf{Z}/n\mathbf{Z}$ is defined as a pair $a, b \in \mathbf{Z}/n\mathbf{Z}$ with $4a^3 + 27b^2 \in (\mathbf{Z}/n\mathbf{Z})^*$. The set of points, defined in a way similar to 3.7.1, is again an abelian group. The group law so far has limited applications in cryptology. Instead, for cryptanalytic purposes it is more useful to define a *partial addition* on $E(\mathbf{Z}/n\mathbf{Z})$, simply by performing the group law defined in 3.7.1 in $\mathbf{Z}/n\mathbf{Z}$ as opposed to $\mathbf{F}_{p^\ell}$. Due to the existence of zero divisors in $\mathbf{Z}/n\mathbf{Z}$ the resulting operation can break down (which is the reason that it is called partial addition) and produce a non-trivial divisor of n instead of the desired sum $P + Q$. As will be shown in 4.2.3, this is precisely the type of 'accident' one is hoping for in cryptanalytic applications.

3.7.5. *Elliptic curve modulo a composite taken modulo a prime*

Let p be any prime dividing n. The elliptic curve $a, b \in \mathbf{Z}/n\mathbf{Z}$ as in 3.7.4 and the set of points $E_{a,b}(\mathbf{Z}/n\mathbf{Z})$ can be mapped to an elliptic curve $\bar{a} = a \bmod p, \bar{b} = b \bmod p$ over $\mathbf{F}_p$ and set of points $E_{\bar{a},\bar{b}}(\mathbf{F}_p)$. The latter is done by reducing the coordinates of a point $P \in E_{a,b}(\mathbf{Z}/n\mathbf{Z})$ modulo p resulting in $P_p \in E_{\bar{a},\bar{b}}(\mathbf{F}_p)$. Let $P, Q \in E(\mathbf{Z}/n\mathbf{Z})$. If $P + Q \in E(\mathbf{Z}/n\mathbf{Z})$ is successfully computed using the partial addition, then $P_p + Q_p$ (using the group law in $E(\mathbf{F}_p)$) equals $(P + Q)_p$. Furthermore, $P = \mathcal{O}$ if and only if $P_p = \mathcal{O}_p$.

3.7.6. *Generating elliptic curves for cryptographic applications*

For cryptographic applications one wants elliptic curves for which the cardinality $\#E(\mathbf{F}_{p^\ell})$ of the group of points is either prime or the product of a small number and a prime. It is known that $\#E(\mathbf{F}_{p^\ell})$ equals $p^\ell + 1 - t$ for some integer t with $|t| \leq 2\sqrt{p^\ell}$ (Hasse's theorem [110]). Furthermore, if E is uniformly distributed over the elliptic curves over $\mathbf{F}_{p^\ell}$, then $\#E(\mathbf{F}_{p^\ell})$ is approximately uniformly distributed over the integers close to $p^\ell + 1$. Thus, if sufficiently many elliptic curves are selected at random over some fixed finite field, it may be expected (3.5.1) that a suitable one will be found after a reasonable number of attempts.

It follows that for cryptographic applications of elliptic curves it is desirable to have an efficient and, if at all possible, simple way to determine $\#E(\mathbf{F}_{p^\ell})$, in order to check if $\#E(\mathbf{F}_{p^\ell})$ satisfies the condition of being prime or almost prime. The first polynomial-time algorithm solving this so-called elliptic curve point counting problem was due to Schoof [106]. It has been improved by many different authors (see [10]). Although enormous progress has been made, the point counting problem is a problem one tries to avoid in cryptographic applications. This may be done by using specific curves with known properties (and group cardinalities), or by using curves from a prescribed list prepared by some trusted party. However, if one insists on a randomly generated elliptic curve (over, say, a randomly generated finite field), one will have to deal with the point counting. As a result, parameter selection in its full generality for elliptic curve cryptography must still considered to be a nuisance. This is certainly the case when compared to the ease of parameter selection in RSA or systems relying on the discrete logarithm problem in ordinary multiplicative groups of finite fields. Refer to the proceedings of recent cryptographic conferences for progress on the subject of point counting and elliptic curve parameter initialization.

Let E be a suitable elliptic curve, i.e., $\#E(\mathbf{F}_{p^\ell}) = sq$ for a small integer s and prime q. A generator of the order q subgroup of $E(\mathbf{F}_{p^\ell})$ can be found in a way similar to 3.6.5. Just pick random points $P \in E(\mathbf{F}_{p^\ell})$ until $sP \neq \mathcal{O}$ (Remark 3.7.2). The desired generator is then given by sP.

Given an elliptic curve E over $\mathbf{F}_{p^\ell}$, the set of points $E(\mathbf{F}_{p^{\ell k}})$ can be considered over an extension field $\mathbf{F}_{p^{\ell k}}$ of $\mathbf{F}_{p^\ell}$. In particular the case $\ell = 1$ is popular in cryptographic applications. This approach facilitates the computation of $\#E(\mathbf{F}_{p^{\ell k}})$ (based on Weil's theorem [78]). However, it is frowned upon by some because of potential security risks. Also, elliptic curves E over $\mathbf{F}_{p^\ell}$ and their corresponding group $E(\mathbf{F}_{p^\ell})$ with ℓ composite are not considered to be a good choice. See [112] and the references given there.

4. Factoring and Computing Discrete Logarithms

Let n be an odd positive composite integer, let $g \in G$ be an element of known order of a group of known cardinality $\#G$, and let $h \in \langle g \rangle$. If required, the probabilistic compositeness test from 3.5.2 can be used to ascertain that n is composite. Checking that $h \in \langle g \rangle$ can often be done by verifying that $h \in G$ and that $h^{\operatorname{order}(g)} = 1$. It is assumed that the elements of G are uniquely represented. The group law in G is referred to as 'multiplication'. If $G = E(\mathbf{F}_p)$ then this multiplication is actually elliptic curve addition (Remark 3.7.2).

This section reviews the most important methods to solve the following two problems:

Factoring
Find a not necessarily prime factor p of n with $1 < p < n$.
Computing discrete logarithms
Find $\log_g(h)$, i.e., the integer $t \in \{0, 1, \dots, \operatorname{order}(g) - 1\}$ such that $g^t = h$.

For generic G and prime order(g) it has been proved that finding $\log_g(h)$ requires, in general, $c\sqrt{\text{order}(g)}$ group operations [88, 108], for a constant $c > 0$. However, this generic model does not apply to any practical situation. For integer factoring no lower bound has been published.

4.1. Exponential-time methods

In this subsection the basic exponential-time methods are sketched:

Factoring: methods that may take n^c operations on integers of size comparable to n,

Computing discrete logarithms: methods that may take order$(g)^c$ multiplications in G,

where c is a positive constant. These runtimes are called exponential-time because $n^c = e^{c\log(n)}$, so that the runtime is an exponential function of the input length $\log(n)$.

4.1.1. Exhaustive search

Factoring n by exhaustive search is referred to as *trial division*: for all primes in succession check if they divide n, until a proper divisor is found. Because n has a prime divisor $\leq \sqrt{n}$, the number of division attempts is bounded by $\pi(\sqrt{n}) \approx \frac{\sqrt{n}}{\log(\sqrt{n})}$ (see 3.5.1). More than 91% of all positive integers have a factor < 1000. For randomly selected composite n trial division is therefore very effective. It is useless for composites as used in RSA

Similarly, $\log_g(h)$ can be found by comparing g^t to h for $t = 0, 1, 2, \ldots$ in succession, until $g^t = h$. This takes at most order(g) multiplications in G. There are no realistic practical applications of this method, unless order(g) is very small.

4.1.2. Pollard's $p - 1$ method [92]

According to Fermat's little theorem (see 3.5.2), $a^{p-1} \equiv 1 \bmod p$ for prime p and any integer a not divisible by p. It follows that $a^k \equiv 1 \bmod p$ if k is an integer multiple of $p - 1$. Furthermore, if p divides n, then p divides $\gcd(a^k - 1, n)$. This may make it possible to find a prime factor p of n by computing $\gcd(a^k - 1, n)$ for an arbitrary integer a with $1 < a < n$ and a coprime to n (assuming one is not so lucky to pick an a with $\gcd(a, n) \neq 1$). This works if one is able to find a multiple k of $p - 1$.

The latter may be possible if n happens to have a prime factor p for which $p-1$ consists of the product of some small primes. If that is the case then k can be chosen as a product of many small prime powers. Obviously, only some of the primes dividing k actually occur in $p-1$. But $p-1$ is not known beforehand, so one simply includes as many small prime powers in k as feasible, and hopes for the best, i.e., that the resulting k is a multiple of $p-1$. The resulting k may be huge, but the number $a^k - 1$ has to be computed only modulo n. Refer to [83] for implementation details.

As a result a prime factor p of n can be found in time proportional to the largest prime factor in $p-1$. This implies that the method is practical only if one is lucky and the largest prime factor in $p-1$ happens to be sufficiently small. For reasonably sized p, such as used as factors of RSA moduli, the probability that p can be found using Pollard's $p-1$ method is negligible. Nevertheless, the existence of Pollard's $p-1$ method is the reason that many cryptographic standards require RSA moduli consisting of primes p for which $p-1$ has a large prime factor. Because a variation using $p+1$ (the 'next' cyclotomic polynomial) follows in a straightforward fashion [121], methods have even been designed to generate primes p that can withstand both a $p-1$ and a $p+1$ attack. This conveniently overlooks the fact that there is an attack for each cyclotomic polynomial [8]. However, in view of the elliptic curve factoring method (4.2.3) and as argued in [103], none of these precautions makes sense.

4.1.3. *The Silver-Pohlig-Hellman method* [91]

Just as p dividing n can be found easily if $p-1$ is the product of small primes, discrete logarithms in $\langle g \rangle$ can be computed easily if $\mathrm{order}(g)$ has just small factors. Assume that $\mathrm{order}(g)$ is composite and that q is a prime dividing $\mathrm{order}(g)$. From $g^t = h$ it follows that

$$\left(g^{\frac{\mathrm{order}(g)}{q}} \right)^{t \bmod q} = h^{\frac{\mathrm{order}(g)}{q}}.$$

Thus, t modulo each of the primes q dividing $\mathrm{order}(g)$ can be found by solving the discrete logarithm problems for the $\frac{\mathrm{order}(g)}{q}$-th powers of g and h. If $\mathrm{order}(g)$ is a product of distinct primes, then t follows using the Chinese remainder theorem (3.3.6). If $\mathrm{order}(g)$ is not squarefree, then use the extension as described in [67].

The difficulty of a discrete logarithm problem therefore depends mostly on the size of the largest prime factor in $\mathrm{order}(g)$. Because the factorization of $\#G$ is generally assumed to be known (and in general believed to be easier to find than discrete logarithms in G), the order of g is typically assumed to be a sufficiently large prime in cryptographic applications.

4.1.4. *Shanks' baby-step-giant-step* [56, Exercise 5.17]

Let $s = \lfloor\sqrt{\text{order}(g)}\rfloor + 1$. Then for any $t \in \{0, 1, \ldots, \text{order}(g) - 1\}$ there are non-negative integers $t_0, t_1 \leq s$ such that $t = t_0 + t_1 s$. From $g^t = h$ it follows that $hg^{-t_0} = g^{t_1 s}$. The values g^{is} for $i \in \{0, 1, \ldots, s\}$ are computed (the 'giant steps') at the cost of about s multiplications in G, and put in a hash table. Then, for $j = 0, 1, \ldots, s$ in succession hg^{-j} is computed (the 'baby steps') and looked up in the hash table, until a match is found. The value t can be derived from j and the location of the match. This (deterministic) method requires at most about $2\sqrt{\text{order}(g)}$ multiplications in G. This closely matches the lower bound mentioned above. The greatest disadvantage of this method is that it requires storage for $\sqrt{\text{order}(g)}$ elements of G. Pollard's rho method, described in 4.1.5 below, requires hardly any storage and achieves essentially the same speed.

4.1.5. *Pollard's rho and lambda methods* [93]

The probability that among a group of 23 randomly selected people at least two people have the same birthday is more than 50%. This probability is much higher than most people would expect. It is therefore referred to as the birthday paradox. It lies at the heart of the most effective general purpose discrete logarithm algorithms, Pollard's rho and lambda methods. If elements are drawn at random from $\langle g \rangle$ (with replacement) then the expected number of draws before an element is drawn twice (a so-called 'collision') is $\sqrt{\text{order}(g)\pi/2}$. Other important cryptanalytic applications are the use of 'large primes' in subexponential-time factoring and discrete logarithm methods (Section 4.2) and collision search for hash functions (not treated here).

Application to computing discrete logarithms. A collision of randomly drawn elements from $\langle g \rangle$ is in itself not useful to solve the discrete logarithm problem. The way this idea is made to work to compute discrete logarithms is as follows. Define a (hopefully) random walk on $\langle g \rangle$ consisting of elements of the form $g^e h^d$ for known e and d, wait for a collision, i.e., e, d and e', d' such that $g^e h^d = g^{e'} h^{d'}$, and compute $\log_g(h) = \frac{e-e'}{d'-d} \bmod \text{order}(g)$. A 'random' walk $(w_i)_{i=1}^{\infty}$ on $\langle g \rangle$ can, according to [93], be achieved as follows. Partition G into three subsets G_1, G_2, and G_3 of approximately equal cardinality. This can usually be done fairly accurately based on the representation of the elements of G. One may expect that this results in three sets $G_j \cap \langle g \rangle$, for $j = 1, 2, 3$, of about the same size. Take $w_1 = g$ (so $e = 1$, $d = 0$), and define w_{i+1} as a function of w_i:

$$w_{i+1} = \begin{cases} hw_i & \text{if } w_i \in G_1 \ ((e, d) \to (e, d+1)) \\ w_i^2 & \text{if } w_i \in G_2 \ ((e, d) \to (2e, 2d)) \\ gw_i & \text{if } w_i \in G_3 \ ((e, d) \to (e+1, d)). \end{cases}$$

This can be replaced by any other function that allows easy computation of the exponents e and d and that looks sufficiently random. It is not necessary to compare each new w_i to all previous ones (which would make the method as slow as exhaustive search). According to *Floyd's cycle-finding algorithm* it suffices to compare w_i to w_{2i} for $i = 1, 2, \ldots$ (see [56]). A pictorial description of the sequence $(w_i)_{i=0}^{\infty}$ is given by a ρ (the Greek character rho): starting at the tail of the ρ it iterates until it bites in its own tail, and cycles from there on.

As shown in [115] partitioning G into only three sets does in general not lead to a truly random walk. In practice that means that the collision occurs somewhat later than it should. Unfortunately a truly random walk is hard to achieve. However, as also shown in [115], if G is partitioned into substantially more than 3 sets, say about 15 sets, then the collision occurs on average almost as fast as it would for a truly random walk. An improvement of Floyd's cycle-finding algorithm is described in [16].

Parallelization. If m processors run the above method independently in parallel, each starting at its own point $w_1 = g^e h^d$ for random e, d, a speed-up of a factor $\sqrt{m}$ can be expected. A parallelization that achieves a speed-up of a factor m when run on m processors is described in [118]. Define *distinguished points* as elements of $\langle g \rangle$ that occur with relatively low probability θ and that have easily recognizable characteristics. Let each processor start at its own randomly selected point $w_1 = g^e h^d$. As soon as a processor hits a distinguished point the processor reports the distinguished point to a central location and starts afresh. In this way m processors generate $\geq m$ independent 'trails' of average length $1/\theta$. Based on the birthday paradox, one may expect that $\sqrt{\text{order}(g)\pi/2}$ trail points have to be generated before a collision occurs among the trails, at an average cost of $(\sqrt{\text{order}(g)\pi/2})/m$ steps per processor. However, this collision itself goes undetected. It is only detected, at the central location, at the first distinguished point after the collision. Each of the two contributing processors therefore has to do an additional expected $1/\theta$ steps to reach that distinguished point. For more details, also on the choice of θ, see [118].

Pollard's parallelized rho is currently the method of choice to attack the discrete logarithm problem in groups of elliptic curves. Groups of well over 2^{100} elements can successfully be attacked. For the most recent results, consult [23].

Pollard's lambda method for catching kangaroos. There is also a non-parallelized discrete logarithm method that is based on just two trails that collide, thus resembling a λ (the Greek character lambda). It finds t in about $2\sqrt{w}$ applications of the group law if t is known to lie in a specified interval of width w. See [93] for a description of this method in terms of tame and wild kangaroos and [94] for a speed-up based on the methods from [118].

Application to factoring. The collision idea can also be applied in the context of factoring. If elements are drawn at random from $\mathbf{Z}/n\mathbf{Z}$ (with replacement) then the expected number of draws before an element is drawn that is identical modulo p to some element drawn earlier is $\sqrt{p\pi/2}$. Exponents do not have to be carried along, a random walk in $\mathbf{Z}/n\mathbf{Z}$ suffices. According to [93] this can be achieved by picking $w_1 \in \mathbf{Z}/n\mathbf{Z}$ at random and by defining

$$w_{i+1} = (w_i^2 + 1) \bmod n.$$

With Floyd's cycle-finding algorithm one may expect that $\gcd(w_i - w_{2i}, n) > 1$ after about $\sqrt{p}$ iterations. In practice products of $|w_i - w_{2i}|$ for several consecutive i's are accumulated (modulo n) before a gcd is computed.

One of the earliest successes of Pollard's rho method was the factorization of the eighth Fermat number $F_8 = 2^{2^8} + 1$. It was found because the factor happened to be unexpectedly small [18]. See also 4.2.6.

4.2. Subexponential-time methods

4.2.1. The L function

For $t, \gamma \in \mathbf{R}$ with $0 \le t \le 1$ the notation $L_x[t, \gamma]$ introduced in [67] is used for any function of x that equals

$$e^{(\gamma + o(1))(\log x)^t (\log \log x)^{1-t}}, \text{ for } x \to \infty.$$

For $t = 0$ this equals $(\log x)^\gamma$ and for $t = 1$ it equals x^γ (up to the $o(1)$ in the exponent). It follows that for $0 \le t \le 1$ the function $L_x[t, \gamma]$ interpolates between polynomial-time ($t = 0$) and exponential-time ($t = 1$). Runtimes equal to $L_x[t, \gamma]$ with $0 < t < 1$ are called subexponential-time in x because they are asymptotically less than $e^{c \log(x)}$ for any constant c.

In this subsection the basic subexponential-time methods for factoring and computing discrete logarithms in $\mathbf{F}_{p^\ell}^*$ are sketched:

Factoring: methods for which the number of operations on integers of size comparable to n is expected to be $L_n[t, \gamma]$ for $n \to \infty$,

Computing discrete logarithms in $\mathbf{F}_{p^\ell}^*$: methods for which the number of multiplications in $\mathbf{F}_{p^\ell}$ is expected to be $L_{p^\ell}[t, \gamma]$ for $p \to \infty$ and fixed ℓ, or fixed p and $\ell \to \infty$,

where $\gamma > 0$ and t are constants with $0 < t < 1$. For most methods presented below these (probabilistic) runtimes cannot rigorously be proved. Instead, they are based on heuristic arguments.

The discrete logarithm methods presented below work only for $G = \mathbf{F}_{p^\ell}^*$, because explicit assumptions have to be made about properties of the representation of group elements. Thus, unlike Section 4.1 it is explicitly assumed that $G = \mathbf{F}_{p^\ell}^*$, and that g generates $\mathbf{F}_{p^\ell}^*$.

4.2.2. *Smoothness*

Integers. A positive integer is B-smooth (or simply smooth if B is clear from the context) if all its prime factors are $\leq B$. Let $\alpha, \beta, r, s \in \mathbf{R}_{>0}$ with $s < r \leq 1$. It follows from [20, 37] that a random positive integer $\leq L_x[r, \alpha]$ is $L_x[s, \beta]$-smooth with probability

$$L_x[r - s, -\alpha(r - s)/\beta], \text{ for } x \to \infty.$$

Polynomials over $\mathbf{F}_p$. Assume that, as in 3.6.1, elements of $\mathbf{F}_{p^\ell}^*$ with p prime and $\ell > 1$ are represented as non-zero polynomials in $\mathbf{F}_p[X]$ of degree $< \ell$. The *norm* of $h \in \mathbf{F}_{p^\ell}^*$ in this representation is defined as $p^{\text{degree}(h)}$.

A polynomial in $\mathbf{F}_p[X]$ is B-smooth if it factors as a product of irreducible polynomials in $\mathbf{F}_p[X]$ of norm $< B$. Let $\alpha, \beta, r, s \in \mathbf{R}_{>0}$ with $r \leq 1$ and $\frac{r}{100} < s < \frac{99r}{100}$. It follows from [90] that a random polynomial in $\mathbf{F}_p[X]$ of norm $\leq L_x[r, \alpha]$ is $L_x[s, \beta]$-smooth with probability

$$L_x[r - s, -\alpha(r - s)/\beta], \text{ for } x \to \infty.$$

Note the similarity with integer smoothness probability.

4.2.3. *Elliptic curve method* [74]

Rephrasing Pollard's $p-1$ method 4.1.2 in terms of 4.2.2, it attempts to find p dividing n for which $p-1$ is B-smooth by computing $\gcd(a^k - 1, n)$ for an integer k that consists of the primes $\leq B$ and some of their powers. Most often the largest prime in $p-1$ is too large to make this practical. The elliptic curve method is similar to Pollard's $p-1$ method in the sense that it tries to take advantage of the smoothness of a group order ($\#\mathbf{Z}/p\mathbf{Z}^*$ in Pollard's $p-1$ method): if the group order is smooth a randomly generated unit in the group (a^k) may lead to a factorization. In Pollard's $p-1$ method the groups are fixed as the groups $\mathbf{Z}/p\mathbf{Z}^*$ for the primes p dividing the number one tries to factor. The elliptic curve method randomizes the choice of the groups (and their orders). Eventually, if one tries often enough, a group of smooth order will be encountered and a factorization found.

Let $a, b \in \mathbf{Z}/n\mathbf{Z}$ be randomly selected so that they define an elliptic curve E over $\mathbf{Z}/n\mathbf{Z}$ (see 3.7.4). According to 3.7.5 an elliptic curve $E_p = E_{a \bmod p, b \bmod p}$ over $\mathbf{F}_p$ is defined for each prime p dividing n, and $\#E_p(\mathbf{F}_p)$ behaves as a random integer close to $p+1$ (see 3.7.6). Based on 4.2.2 with $r = 1$, $\alpha = 1$, $s = 1/2$, and $\beta = \sqrt{1/2}$ it is not unreasonable to assume that $\#E_p(\mathbf{F}_p) = L_p[1, 1]$ is $L_p[1/2, \sqrt{1/2}]$-smooth with probability $L_p[1/2, -\sqrt{1/2}]$. Thus, for a fixed p, once every $L_p[1/2, \sqrt{1/2}]$ random elliptic curves over $\mathbf{Z}/n\mathbf{Z}$ one expects to find a curve for which the group order $\#E_b(\mathbf{F}_p)$ is $L_p[1/2, \sqrt{1/2}]$-smooth.

Assume that $\#E_p(\mathbf{F}_p)$ is $L_p[1/2, \sqrt{1/2}]$-smooth, let k be the product of the primes $\leq L_p[1/2, \sqrt{1/2}]$ and some of their powers, and let P be a random element of $E(\mathbf{Z}/n\mathbf{Z})$. If one attempts to compute kP in $E(\mathbf{Z}/n\mathbf{Z})$ using the partial addition defined in 3.7.4 and the computation does not break down, then the result is some point $R \in E(\mathbf{Z}/n\mathbf{Z})$. According to 3.7.5 the point $R_p \in E_p(\mathbf{F}_p)$ would have been obtained by computing the elliptic curve scalar product of k and the point $P_p \in E_p(\mathbf{F}_p)$ as defined in 3.7.5. If enough prime powers are included in k, then the order of $P_p \in E_p(\mathbf{F}_p)$ divides k, so that $R_p = \mathcal{O}_p \in E_p(\mathbf{F}_p)$, where $\mathcal{O}_p$ is the zero element in $E_p(\mathbf{F}_p)$. But, according to 3.7.5, $R_p = \mathcal{O}_p$ implies $R = \mathcal{O}$. The latter implies that $R_q = \mathcal{O}_q$ for any prime q dividing n. It follows that, if R has been computed successfully, k must be a multiple of the order of P when taken modulo any prime dividing n. Given how much luck is already involved in picking E such that $\#E_p(\mathbf{F}_p)$ is smooth for one particular p dividing n, it is unlikely that this would happen for all prime factors of n simultaneously. Thus if E was randomly selected in such a way that

$\#E_p(\mathbf{F}_p)$ is $L_p[1/2, \sqrt{1/2}]$-smooth, it is much more likely that R has not been computed to begin with, but that the partial addition broke down, i.e., produced a non-trivial factor of n. From $\mathcal{O}_p = (0:1:0)$ it follows that p divides that factor.

Since one in every $L_p[1/2, \sqrt{1/2}]$ elliptic curves over $\mathbf{Z}/n\mathbf{Z}$ can be expected to be lucky, the total expected runtime is $L_p[1/2, \sqrt{1/2}]$ times the time required to compute kP, where k is the product of powers of primes $\leq L_p[1/2, \sqrt{1/2}]$. The latter computation requires $L_p[1/2, \sqrt{1/2}]$ partial additions, i.e., has cost proportional to $\log(n)^2 L_p[1/2, \sqrt{1/2}]$. The total cost is proportional to

$$L_p[1/2, \sqrt{1/2}] \cdot \log(n)^2 L_p[1/2, \sqrt{1/2}] = \log(n)^2 L_p[1/2, \sqrt{2}].$$

It follows that using the elliptic curve method small factors can be found faster than large factors. For $p \approx \sqrt{n}$, the worst case, the expected runtime becomes $L_n[1/2, 1]$. For RSA moduli it is known that the worst case applies. For composites without known properties and, in particular, a smallest factor of unknown size, one generally starts off with a relatively small k aimed at finding small factors. This k is gradually increased for each new attempt, until a factor is found or until the factoring attempt is aborted. See [14, 83] for implementation details of the elliptic curve method. The method is ideally parallelizable: any number of attempts can be run independently on any number of processors in parallel, until one of them is lucky.

The reason that the runtime argument is heuristic is that $\#E_p(\mathbf{F}_p)$ is contained in an interval of short length (namely, about $4\sqrt{p}$) around $p+1$. Even though $\#E_p(\mathbf{F}_p)$ cannot be distinguished from a truly random integer in that interval, intervals of short length around $p+1$ cannot be proved (yet) to be smooth with approximately the same probability as random integers $\leq p$. The heuristics are, however, confirmed by experiments and the elliptic curve method so far behaves as heuristically predicted.

Remarkable successes of the elliptic curve method were the factorizations of the tenth and eleventh Fermat numbers, $F_{10} = 2^{2^{10}} + 1$ and $F_{11} = 2^{2^{11}} + 1$; see [17]. Factors of over 50 decimal digits have occasionally been found using the elliptic curve method. The method is not considered to be a threat against RSA, but its existence implies that care should be taken when using RSA moduli consisting of more than two prime factors.

A variant of the elliptic curve method suitable for the computation of discrete logarithms has never been published.

4.2.4. *The Morrison-Brillhart approach*

The expected runtimes of all factoring methods presented so far depend strongly on properties of the factor to be found, and only polynomially on the size of the number to be factored. For that reason they are referred to as *special purpose* factoring methods. All factoring methods described in the sequel are *general purpose* methods: their expected runtimes depend just on the size of the number n to be factored. They are all based, in some way or another, on Fermat's factoring method of solving a congruence of squares modulo n: try to find integers x and y such that their squares are equal modulo n, i.e., $x^2 \equiv y^2 \bmod n$. Such x and y are useful for factoring purposes because it follows from $x^2 - y^2 \equiv 0 \bmod n$ that n divides $x^2 - y^2 = (x-y)(x+y)$, so that

$$n = \gcd(n, x-y) \gcd(n, x+y).$$

This may yield a non-trivial factorization of n. There is a probability of at least 50% that this produces a non-trivial factor of n if n is composite and not a prime power and x and y are random solutions to $x^2 \equiv y^2 \bmod n$.

Fermat's method to find x and y consists of trying $x = [\sqrt{n}] + 1, [\sqrt{n}] + 2, \dots$ in succession, until $x^2 - n$ is a perfect square. In general this cannot be expected to be competitive with any of the methods described above, not even trial division. Morrison and Brillhart [87] proposed a faster way to find x and y. Their general approach is not to solve the identity $x^2 \equiv y^2 \bmod n$ right away, but to construct x and y based on a number of other identities modulo n which are, supposedly, easier to solve. Thus, the Morrison-Brillhart approach consists of two stages: a first stage where a certain type of identities modulo n are found, and a second stage where those identities are used to construct a solution to $x^2 \equiv y^2 \bmod n$. This approach applies to all factoring and discrete logarithm algorithms described below.

Let B be a smoothness bound (4.2.2) and let P be the *factor base*: the set of primes $\le B$ of cardinality $\#P = \pi(B)$ (see 3.5.1). In the first stage, one collects $> \#P$ integers v such that $v^2 \bmod n$ is B-smooth:

$$v^2 \equiv \left(\prod_{p \in P} p^{e_{v,p}} \right) \bmod n.$$

These identities are often referred to as *relations* modulo n, and the first stage is referred to as the *relation collection stage*. Morrison and Brillhart determined relations using continued fractions (4.2.5). Dixon [39] proposed a simpler, but slower, method to find relations: pick $v \in \mathbf{Z}/n\mathbf{Z}$ at random and keep the ones for which $v^2 \in \mathbf{Z}/n\mathbf{Z}$ is B-smooth. Let V be the resulting set of cardinality $\#V > \#P$ consisting of the 'good' v's, i.e., the relations.

Each $v \in V$ gives rise to a $\#P$-dimensional vector $(e_{v,p})_{p\in P}$. Because $\#V > \#P$, the vectors $\{(e_{v,p})_{p\in P} : v \in V\}$ are linearly dependent. This implies that there exist at least $\#V - \#P$ linearly independent subsets S of V for which

$$\sum_{v\in S} e_{v,p} = 2(s_p)_{p\in P} \text{ with } s_p \in \mathbf{Z} \text{ for } p \in P.$$

These can, in principle, be found using Gaussian elimination modulo 2 on the matrix having the vectors $(e_{v,p})_{p\in P}$ as rows. The second stage is therefore referred to as the *matrix step*.

Given such a subset S of V and corresponding integer vector $(s_p)_{p\in P}$, the integers

$$x = \left(\prod_{v\in S} v\right) \bmod n, \quad y = \left(\prod_{p\in P} p^{s_p}\right) \bmod n$$

solve the congruence $x^2 \equiv y^2 \bmod n$. Each of the $\#V - \#P$ independent subsets leads to an independent chance of at least 50% to factor n.

There are various ways to analyse the expected runtime of Dixon's method, depending on the way the candidate v's are tested for smoothness. If trial division is used, then the best choice for B turns out to be $B = L_n[1/2, 1/2]$. For each candidate v, the number $v^2 \bmod n$ is assumed to behave as a random number $\le n = L_n[1,1]$, and therefore B-smooth with probability $L_n[1/2, -1]$ (see 4.2.2). Testing each candidate takes time $\#P = \pi(B) = L_n[1/2, 1/2]$ (all $\log(n)$ factors 'disappear' in the $o(1)$), so collecting somewhat more than $\#P$ relations can be expected to take time

$$\overbrace{L_n[1/2, 1/2]}^{\substack{\text{number} \\ \text{of relations} \\ \text{to be collected}}} \cdot \overbrace{L_n[1/2, 1/2]}^{\substack{\text{trial} \\ \text{division}}} \cdot \overbrace{(L_n[1/2, -1])^{-1}}^{\substack{\text{inverse of} \\ \text{smoothness} \\ \text{probability}}} = L_n[1/2, 2].$$

Gaussian elimination on the $\#V \times \#P$ matrix takes time

$$L_n[1/2, 1/2]^3 = L_n[1/2, 1.5].$$

It follows that the total time required for Dixon's method with trial division is $L_n[1/2, 2]$. The runtime is dominated by the relation collection stage.

If trial division is replaced by the elliptic curve method (4.2.3), the time to test each candidate for B-smoothness is reduced to $L_n[1/2, 0]$: the entire cost disappears in the $o(1)$. As a result the two stages can be seen to require time $L_n[1/2, 1.5]$ each. This can be further reduced as follows. In the first place, redefine B as $L_n[1/2, \sqrt{1/2}]$ so that the entire relation collection stage takes time

$$L_n[1/2, \sqrt{1/2}] \cdot L_n[1/2, 0] \cdot (L_n[1/2, -\sqrt{1/2}])^{-1} = L_n[1/2, \sqrt{2}].$$

Secondly, observe that at most $\log_2(n) = L_n[1/2, 0]$ entries are non-zero for each vector $(e_{v,p})_{p \in P}$, so that the total number of non-zero entries of the matrix (the 'weight') is $\#V \times L_n[1/2, 0] = L_n[1/2, \sqrt{1/2}]$. The number of operations required by sparse matrix techniques is proportional to the product of the weight and the number of columns, so the matrix step can be done in time $L_n[1/2, \sqrt{1/2}]^2 = L_n[1/2, \sqrt{2}]$. Thus, using the elliptic curve method to test for smoothness and using sparse matrix techniques for the second stage, the overall runtime of Dixon's method becomes

$$L_n[1/2, \sqrt{2}] + L_n[1/2, \sqrt{2}] = L_n[1/2, \sqrt{2}].$$

Asymptotically the relation collection stage and the matrix step take the same amount of time. Sparse matrix methods are not further treated here; see [29, 61, 84, 100, 120] for various methods that can be applied.

Dixon's method has the advantage that its expected runtime can be rigorously analysed and does not depend on unproved hypotheses or heuristics. In practice, however, it is inferior to the original Morrison-Brillhart continued fraction approach (see 4.2.5) and to the other methods described below. The Morrison-Brillhart method was used to factor the seventh Fermat number $F_7 = 2^{2^7} + 1$ (see [87]).

4.2.5. *Quadratic sieve*

The most obvious way to speed-up Dixon's method is by generating the integers v in such a way that $v^2 \bmod n$ is substantially smaller than n, thereby improving the smoothness probability. In CFRAC, Morrison and Brillhart's continued fraction method, the residues to be tested for smoothness are of the form $a_i^2 - nb_i^2$, where a_i/b_i is the i-th continued fraction convergent to $\sqrt{n}$. Thus, $v = a_i$ for $i = 1, 2, \ldots$ and the residues $v^2 \bmod n = a_i^2 - nb_i^2$ to be tested for smoothness are only about $2\sqrt{n}$. However, each residue has to be processed separately (using trial division or the elliptic curve method).

A simpler way to find residues that are almost as small but that can, in practice, be tested much faster was proposed by Pomerance [95, 96]. Let

$v(i) = i + [\sqrt{n}]$ for small i, then

$$v(i)^2 \bmod n = (i + [\sqrt{n}])^2 - n \approx 2i\sqrt{n}.$$

Compared to Dixon's method this approximately halves the size of the residues to be tested for B-smoothness. The important advantage – in practice, not in terms of L_n – compared to the continued fraction method is that a sieve can be used to test the values $(v(i)^2 \bmod n)$ for smoothness for many consecutive i's simultaneously, in a manner similar to 3.5.8. This is based on the observation that if p divides $(v(r)^2 \bmod n)$, i.e., r is a root modulo p of $f(X) = (X + [\sqrt{n}])^2 - n$, then p divides $(v(r + kp)^2 \bmod n)$ for any integer k, i.e., $r + kp$ is a root modulo p of $f(X)$ for any integer k.

This leads to the following sieving step. Let $(s(i))_{i=0}^{L-1}$ be a sequence of L locations, corresponding to $(v(i)^2 \bmod n)$ for $0 \le i < L$, with initial values equal to 0. For all $p \le B$ find all roots r modulo p of $(X + [\sqrt{n}])^2 - n$. For all resulting pairs (p, r) replace $s(r + kp)$ by $s(k + rp) + \log_2(p)$ for all integers k such that $0 \le r + kp < L$. As a result, values $s(i)$ that are close to the 'report bound' $\log_2((i + [\sqrt{n}])^2 - n)$ are likely to be B-smooth. Each such value is tested separately for smoothness. In practice the sieve values $s(i)$ and (rounded) $\log_2(p)$ values are represented by bytes.

With $B = L_n[1/2, 1/2]$ and assuming that $(v(i)^2 \bmod n)$ behaves as a random number close to $\sqrt{n} = L_n[1, 1/2]$ it is B-smooth with probability $L_n[1/2, -1/2]$. This assumption is obviously incorrect: if an odd prime p divides $(v(i)^2 \bmod n)$ (and does not divide n) then $(i + [\sqrt{n}])^2 \equiv n \bmod p$, so that n is a quadratic residue modulo p. As a consequence, one may expect that half the primes $\le B$ cannot occur in $(v(i)^2 \bmod n)$, so that $\#P \approx \pi(B)/2$. On the other hand, for each prime p that may occur, one may expect two roots of $f(X)$ modulo p. In practice the smoothness probabilities are very close to what is naïvely predicted based on 4.2.2.

To find $> \#P = L_n[1/2, 1/2]$ relations,

$$L_n[1/2, 1/2] \cdot (L_n[1/2, -1/2])^{-1} = L_n[1/2, 1]$$

different i's have to be considered, so that the sieve length L equals $L_n[1/2, 1]$. This justifies the implicit assumption made above that i is small. The sieving time consists of the time to find the roots (i.e., $\#P = L_n[1/2, 1/2]$ times an effort that is polynomial-time in n and p) plus the actual sieving time. The latter can be expressed as

$$\sum_{\{(p,r):\, p \in P\}} \frac{L}{p} \approx 2L \sum_{p \in P} \frac{1}{p} = L_n[1/2, 1]$$

because $\sum_{p \in P} \frac{1}{p}$ is proportional to $\log \log(L_n[1/2, 1/2])$ (see [48]) and disappears in the $o(1)$. The matrix is sparse again. It can be processed in time $L_n[1/2, 1/2]^2 = L_n[1/2, 1]$. The total (heuristic) expected runtime of Pomerance's quadratic sieve factoring method becomes

$$L_n[1/2, 1] + L_n[1/2, 1] = L_n[1/2, 1].$$

The relation collection and matrix steps take, when expressed in L_n, an equal amount of time. For all numbers factored so far with the quadratic sieve, however, the actual runtime spent on the matrix step is only a small fraction of the total runtime.

The runtime of the quadratic sieve is the same as the worst case runtime of the elliptic curve method applied to n. The sizes of the various polynomial-time factors involved in the runtimes (which all disappear in the $o(1)$'s) make the quadratic sieve much better for numbers that split into two primes of about equal size. In the presence of small factors the elliptic curve method can be expected to outperform the quadratic sieve.

Because of its practicality many enhancements of the basic quadratic sieve as described above have been proposed and implemented. The most important ones are listed below. The runtime of quadratic sieve when expressed in terms of L_n is not affected by any of these improvements. In practice, though, they make quite a difference.

Multiple polynomials Despite the fact that 'only' $L_n[1/2, 1]$ different i's have to be considered, in practice the effect of the rather large i's is quite noticeable: the larger i gets, the smaller the 'yield' becomes. Davis and Holdridge were the first to propose the use of more polynomials [36]. A somewhat more practical but similar solution was independently suggested by Montgomery [96, 111]. As a result a virtually unlimited amount of equally useful polynomials can be generated, each playing the role of $f(X)$ in the description above. As soon as one would be sieving too far away from the origin ($i = 0$), sieving continues with a newly selected polynomial. See [67, 96, 111] for details.

Self-initializing For each polynomial all roots modulo all primes $\leq B$ have to be computed. In practice this is a time consuming task. In [99] it is shown how large sets of polynomials can be generated in such a way that the most time consuming part of the root computation has to be carried out only once per set.

Large primes In the above description sieve values $s(i)$ are discarded if they are not sufficiently close to the report bound. By relaxing the report bound somewhat, relations of the form

$$v^2 = q \cdot \left(\prod_{p \in P} p^{e_{v,p}} \right) \bmod n$$

can be collected at hardly any extra cost. Here q is a prime larger than B, the *large prime*. Relations involving large primes are referred to as *partial relations*. Two partial relations with the same large prime can be multiplied (or divided) to yield a relation that is, for factoring purposes, just as useful as any ordinary relation. The latter are, in view of the partial relations, often referred to as *full relations*. However, combined partial relations make the matrix somewhat less sparse. It is a consequence of the birthday paradox (4.1.5) and of the fact that smaller large primes occur more frequently than larger ones, that matches between large primes occur often enough to make this approach worthwhile. Actually, it more than halves the sieving time. The obvious extension is to allow more than a single large prime. Using two large primes again more than halves the sieving time [71]. Three large primes have, yet again, almost the same effect, according to the experiment reported in [75]. Large primes can be seen as a cheap way to extend the size of the factor base P – cheap because the large primes are not sieved with.

Parallelization The multiple polynomial variation of the quadratic sieve (or its self-initializing variant) allows straightforward parallelization of the relation collection stage on any number of independent processors [21]. Communication is needed only to send the initial data, and to report the resulting relations back to the central location (where progress is measured and the matrix step is carried out). This can be done on any loosely coupled network of processors, such as the Internet [70].

So far the largest number factored using the quadratic sieve is the factorization of the RSA challenge number, a number of 129 decimal digits [42], reported in [7] (but see [75]). Since that time the number field sieve (4.2.7) has overtaken the quadratic sieve, and the method is no longer used to pursue record factorizations.

4.2.6. *Historical note*

In the late 1970s Schroeppel invented the *Linear sieve* factoring algorithm, based on the Morrison-Brillhart approach. He proposed to look for pairs of small integers i, j such that

$$(i + [\sqrt{n}])(j + [\sqrt{n}]) - n \approx (i + j)\sqrt{n}$$

is smooth. Compared to Morrison and Brillhart's continued fraction method this had the advantage that smoothness could be tested using a sieve. This led to a much faster relation collection stage, despite the fact that $(i+j)\sqrt{n}$ is larger than $2\sqrt{n}$, the order of magnitude of the numbers generated by Morrison and Brillhart. A disadvantage was, however, that $(i + [\sqrt{n}])(j + [\sqrt{n}])$ is not a square. This implied that all values $i + [\sqrt{n}]$ occurring in a relation had to be carried along in the matrix as well, to combine them into squares. This led to an unusually large matrix, for that time at least.

The runtime of the linear sieve was fully analysed by Schroeppel in terms equivalent to the L function defined in 4.2.1. It was the first factoring method for which a (heuristic) subexponential expected runtime was shown. The runtime of Morrison and Brillhart's continued fraction method, though also subexponential, had up to that time never been analysed.

Schroeppel implemented his method and managed to collect relations for the factorization of the eighth Fermat number $F_8 = 2^{2^8} + 1$. Before he could embark on the matrix step, however, F_8 was factored by a stroke of luck using Pollard's rho method (4.1.5 and [18]). As a result of this fortunate – or rather, unfortunate – factorization the linear sieve itself and its runtime analysis never got the attention it deserved. Fortunately, however, it led to the quadratic sieve when Pomerance, attending a lecture by Schroeppel, realized that it may be a good idea to take $i = j$ in the linear sieve.

4.2.7. Number field sieve [68]

At this point the number field sieve is the fastest general purpose factoring algorithm that has been published. It is based on an idea by Pollard in 1988 to factor numbers of the form $x^3 + k$ for small k (see his first article in [68]). This method was quickly generalized to a factoring method for numbers of the form $x^d + k$, a method that is currently referred to as the special number field sieve. It proved to be practical by factoring the ninth Fermat number $F_9 = 2^{2^9} + 1$. This happened in 1990, long before F_9 was expected to 'fall' [69]. The heuristic expected runtime of the special number field sieve is $L_n[1/3, 1.526]$, where $1.526 \approx (32/9)^{1/3}$. It was the first factoring algorithm with runtime substantially below $L_n[1/2, c]$ (for constant c), and as such an enormous breakthrough. It was also an unpleasant surprise for factoring based cryptography, despite the fact that the method was believed to have very limited applicability.

This hope was, however, destroyed by the development of the general number field sieve, as it was initially referred to. Currently it is referred to as the number field sieve. In theory the number field sieve should be able to factor any number in heuristic expected time $L_n[1/3, 1.923]$, with $1.923 \approx (64/9)^{1/3}$. It took a few years, and the dogged determination of a handful of true believers, to show that this algorithm is actually practical. Even for numbers having fewer than 100 decimal digits it already beats the quadratic sieve. The crossover point is much lower than expected and reported in the literature [98].

The number field sieve follows the traditional Morrison-Brillhart approach of collecting relations, based on some concept of smoothness, followed by a matrix step. The reason that it is so much faster than previous smoothness based approaches, is that the numbers that are tested for smoothness are of order $n^{o(1)}$, for $n \to \infty$, as opposed to n^c for a (small) constant c for the older methods. One of the reasons of its practicality is that it allows relatively easy use of more than two large primes, so that relatively small factor bases can be used during sieving. The relation collection stage can be distributed over almost any number of loosely coupled processors, similar to quadratic sieve.

For an introductory description of the number field sieve, refer to [66, 69, 97]. For complete details see [68] and the references given there. The latest developments are described in [22, 85].

The largest 'special' number factored using the special number field sieve is $2^{773} + 1$ (see [35]). This was done by the same group that achieved the current general number field sieve record by factoring a 512-bit RSA modulus [22]. Neither of these records can be expected to stand for a long time. Consult [35] for the most recent information. Such 'public domain' factoring records should not be confused with factorizations that could, in principle or in practice, be obtained by well funded agencies or other large organizations. Also, it should be understood that the computational effort involved in a 512-bit RSA modulus factorization is dwarfed by DES cracking projects. See [72] for a further discussion on these and related issues.

4.2.8. *Index calculus method*

As was first shown in [1], a variation of the Morrison-Brillhart approach can be used to compute discrete logarithms in $\mathbf{F}_{p^\ell}$. First use the familiar two stage approach to compute the discrete logarithms of many 'small' elements of $\mathbf{F}_{p^\ell}^*$. Next use this information to compute the discrete logarithm of arbitrary elements of $\mathbf{F}_{p^\ell}^*$. The outline below applies to any finite field $\mathbf{F}_{p^\ell}$. The expected runtimes are for $p \to \infty$ and ℓ fixed or p fixed and $\ell \to \infty$, as in 4.2.2. See also [2]. If $\ell = 1$ the 'norm' of a field element is simply the integer representing the field element, and an element is 'prime' if that integer is prime. If $\ell > 1$ the 'norm' is as in 4.2.2, and an element is 'prime' if its representation is an irreducible polynomial over $\mathbf{F}_p$ (see 3.6.1).

Let B be a smoothness bound, and let the factor base P be the subset of $\mathbf{F}_{p^\ell}^*$ of primes of norm $\leq B$. Relations are defined as identities of the form

$$g^v = \prod_{p \in P} p^{e_{v,p}},$$

with $v \in \{0, 1, \ldots, p^\ell - 2\}$ and where g generates $\mathbf{F}_{p^\ell}^*$. This implies that

$$v \equiv \left(\sum_{p \in P} e_{v,p} \log_g(p) \right) \bmod (p^\ell - 1).$$

It follows that with more than $\#P$ distinct relations, the values of $\log_g(p)$ for $p \in P$ can be found by solving the system of linear relations defined by the vectors $(e_{v,p})_{p \in P}$.

Given an arbitrary $h \in \mathbf{F}_{p^\ell}^*$ and $\log_g(p)$ for $p \in P$, the value of $\log_g(h)$ can be found by selecting an integer u such that hg^u is B-smooth, i.e.,

$$hg^u = \prod_{p \in P} p^{e_{u,p}},$$

because it leads to

$$\log_g(h) = \left(\left(\sum_{p \in P} e_{u,p} \log_g(p) \right) - u \right) \bmod (p^\ell - 1).$$

With $B = L_{p^\ell - 1}[1/2, \sqrt{1/2}]$ and Dixon's approach to find relations (i.e., pick v at random and test g^v for B-smoothness) the relation collection stage takes expected time $L_{p^\ell}[1/2, \sqrt{2}]$. This follows from the smoothness probabilities in 4.2.2, the runtime of the elliptic curve method (4.2.3) if $\ell = 1$, and, if $\ell > 1$, the fact that polynomials over $\mathbf{F}_p$ of degree k can be factored in expected time polynomial in k and $\log(p)$ (see [56]). Solving the system of linear equations takes the same expected runtime $L_{p^\ell}[1/2, \sqrt{2}]$ because the relations are sparse, as in 4.2.4. Finally, an appropriate u can be found using the same Dixon approach. This results in an expected runtime $L_{p^\ell}[1/2, \sqrt{1/2}]$ per discrete logarithm to be computed (given the $\log_g(p)$ for $p \in P$).

Variations. Various asymptotically faster variants of the same basic idea have been proposed that reduce the heuristic expected runtime to $L_{p^\ell}[1/2, 1]$ for the preparatory stage and $L_{p^\ell}[1/2, 1/2]$ per individual discrete logarithm; see [33, 62, 90] for details. One of these methods, the *Gaussian integers method* for the computation of discrete logarithms in $\mathbf{F}_p$ is of particular interest. It not only gave Pollard the inspiration for the (special) number field sieve integer factoring method, but it is also still of practical interest despite asymptotically faster methods that have in the mean time been found (see below).

Coppersmith's method. Another important variant of the index calculus method is Coppersmith's method. It applies only to finite fields of small fixed characteristic, such as $\mathbf{F}_{2^\ell}$. It was the first cryptanalytic method to break through the $L_x[1/2, c]$ barrier, with an expected runtime $L_{2^\ell}[1/3, 1.588]$. This is similar to the runtime of the number field sieve, but the method was found much earlier. Refer to [28] for details or to [67] for a simplified description.

Coppersmith's method has proved to be very practical. Also, the constant 1.588 is substantially smaller than the constant 1.923 in the runtime of the number field sieve integer factoring method. For those reasons RSA moduli of a given size s are believed to offer more security than the multiplicative group of a finite field $\mathbf{F}_{p^\ell}$ with small constant p and $p^\ell \approx s$. If such fields are used, then p^ℓ must be chosen considerably larger than s to achieve the same level of security.

In [47] an (incomplete) application of Coppersmith's method to the finite field $\mathbf{F}_{2^{503}}$ is described. A description of a completed discrete logarithm computation in $\mathbf{F}_{2^{607}}$ can be found in [116] and [117].

Discrete logarithm variant of the number field sieve. For $p \to \infty$ and ℓ fixed discrete logarithms in $\mathbf{F}_{p^\ell}$ can be computed in heuristic expected runtime $L_{p^\ell}[1/3, 1.923]$. For small fixed p this is somewhat slower than Coppersmith's method. However, the method applies to finite fields of arbitrary characteristic. The method is based on the number field sieve. For details refer to [45, 104]. See also [54].

The current record for finite field discrete logarithm computation is a 399-bit (120 decimal digit) prime field, reported in [53].

Acknowledgment

These notes were written on the occasion of the tutorial on mathematical foundations of coding theory and cryptology, held from July 23 to 26, 2001, at the Institute for Mathematical Sciences (IMS) of the National University of Singapore, as part of the program on coding theory and data integrity. The author wants to thank the IMS and in particular Professor Harald Niederreiter for their invitation to participate in the program.

References

1. L.M. Adleman, *A subexponential-time algorithm for the discrete logarithm problem with applications*, Proceedings 20th Ann. IEEE symp. on foundations of computer science (1979) 55-60.
2. L.M. Adleman, J. DeMarrais, *A subexponential algorithm for discrete logarithms over all finite fields*, Proceedings Crypto'93, LNCS 773, Springer-Verlag 1994, 147-158.
3. L.M. Adleman, M.A. Huang, *Primality testing and abelian varieties over finite fields*, Lecture Notes in Math. **1512**, Springer-Verlag 1992.
4. M. Agrawal, N. Kayal, N. Saxena, *PRIMES is in P*, preprint, Department of Computer Science & Engineering, Indian Institute of Technology Kanpur, India, August 2002.
5. W.R. Alford, A. Granville, C. Pomerance, *There are infinitely many Carmichael numbers*, Ann. of Math. **140** (1994) 703-722.
6. I. Anshel, M. Anshel, D. Goldfeld, *An algebraic method for public-key cryptography*, Mathematical Research Letters **6** (1999) 287-291.
7. D. Atkins, M. Graff, A.K. Lenstra, P.C. Leyland, *THE MAGIC WORDS ARE SQUEAMISH OSSIFRAGE*, Proceedings Asiacrypt'94, LNCS 917, Springer-Verlag 1995, 265-277.
8. E. Bach, J. Shallit, *Cyclotomic polynomials and factoring*, Math. Comp. **52** (1989) 201-219.
9. E. Bach, J. Shallit, *Algorithmic number theory, Volume 1, Efficient Algorithms*, The MIT press, 1996.
10. I. Blake, G. Seroussi, N. Smart, *Elliptic curves in cryptography*, Cambridge University Press, 1999.
11. D. Boneh, R.A. DeMillo, R.J. Lipton, *On the importance of checking cryptographic protocols for faults*, Proceedings Eurocrypt'97, LNCS 1233, Springer-Verlag 1997, 37-51.
12. D. Boneh, G. Durfee, *Cryptanalysis of RSA with private key d less than* $N^{0.292}$, Proceedings Eurocrypt'99, LNCS 1592, Springer-Verlag 1999, 1-11.
13. D. Boneh, R.J. Lipton, *Algorithms for black-box fields and their application to cryptography*, Proceedings Crypto'96, LNCS 1109, Springer-Verlag 1996, 283-297.
14. W. Bosma, A.K. Lenstra, *An implementation of the elliptic curve integer factorization method*, chapter 9 in *Computational algebra and number theory* (W. Bosma, A. van der Poorten, eds.), Kluwer Academic Press (1995).
15. W. Bosma, M.P.M van der Hulst, *Primality proving with cyclotomy*, PhD. thesis, Universiteit van Amsterdam (1990).
16. R.P. Brent, *An improved Monte Carlo factorization algorithm*, BIT **20** (1980) 176-184.

17. R.P. Brent, *Factorization of the tenth and eleventh Fermat numbers*, manuscript, 1996.
18. R.P. Brent, J.M. Pollard, *Factorization of the eighth Fermat number*, Math. Comp. **36** (1980) 627-630.
19. E.F. Brickell, D.M. Gordon, K.S. McCurley, D.B. Wilson, *Fast exponentiation with precomputation*, Proceedings Eurocrypt'92, LNCS 658, Springer-Verlag, 1993, 200-207.
20. E.R. Canfield, P. Erdös, C. Pomerance, *On a problem of Oppenheim concerning "Factorisatio Numerorum"*, J. Number Theory **17** (1983) 1-28.
21. T.R. Caron, R.D. Silverman, *Parallel implementation of the quadratic sieve*, J. Supercomput. **1** (1988) 273-290.
22. S. Cavallar, B. Dodson, A.K. Lenstra, W. Lioen, P.L. Montgomery, B. Murphy, H. te Riele, et al., *Factorization of a 512-bit RSA modulus*, Proceedings Eurocrypt 2000, LNCS 1807, Springer-Verlag 2000, 1-18.
23. www.certicom.com.
24. H. Cohen, *Analysis of the flexible window powering algorithm*, available from http://www.math.u-bordeaux.fr/~cohen, 2001.
25. H. Cohen, A.K. Lenstra, *Implementation of a new primality test*, Math. Comp. **48** (1986) 187-237.
26. H. Cohen, H.W. Lenstra, Jr., *Primality testing and Jacobi sums*, Math. Comp. **42** (1984) 297-330.
27. H. Cohen, A. Miyaji, T. Ono, *Efficient elliptic curve exponentiation using mixed coordinates*, Proceedings Asiacrypt'98, LNCS 1514, Springer-Verlag 1998, 51-65.
28. D. Coppersmith, *Fast evaluation of logarithms in fields of characteristic two*, IEEE Trans. Inform. Theory 30 (1984) 587-594.
29. D. Coppersmith, *Solving linear equations over GF(2) using block Wiedemann algorithm*, Math. Comp. **62** (1994) 333-350.
30. D. Coppersmith, *Finding a small root of a univariate modular equation*, Proceedings Eurocrypt'96, LNCS 1070, Springer-Verlag 1996, 155-165.
31. D. Coppersmith, *Small solutions to polynomial equations, and low exponent RSA vulnerabilities*, Journal of Cryptology **10** (1997) 233-260.
32. D. Coppersmith, M. Franklin, J. Patarin, M. Reiter, *Low-exponent RSA with related messages*, Proceedings Eurocrypt'96, LNCS 1070, Springer-Verlag 1996, 1-9.
33. D. Coppersmith, A.M. Odlyzko, R. Schroeppel, *Discrete logarithms in GF(p)*, Algorithmica **1** (1986) 1-15.
34. D. Coppersmith, A. Shamir, *Lattice attacks on NTRU*, Proceedings Eurocrypt'97, LNCS 1233, Springer Verlag 1997, 52-61.
35. www.cwi.nl.
36. J.A. Davis, D.B. Holdridge, *Factorization using the quadratic sieve algorithm*, Tech. Report SAND 83-1346, Sandia National Laboratories, Albuquerque, NM, 1983.
37. N.G. De Bruijn, *On the number of positive integers $\leq x$ and free of prime factors $> y$, II*, Indag. Math. **38** (1966) 239-247.

38. W. Diffie, M.E. Hellman, *New directions in cryptography*, IEEE Trans. Inform. Theory 22 (1976) 644-654.
39. J.D. Dixon, *Asymptotically fast factorization of integers*, Math. Comp. **36** (1981) 255-260.
40. *Quantum dreams*, The Economist, March 10, 2001, 81-82.
41. T. ElGamal, *A public key cryptosystem and a signature scheme based on discrete logarithms*, IEEE Transactions on Information Theory 31(4) (1985) 469-472.
42. M. Gardner, *Mathematical games, a new kind of cipher that would take millions of years to break*, Scientific American, August 1977, 120-124.
43. C. Gentry, J. Johnson, J. Stern, M, Szydlo, *Cryptanalysis of the NSS signature scheme* and *Exploiting several flaws in the NSS signature scheme*, presentations at the Eurocrypt 2001 rump session, May 2001.
44. S. Goldwasser, J. Kilian, *Almost all primes can be quickly verified*, Proc. 18th Ann. ACM symp. on theory of computing (1986) 316-329.
45. D. Gordon, *Discrete logarithms in GF(p) using the number field sieve*, SIAM J. Discrete Math. **6** (1993) 312-323.
46. D.M. Gordon, *A survey on fast exponentiation methods*, Journal of Algorithms **27** (1998) 129-146.
47. D.M. Gordon, K.S. McCurley, *Massively parallel computation of discrete logarithms*, Proceedings Crypto'92, LNCS 740, Springer-Verlag 1993, 312-323.
48. G.H. Hardy, E.M. Wright, *An introduction to the theory of numbers*, Oxford Univ. Press, Oxford, 5th ed., 1979.
49. J. Hastad, *Solving simultaneous modular equations of low degree*, SIAM J. Comput. **17** (1988) 336-341.
50. J. Hoffstein, J. Pipher, J.H. Silverman, *NTRU: a ring-based public key cryptosystem*, Proceedings ANTS III, LNCS 1423, Springer-Verlag 1998, 267-288.
51. J. Hoffstein, J. Pipher, J.H. Silverman, *NSS: an NTRU lattice-based signature scheme*, Proceedings Eurocrypt 2001, LNCS 2045, Springer-Verlag 2001, 211-228.
52. T. Itoh, S, Tsujii, *A fast algorithm for computing multiplicative inverses in $GF(2^k)$ using normal bases*, Information and computation **78** (1988) 171-177.
53. A. Joux, R. Lercier, *Discrete logarithms in GF(p) (120 decimal digits)*, available from http://listserv.nodak.edu/archives/nmbrthry.html, April 2001.
54. A. Joux, R. Lercier, *The function field sieve is quite special*, Proceedings ANTS V, LNCS 2369, Springer-Verlag 2002, 431-445.
55. A. Joux, K. Nguyen, *Separating decision Diffie-Hellman from Diffie-Hellman in cryptographic groups*, available from eprint.iacr.org, 2001.
56. D.E. Knuth, *The art of computer programming, Volume 2, Seminumerical Algorithms*, third edition, Addison-Wesley, 1998.
57. K.H. Ko, S.J. Lee, J.H. Cheon, J.W. Han, J. Kang, C. Park, *New public-key cryptosystem using braid groups*, Proceedings Crypto 2000, LNCS 1880, Springer-Verlag 2000, 166-183.

58. T. Kobayashi, H. Morita, K. Kobayashi, F. Hoshino, *Fast elliptic curve algorithm combining Frobenius map and table reference to adapt to higher characteristic*, Proceedings Eurocrypt'99, LNCS 1592, Springer-Verlag (1999) 176-189.

59. N. Koblitz, *Elliptic curve cryptosystems*, Math. Comp. **48** (1987) 203-209.

60. P.C. Kocher, *Timing attacks on implementations of Diffie-Hellman, RSA, DSS, and other systems*, Proceedings Crypto'96, LNCS 1109, Springer-Verlag 1996, 104-113.

61. B.A. LaMacchia, A.M. Odlyzko, *Solving large sparse linear systems over finite fields*, Proceedings Crypto'90, LNCS 537, Springer-Verlag 1990, 109-133.

62. B.A. LaMacchia, A.M. Odlyzko, *Computation of discrete logarithms in prime fields*, Design, Codes and Cryptography **1** (1991) 47-62.

63. A.K. Lenstra, *The long integer package FREELIP*, available from www.ecstr.com or directly from the author.

64. A.K. Lenstra, *Using cyclotomic polynomials to construct efficient discrete logarithm cryptosystems over finite fields*, Proceedings ACISP'97, LNCS 1270, Springer-Verlag 1997, 127-138.

65. A.K. Lenstra, *Generating RSA moduli with a predetermined portion*, Proceedings Asiacrypt'98, LNCS 1514, Springer-Verlag 1998, 1-10.

66. A.K. Lenstra, *Integer factoring*, Designs, codes and cryptography **19** (2000) 101-128.

67. A.K. Lenstra, H.W. Lenstra, Jr., *Algorithms in number theory*, chapter 12 in *Handbook of theoretical computer science, Volume A, algorithms and complexity* (J. van Leeuwen, ed.), Elsevier, Amsterdam (1990).

68. A.K. Lenstra, H.W. Lenstra, Jr., (eds.), *The development of the number field sieve*, Lecture Notes in Math. **1554**, Springer-Verlag 1993.

69. A.K. Lenstra, H.W. Lenstra, Jr., M.S. Manasse, J.M. Pollard, *The factorization of the ninth Fermat number*, Math. Comp. **61** (1993) 319-349.

70. A.K. Lenstra, M.S. Manasse, *Factoring by electronic mail*, Proceedings Eurocrypt'89, LNCS 434, Springer-Verlag 1990, 355-371.

71. A.K. Lenstra, M.S. Manasse, *Factoring with two large primes*, Proceedings Eurocrypt'90, LNCS 473, Springer-Verlag 1990, 72-82; Math. Comp. **63** (1994) 785-798.

72. A.K. Lenstra, E.R. Verheul, *Selecting cryptographic key sizes*, Journal of Cryptology **14** (2001) 255-293; available from www.cryptosavvy.com.

73. A.K. Lenstra, E.R. Verheul, *The XTR public key system*, Proceedings Crypto 2000, LNCS 1880, Springer-Verlag, 2000, 1-19; available from www.ecstr.com.

74. H.W. Lenstra, Jr., *Factoring integers with elliptic curves*, Ann. of Math. **126** (1987) 649-673.

75. P. Leyland, A.K. Lenstra, B. Dodson, A. Muffett, S.S. Wagstaff, Jr., *MPQS with three large primes*, Proceedings ANTS V, LNCS 2369, Springer-Verlag 2002, 446-460.

76. R. Lidl, H, Niederreiter, *Introduction to finite fields and their applications*, Cambridge University Press, 1994.

77. U. Maurer, *Fast generation of prime numbers and secure public-key cryptographic parameters*, Journal of Cryptology **8** (1995) 123-155.

78. A.J. Menezes (ed.), *Elliptic curve public key cryptosystems*, Kluwer academic publishers, 1993.

79. A.J. Menezes (ed.), *Applications of finite fields*, Kluwer academic publishers, 1993.

80. A.J. Menezes, P.C. van Oorschot, S.A. Vanstone, *Handbook of applied cryptography*, CRC Press, 1997.

81. V. Miller *Use of elliptic curves in cryptography*, Proceedings Crypto'85, LNCS 218, Springer-Verlag 1986, 417-426.

82. P.L. Montgomery, *Modular multiplication without trial division*, Math. Comp. **44** (1985) 519-521.

83. P.L. Montgomery, *Speeding the Pollard and elliptic curve methods of factorization*, Math. Comp. **48** (1987) 243-264.

84. P.L. Montgomery, *A block Lanczos algorithm for finding dependencies over GF(2)*, Proceedings Eurocrypt'95, LNCS 925, Springer-Verlag 1995, 106-120.

85. P.L. Montgomery, B. Murphy, *Improved polynomial selection for the number field sieve*, extended abstract for the conference on the mathematics of public-key cryptography, June 13-17, 1999, the Fields institute, Toronto, Ontario, Canada.

86. F. Morain, *Implementation of the Goldwasser-Kilian-Atkin primality testing algorithm*, INRIA Report 911, INRIA-Rocquencourt, 1988.

87. M.A. Morrison, J. Brillhart, *A method of factorization and the factorization of F_7*, Math. Comp. **29** (1975) 183-205.

88. V.I. Nechaev, *Complexity of a determinate algorithm for the discrete logarithm*, Mathematical Notes, 55(2) (1994) 155-172. Translated from Matematicheskie Zametki, 55(2) (1994) 91-101. This result dates from 1968.

89. P.Q. Nguyen, J. Stern, *Lattice reduction in cryptology: an update*, Proceedings ANTS IV, LNCS 1838, Springer-Verlag 2000, 85-112.

90. A.M. Odlyzko, *Discrete logarithms and their cryptographic significance*, Proceedings Eurocrypt'84, LNCS 209, Springer-Verlag 1985, 224-314.

91. S.C. Pohlig, M.E. Hellman, *An improved algorithm for computing logarithms over $GF(p)$ and its cryptographic significance*, IEEE Trans. on Inform Theory 24 (1978) 106-110.

92. J.M. Pollard, *Theorems on factorization and primality testing*, Proceedings of the Cambridge philosophical society, **76** (1974) 521-528.

93. J.M. Pollard, *Monte Carlo methods for index computation (mod p)*, Math. Comp., 32 (1978) 918-924.

94. J.M. Pollard, *Kangaroos, monopoly and discrete logarithms*, Journal of Cryptology **13** (2000) 437-447.

95. C. Pomerance, *Analysis and comparison of some integer factoring algorithms*, in *Computational methods in number theory* (H.W. Lenstra, Jr., R. Tijdeman, eds.), Math. Centre Tracts **154**, **155**, Mathematisch Centrum, Amsterdam (1983) 89-139.

96. C. Pomerance, *The quadratic sieve factoring algorithm*, Proceedings Euro-crypt'84, LNCS 209, Springer-Verlag 1985, 169-182.
97. C. Pomerance, *The number field sieve*, Proc. Symp. Appl. Math. **48** (1994) 465-480.
98. C. Pomerance, *A tale of two sieves*, Notices of the AMS (1996) 1473-1485.
99. C. Pomerance, J.W. Smith, R. Tuler, *A pipeline architecture for factoring large integers with the quadratic sieve algorithm*, SIAM J. Comput. **17** (1988) 387-403.
100. C. Pomerance, J.W. Smith, *Reduction of huge, sparse matrices over finite fields via created catastrophes*, Experimental Math. **1** (1992) 89-94.
101. M.O. Rabin, *Probabilistic algorithms for primality testing*, J. Number Theory, **12** (1980) 128-138.
102. R.L. Rivest, A. Shamir, L.M. Adleman, *A method for obtaining digital signatures and public key cryptosystems*, Comm. of the ACM, **21** (1978) 120-126.
103. R.L. Rivest, R.D. Silverman, *Are 'strong' primes needed for RSA?*, manuscript, April 1997, available from www.iacr.org.
104. O. Schirokauer, *Discrete logarithms and local units*, Phil. Trans. R. Soc. Lond. A 345 (1993) 409-423.
105. C.P. Schnorr, *Efficient signature generation by smart cards*, Journal of Cryptology **4** (1991) 161-174.
106. R.J. Schoof, *Elliptic curves over finite fields and the computation of square roots mod p*, Math. Comp. **44** (1985) 483-494.
107. P.W. Shor, *Algorithms for quantum computing: discrete logarithms and factoring*, Proceedings of the IEEE 35th Annual Symposium on Foundations of Computer Science (1994) 124-134.
108. V. Shoup, *Lower bounds for discrete logarithms and related problems*, Proceedings Eurocrypt'97, LNCS 1233, 256-266, Springer 1997.
109. V. Shoup, *NTL*, available from www.shoup.net/ntl.
110. J.H. Silverman, *The arithmetic of elliptic curves*, Springer-Verlag, 1986.
111. R.D. Silverman, *The multiple polynomial quadratic sieve*, Math. Comp. **46** (1987) 327-339.
112. N.P. Smart, *How secure are elliptic curves over composite extension fields?*, Proceedings Eurocrypt 2001, LNCS 2045, Springer-Verlag 2001, 30-39.
113. P. Smith, C. Skinner, *A public-key cryptosystem and a digital signature system based on the Lucas function analogue to discrete logarithms*, Proceedings Asiacrypt'94, LNCS 917, Springer-Verlag 1995, 357-364.
114. J. Stern, *Cryptanalysis of the NTRU signature scheme (NSS)*, manuscript, April 2001.
115. E. Teske, *Speeding up Pollard's rho methods for computing discrete logarithms*, Proceedings ANTS III, LNCS 1423, Springer-Verlag 1998, 541-554.
116. E. Thomé, *Computation of discrete logarithms in $F_{2^{607}}$*, Proceedings Asiacrypt 2001, LNCS 2248, Springer-Verlag 2001, 107-124.
117. E. Thomé, *Discrete logarithms in $F_{2^{607}}$*, available from http://listserv.nodak.edu/archives/nmbrthry.html, February 2002.
118. P.C. van Oorschot, M.J. Wiener, *Parallel collision search with cryptanalytic applications*, Journal of Cryptology **12** (1999) 1-28.

119. M. Wiener, *Cryptanalysis of short RSA secret exponents*, IEEE Trans. Inform. Theory 36 (1990) 553-558.
120. D.H. Wiedemann, *Solving sparse linear equations over finite fields*, IEEE Trans. Inform. Theory 32 (1986) 54-62.
121. H.C. Williams, *A p+1 method of factoring*, Math. Comp. 39 (1982) 225-234.

DETECTING AND REVOKING COMPROMISED KEYS

Tsutomu Matsumoto

Graduate School of Environment and Information Sciences
Yokohama National University
79-7, Tokiwadai, Hodogaya, Yokohama, 240-8501, Japan
E-mail: tsutomu@mlab.jks.ynu.ac.jp

This note describes two correlated topics in cryptography. The first topic is the entity exclusion, or how to distribute a cryptographic key over a broadcasting shared by n entities so that all but d excluded entities can get a group key. In a system such as Pay-TV, Internet multicasting and group mobile telecommunication, a dishonest user or an unauthorized terminal should be excluded as quickly as possible. We discuss the points of evaluation and history of the field followed by concrete schemes smartly enabling the entity exclusion. The second topic is how to discover the existence of a "clone," that is, another entity with the same ID and the same secret key as the original. This problem is rather hard to solve in general. However, depending on environmental conditions there are approaches for solving the problem. We suggest some effective ways for the clone discovery.

1. Introduction

Imagine a system consisting of a lot of entities. An entity is what creates, sends, receives, processes, modifies, or uses data. The system employs cryptography to maintain the confidentiality, integrity, or authenticity of data exchanged among the entities. A typical scenario is that each entity is assigned a unique identifier (ID) and an individualized private key as well as other parameters. Security attained by cryptography is based on the assumption that every private key is kept properly. Therefore two natural questions arise. How can the system know that the assumption is satisfied? How can the system revoke a particular key if there is a need to do so? Revoking a key may be restated as excluding an entity possessing the corresponding key. This note discusses the latter question in detail in Sections 2-4 and then the former question briefly in the final section.

2. Entity Exclusion in Group Communication

2.1. *Theme*

A *broadcast encryption* allows a distributor to send the same message simultaneously over a broadcast channel to all authorized users with confidentiality. Pay-TV via cable and satellite networks, Internet multicasts, and mobile group telecommunication such as a private mobile radio or a taxi radio, are typical examples. A secure and fast method to distribute a shared key (which is called a *group key* in this note) to all the proper users is required.

The main part of this note focuses on the *entity exclusion*, or how to transmit a group key over a broadcast channel shared by n entities so that all but d excluded entities can get the group key. For example, entity exclusion can prevent a lost or stolen mobile terminal to be used to eavesdrop the secret broadcasting. Entity exclusion can also prevent unauthorized access to Pay-TV and Internet.

2.2. *Development*

A simple method of entity exclusion is that a distributor distributes a new group key to each entity except the excluded users, in encrypted form by a secret key of each user. This method requires each entity to keep only one secret key, while the distributor should transmit $n - d$ encrypted new group keys.

Another simple method is that each entity has common keys for all subsets of n users. This method does not require the distributor to transmit any message, while each entity should keep a lot of keys.

To improve this trade-off between the amount of transmission and the key storage of each user, many ideas [3, 5, 6, 7, 8, 9, 13, 17, 20, 18, 26, 27] have been proposed. Criteria for evaluation may be listed as

flexibility a fixed and privileged distributor is required or not.
reusability a secret key of each entity can be reused or not.
scalability in complexity the amount of transmission and key storage
     of each entity is independent of the group scale n or not.

Berkovits [5] proposed a scheme using secret sharing. Mambo, Nishikawa, Tsujii and Okamoto [18] proposed a broadcast communication scheme, which is efficient in terms of computation of each entity and the amount of transmission. These two works can be applied to key distribution for a pre-determined privileged subset, but not to a dynamically changing subset.

A major step in key distribution with entity exclusion was marked when Fiat and Naor proposed a scheme [13]. The scheme is resilient to any coalition of d users, by extending an original scheme, which excludes a single user, using multi-layered hashing techniques. In the scheme, each entity stores $O(d \log d \log n)$ keys and the distributor broadcasts $O(d^2 (\log d)^2 \log n)$ messages. Blundo, Mattos and Stinson extend this basic work in the papers [6, 7], and Luby and Staddon studied the trade-off in the paper [17].

A second major step is a hierarchical key distribution scheme (called HKDS in the current note) using a balanced binary tree. This was done by two research groups of Wallner, Harder and Agee [26] and Wong, Gouda and Lam [27]. In this scheme, the amount of transmission is $O((degree - 1) \times \log n)$ and the number of keys for each entity is $O(\log n)$, where n is the number of users on the broadcast channel and $degree$ is the number of users in the bottom subgroup of the binary tree. Canetti, Garay, Itkis, Micciancio, Naor and Pinkas proposed an extended method that reduces the amount of transmission in the paper [8]. Canetti, Malkin and Nissim studied the trade-off in the paper [9].

Recently, two works on the entity exclusion problem have been presented: one by Kumar, Rajagopalan and Sahai [15] and one by Matsuzaki and Anzai [20]. In the Kumar-Rajagopalan-Sahai scheme [15] using algebraic geometric codes, each entity has an individual subset of a key-set. Redundant pieces of message using an error-correcting code are encrypted by keys belonging to users who are not excluded and are broadcast. The amount of transmission is $O(d^2)$ regardless of n and the key storage of each entity is $O(d \times \log n)$. Consequently, the scheme enables an efficient entity exclusion of which the amount of transmission does not depend on the group scale. However, this scheme still requires the key storage that depends on the group scale n, and a fixed and privileged distributor. Matsuzaki and Anzai proposed a scheme [20] using mathematical techniques which are well-known as *RSA common modulus attack* and *RSA low exponent attack*. The Matsuzaki-Anzai scheme can simultaneously exclude up to d users. The amount of transmission is $O(d)$ regardless of n and each entity has only one key. Therefore, the scheme enables an efficient entity exclusion when the group scale n becomes large, while the distributor should pre-send a secret key of each entity for every key distribution, since the secret key cannot be reused, and the scheme requires a fixed and privileged distributor who knows all secret keys of users.

Anzai, Matsuzaki and Matsumoto [3] proposed a scheme, called **MaSK**, which is the abbreviation of **Ma**sked **S**haring of Group **K**eys. They applied the *threshold cryptosystems* given in [12] to achieve good reusability and flexibility and scalability in complexity. Therefore the trick they used may be interesting. The following two sections precisely introduce them.

3. MaSK: An Entity Exclusion Scheme

We describe MaSK, the Anzai-Matsuzaki-Matsumoto scheme [3]. MaSK can be based on an appropriate *Diffie-Hellman Problem* [21] defined over a finite cyclic group, including a subgroup of Jacobian of an elliptic curve and so on. We describe it over a prime field $\mathbf{Z}_p$. MaSK contains two phases: system setup phase and key distribution phase. Before explaining them we should clarify the target system and assumptions.

3.1. *Target system and assumptions*

The target system consists of the following:

System manager: A trusted party which decides system parameters and sets each user's secret key. It manages a public bulletin board.

Public bulletin board: A public bulletin board keeps system parameters and public keys for all users.

User i: An entity labeled i as its ID number is a member of the group. We assume the number of total users is n. Let $\Phi = \{1, 2, \ldots, n\}$ be the set of all users.

Coordinator x: A coordinator decides one or more excluded user(s) and coordinates a group key distribution with entity exclusion. We use the term "coordinator" to distinguish it from the fixed and privileged distributor discussed before. In the scheme, any entity can become the coordinator.

Excluded user j: A user to be excluded by the coordinator. Let $\Lambda(\subset \Phi)$ be a set of excluded users, having d users.

Valid user v: A user who is not an excluded user. The set of all valid users forms the *group*.

In the target system, we make the following system assumptions:

(1) All users trust the system manager. The system manager does not do anything illegal.
(2) All users have simultaneous access to the data that the coordinator broadcasts.
(3) All users can get any data from the public bulletin board at any time.
(4) The broadcast channel is not secure, i.e., anyone can see the data on the broadcast channel.

and the following security assumptions:

(1) The *Diffie-Hellman Problem* is computationally hard to solve.
(2) In $(k, n + k - 1)$ threshold cryptosystems, anyone with less than k shadows cannot get any information about the secret S.
(3) Excluded users may conspire to get a group key.
(4) Excluded users may publish their secret information to damage the system security.
(5) Valid users do not conspire with excluded users. If this assumption is not satisfied, excluded users can get the group key from valid users.
(6) Valid users do not publish their secret keys.
(7) The system manager manages the public bulletin board strictly so as not to change it. Or, the public bulletin board has each parameter with the certificate checked before using the public parameters.

3.2. *System setup phase*

At the beginning, a system setup phase is carried out only once.

(1) A system manager decides a parameter k satisfying

$$0 \le d \le k - 2 < n,$$

where n is the number of users in the group and d is the upper bound of the number of excluded users.

(2) The system manager decides the following system parameters and publishes them to the public bulletin board:

- p: a large prime number
- q: a large prime number such that $q \mid p - 1$ and $n + k - 1 < q$
- g: a qth root of unity over $\mathbf{Z}_p$
- $sign(s, m)$: a secure signature generation function, which outputs a signature Z of the message m using a secret key s. We use the signature scheme based on DLP here, such as DSA [25] or Nyberg-Rueppel message recovery signature scheme [22].

- *verify*(y, Z): a secure signature verification function, which checks the validity of the signature Z using a public key y. The function outputs the original message m if the signature Z is "valid".

(3) The system manager generates a system secret key $S \in \mathbf{Z}_q$ and stores it secretly. And, the system manager divides the system secret key S into $n + k - 1$ shadows with the threshold k, using the well-known Shamir's secret sharing scheme [24] as follows:

  (a) The system manager puts $a_0 = S$.

  (b) The system manager defines the following equation over $\mathbf{Z}_q$:

  $$f(x) = \sum_{f=0}^{k-1} a_f x^f \bmod q \tag{1}$$

  where $a_1, a_2, \ldots, a_{k-1}$ are random integers that satisfy the following conditions:

  $$0 \leq a_i \leq q - 1 \quad \text{for all } 1 \leq i \leq k - 1 \quad \text{and } a_{k-1} \neq 0.$$

  (c) The system manager generates $n + k - 1$ shadows as follows:

  $$s_i = f(i) \quad (1 \leq i \leq n + k - 1) \tag{2}$$

(4) The system manager distributes the shadows $s_1, \ldots, s_n$ to each user $1, \ldots, n$, respectively, in a secure manner. Each user keeps its own shadow as its secret key.

(5) The system manager calculates public keys $y_1, \ldots, y_{n+k-1}$ by the following equation:

$$y_i = g^{s_i} \bmod p \quad (1 \leq i \leq n + k - 1) \tag{3}$$

Then, the system manager publishes $y_1, \ldots, y_n$ on the public bulletin board with the corresponding user's ID numbers. The remaining $y_{n+1}, \ldots, y_{n+k-1}$ and the corresponding ID numbers are also published to the public bulletin board as spare public keys. The system manager may remove the secret keys $s_1, \ldots, s_{n+k-1}$ after the system setup phase. The remaining tasks of the system manager are to maintain the public bulletin board and to generate a secret key and a public key of a new user.

3.3. *Key distribution phase*

3.3.1. *Broadcasting by a coordinator*

First, a coordinator x generates a broadcast data $B(\Lambda, r)$ as follows:

(1) The coordinator x decides excluded users. Let Λ be the set of excluded users and d is the number of the excluded users.

(2) The coordinator x chooses $r \in Z_q$ at random and picks $k-d-1$ integers from the set $\{n+1, \ldots, n+k-1\}$ and let Θ be the set of chosen integers. Then, the coordinator calculates $k-1$ exclusion data as follows:

$$M_j = y_j^r \bmod p \quad (j \in \Lambda \cup \Theta) \tag{4}$$

using the public keys of excluded users and the spare public keys on the public bulletin board.

(3) The coordinator x calculates the following preparation data:

$$X = g^r \bmod p \tag{5}$$

(4) Using its own secret key s_x, the coordinator x generates the signature for the data consisting of the preparation data, own ID number, $k-1$ exclusion data, and the corresponding ID numbers:

$$Z = sign(s_x, X \| x \| \{[j, M_j] \mid j \in \Lambda \cup \Theta\}) \tag{6}$$

where $\|$ indicates *concatenation* of data.

(5) The coordinator x broadcasts the following broadcast data to all users:

$$B(\Lambda, r) = Z \| x \tag{7}$$

Next, the coordinator x calculates a group key U using its own secret key s_x and broadcast data $B(\Lambda, r)$:

$$U = X^{s_x \times L(\Lambda \cup \Theta \cup \{x\}, x)}$$

$$\times \prod_{j \in \Lambda \cup \Theta} M_j^{L(\Lambda \cup \Theta \cup \{x\}, j)} \bmod p \tag{8}$$

where

$$L(\Psi, w) = \sum_{t \in \Psi \setminus \{w\}} \frac{t}{t - w} \bmod q \quad (\forall \Psi : set, \forall w : integer) \tag{9}$$

Since $M_j = g^{s_j \times r} \bmod p$ holds, the system secret key S is recovered in the exponent part of equation (8), gathering k sets of secret keys:

$$U \equiv g^{r \times s_x \times L(\Lambda \cup O \cup \{x\}, x)}$$

$$\times \prod_{j \in \Lambda \cup \Theta} g^{s_j \times r \times L(\Lambda \cup \Theta \cup \{x\}, j)} \pmod{p}$$

$$\equiv g^{r\{s_x \times L(\Lambda \cup \Theta \cup \{x\}, x) + \sum_{j \in \Lambda \cup \Theta} s_j \times L(\Lambda \cup \Theta \cup \{x\}, j)\}} \pmod{p}$$

$$\equiv g^{r \times S} \pmod{p}.$$

3.3.2. Receiving by a user

Receiving the broadcast data, a valid user v calculates the group key U using its own secret key s_v as follows:

(1) The valid user v verifies the signature Z using the public key y_x of the coordinator x. If Z is "valid," the user v gets the preparation data X and the exclusion data M_j and their ID numbers as follows:

$$X\|x\|\{[j, M_j] \mid j \in \Lambda \cup \Theta\} = verify(y_x, Z). \qquad (10)$$

If Z is "invalid," the user v rejects $B(\Lambda, r)$.

(2) The valid user v calculates the group key U using its own secret key s_v:

$$U = X^{s_v \times L(\Lambda \cup \Theta \cup \{v\}, v)}$$
$$\times \prod_{j \in \Lambda \cup \Theta} M_j^{L(\Lambda \cup \Theta \cup \{v\}, j)} \bmod p \qquad (11)$$

The group key U is the same as the group key of the coordinator:

$$U \equiv g^{r \times s_v \times L(\Lambda \cup \Theta \cup \{v\}, v)}$$
$$\times \prod_{j \in \Lambda \cup \Theta} g^{s_j \times r \times L(\Lambda \cup \Theta \cup \{v\}, j)} \pmod{p}$$
$$\equiv g^{r\{s_v \times L(\Lambda \cup \Theta \cup \{v\}, v) + \sum_{j \in \Lambda \cup \Theta} s_j \times L(\Lambda \cup \Theta \cup \{v\}, j)\}} \pmod{p}$$
$$\equiv g^{r \times S} \pmod{p}.$$

Every valid user obtains the same group key $g^{r \times S} \bmod p$, gathering k shadows on the exponent part in equation (11). This technique is similar to the *threshold cryptosystems* proposed by Desmedt and Frankel [12].

On the other hand, an excluded user j cannot calculate the group key U because the exclusion data M_j includes the secret key s_j of the excluded user j, and the user j can gather only $k - 1$ shadows.

3.4. An example

We show an example of MaSK, where $d = 1$, $n = 4$ and $k = 3$.

Step1: First, the coordinator (user 2) decides to exclude user 4 and picks $k - d - 1 (= 1)$ integer 5 from $\{5, 6\}$. Next, he chooses $r \in Z_q$ at random. Then, he calculates $k - 1 (= 2)$ exclusion data $M_4 (= y_4^r \bmod p)$, M_5 $(= y_5^r \bmod p)$, the preparation data $X = g^r \bmod p$, and the signature $Z = sign(s_2, X\|2\|\{[4, M_4], [5, M_5]\})$. He then broadcasts the broadcast data $B(\{4\}, r) = Z\|2$.

The following steps can be executed simultaneously.

Step2-a: The coordinator calculates the group key U $(= g^{r \times S} \bmod p)$ by using its own secret key s_2 and the broadcast data $B(\{4\}, r)$.

Step2-b: User 1 verifies the signature Z using the public key y_2 of the coordinator and calculates the group key U by using its own secret key s_1 and the broadcast data $B(\{4\}, r)$.

Step2-c: User 3 verifies the signature Z using the public key y_2 of the coordinator and calculates the group key U by using its own secret key s_3 and the broadcast data $B(\{4\}, r)$.

Step2-d: User 4 cannot calculate the group key U even if using its own secret key s_4 and the broadcast data $B(\{4\}, r)$, since he cannot collect k shadows.

4. Properties of MaSK

4.1. *Security analysis*

We discuss the passive attacks, where an excluded user or an outsider of the group uses only the public data to get the group key U or the secret parameters:

(1) **Getting r:** The random number r is included in the exponent part of the preparation data $X = g^r \bmod p$ and the exclusion data $M_j = y_j^r \bmod p$ of the broadcast data $B(\Lambda, r)$. Therefore, the difficulty of getting r is the same as that of solving DLP.

(1) **Getting S:** All users can get $g^S \bmod p$ by using k public keys on the public bulletin board. Since the secret key S is included in the exponent part, the difficulty of getting S is the same as that of solving DLP.

(2) **Getting U from $g^r \bmod p$ and $g^S \bmod p$:** All users can get g^r from $B(\Lambda, r)$ and $g^S \bmod p$ from the public bulletin board. Getting the group key U $(= g^{r \times S} \bmod p)$ from $g^r \bmod p$ and $g^S \bmod p$ is as difficult as solving the *Diffie-Hellman Problem*.

Next, we discuss the active attacks:

(1) **Modifying or forging $B(\Lambda, r)$:** The broadcast data $B(\Lambda, r)$ consists of the signature Z and the coordinator's ID number x. If an attacker modifies Z or x, the signature verification function outputs "invalid". Therefore, it is difficult to modify or forge $B(\Lambda, r)$. Moreover, we consider that a time-stamp on the broadcast data is necessary to prevent replay attacks.

(2) **Modifying or forging the public bulletin board:** Modifying or forging the public bulletin board is hard because we assume that the system manager manages the public bulletin board strictly so as not to change it. Or, we assume that the public bulletin board has each parameter with the certificate, and the validity of the certificate shall be checked before using the public parameters.

(3) **Publishing s_j by an excluded user j:** We suppose that all excluded users publish their secret keys s_j. Even if a valid user uses his own secret key together, he cannot calculate S, since he can get $d+1$ shadows that are less than the threshold k.

(4) **Conspiracy:** We assume that the valid user does not make conspiracy. Even if all excluded users conspire, they cannot reconstruct the secret key S since they can get at most d ($\leq k - 2$) shadows s_j of S, which is less than the threshold k.

4.2. Applications

In this section, we describe issues that are necessary for applying the scheme to an actual group communication system.

4.2.1. New user

When a new user wants to join a group communication system, the system manager decides its unique user's ID number c which satisfies $n + k \leq c \leq q - 1$. The system manager calculates its secret key $s_c = f(c)$ and sends it to the new user through a secure way. Then, the system manager calculates the corresponding public key y_c ($= g^{s_c} \bmod p$) and adds it to the public bulletin board. This procedure does not affect the existing users.

4.2.2. How to decide a threshold k

MaSK enables a coordinator to exclude a maximum of $k - 2$ users at one time. Also, the parameter k determines the amount of broadcast data shown in equation (7) and the number of exponentiations in equation (11). Moreover, k is a security parameter since the system secret key S is recovered from k shadows. If k is large, the number of users that the coordinator can exclude at once becomes large and the security becomes high, however, the broadcast data amount becomes large and each user must calculate a lot of exponentiations. Therefore, the system manager should decide the parameter k to fit for an actual system.

4.2.3. *Continuity of exclusion*

In an actual group communication, MaSK is used repeatedly by a different coordinator and excluded users. A coordinator can decide either one of the following:

- the coordinator continues to exclude users that were excluded last time by including excluded users' ID numbers into a set Λ, or
- the coordinator revives users that were excluded last time by excluding their ID numbers from the set Λ.

Also, the coordinator can use the previous group key to make a new group key in order to continue excluding users that were excluded last time.

Next, we show a method that a system manager excludes users from the group communication permanently:

(1) The system manager distributes a random number e to all users other than the excluded users by MaSK shown in Section 3.
(2) Each valid user replaces its own secret key s_i with

$$s_i' = s_i \times e \bmod q \tag{12}$$

(3) The system manager replaces the system parameter g on the public bulletin board with

$$g' = g^{1/e} \bmod p \tag{13}$$

Valid users can continue to join the group using s_i' and g' instead of previous s_i and g. The excluded users cannot join the group permanently, because they do not have the new secret key s_j'. When the system manager wants the excluded user to join the group communication again, the system manager would send its new secret key s_j' through a secure way. With this method, all public keys on the public bulletin board do not change since $y_i = (g')^{s_i'} \bmod p$ is satisfied. So, each user can also use the public keys in his local storage.

4.2.4. *Some modifications*

Using MaSK, a coordinator can change the group key without entity exclusion. The coordinator sets $d = 0$ and uses $k - 1$ sets of spare public keys for the broadcast data.

When a fixed coordinator can be assumed, he may keep all the public keys of users in his local storage. Also, if all users can keep all the public keys of other users, the public bulletin board is not needed. In these cases, even a system manager is not necessary after a system setup phase.

If the broadcast channel is assumed to be secure, no verification of coordinator's signature might be necessary.

For example, when the group contains 1000 users, the number of bits of n is only 10 bits, which is much smaller than $|q|$. Using this condition, we can modify the calculation of equation (11) to speed up, reducing the number of bits of an exponent part in equation (11). The detail modifications can be found in the paper [2].

MaSK is considered a one-pass Diffie-Hellman key exchange scheme with entity exclusion function. The coordinator distributes the broadcast data, and shares a group key U ($= g^{S \times r} \bmod p$) with the other users, where we regard y ($= g^S \bmod p$) as a public key for the group. Thus, the coordinator can select any group key GK and share it, by broadcasting $C = GK \times U \bmod p$ together, similar to ElGamal public key cryptosystem.

Also, if an attacker can use a key calculation mechanism of a user (or terminal) as an oracle, a similar modification like Cramer-Shoup public key cryptosystem [11] would be effective against adaptive chosen ciphertext attacks.

4.3. Comparison

In this section, we evaluate the features and performance of the following five methods:

MaSK: the scheme proposed by Anzai, Matsuzaki and Matsumoto [3]
HKDS: the hierarchical key distribution scheme proposed by Wong, Gouda and Lam [27] and Wallner, Harder and Agee [26].
FN: the scheme proposed by Fiat and Naor [13].
MA: the scheme proposed by Matsuzaki and Anzai [20].
KRS: the scheme proposed by Kumar, Rajagopalan and Sahai [15].

The following table shows the features of MaSK with respect to the criteria described before. Besides MaSK, only HKDS has good flexibility. All methods except MA have good reusability. The amount of transmission of MaSK and MA is independent of the group scale n. The key storage of MaSK and MA is constant regardless of n. Other schemes do not have good scalability in complexity.

In summary, MaSK is effective to implement quick group key distribution with entity exclusion function when the group scale n is very large

Table 1.   Comparison of five schemes.

	Flexibility	Reusability	Scalability	
			Amount of transmission	Key storage
MaSK	OK	OK	OK	OK
HKDS	OK	OK	NG	NG
FN	NG	OK	NG	NG
MA	NG	NG	OK	OK
KRS	NG	OK	OK	NG

compared to the number of excluded users d and it works well for devices with limited storage.

5. How to Discover the Existence of a Clone

5.1. *Theme*

Let us turn to the first question: how can the system check that every private key is kept properly? Or, if there is a clone then how can we know the fact? In this note a clone means an entity, which is not the original but with the same ID and the same secret private key as the original. If a tamper resistant module containing a secret key is not as strong as expected, attackers can make a fake terminal having the same secret key and ID as the original. For example, J.R. Rao *et al.* [23] reported power analysis attacks [14] against SIM cards used in GSM phone networks. Clone discovery is rather hard to solve in general. However, depending on environmental conditions there are approaches for solving the problem.

5.2. *Idea*

A simple idea is used in a cellular phone system [1]. A center can determine the cell to which a terminal belongs. The center can find a clone if there are two or more calls with the same ID from remote cells within a short time difference. However, the probability of finding a clone is very small since such a check can be done only when the terminal makes a call.

Recently, based on a similar but slightly different idea, J. Anzai, N. Matsuzaki and T. Matsumoto [4] suggested a practical method to actively discover the existence of a clone in a system consisting of a center and plural terminals, where the center provides a valuable service consecutively in encrypted form. Anzai-Matsuzaki-Matsumoto's basic idea is as follows:

Round. The center changes a session key periodically, where a session key is used for encrypting valuable data, e.g. copy-righted contents. We call the period a "round". The terminal should get a session key for every round, if the terminal wishes to obtain valuable data consecutively.

Random number. The method uses a random number generated by each terminal that a clone cannot guess, as a difference between the original and the clone. For each round each terminal generates a random number and transmits it to the center in encrypted form. The center can discover the existence of a clone if it gets two or more different random numbers with the same ID.

Forcing. The center forces each terminal to pass the random number in exchange for providing the session key of the round. The center sends the session key to any authorized terminal in the encrypted form by the random number. A clone cannot get the session key since it cannot predict the random number.

5.3. *Outline of the protocol*

Each terminal operates the following steps to get the session key for each round:

(1) The terminal generates a random number.

(2) The terminal transmits to the center an encrypted random number by the center's public key. A clone cannot read the random number.

(3) The center checks if there exists a clone using the log of the random numbers. If the center gets two or more different random numbers with the same ID, the center judges that it can discover clones with the ID.

(4) The center encrypts a session key using the random number, and distributes it. If the center discovers clones, the center does not distribute the session key to the clones nor to the original terminal.

5.4. *The clone discovery protocol*

Let us look at the scheme [4] of clone discovery in detail. Let SK_j be a session key at a round b_j. Here, the round b_j means the valid duration of SK_j.

(1) A terminal T_i generates a random number R_{ij}. Then T_i keeps R_{ij} secret during the round b_j and does not use R_{ij} in other rounds.

(2) The terminal T_i calculates an encrypted data as follows:

$$E_{ij} = ENC_A(Y_c, i \parallel R_{ij} \parallel b_j \parallel Sign(S_i, i \parallel R_{ij} \parallel b_j)) \qquad (14)$$

where $ENC_A(Y_c, message)$ is an encrypted message using a public key cryptosystem and the center's public key Y_c, $Sign(S_i, message)$ is a signature of message using the terminal T_i's private key S_i, and "$\parallel$" indicates the concatenation of data. The terminal transmits the encrypted data E_{ij} to the center.

(3) The center decrypts E_{ij} to get R_{ij} and $Sign(S_i, i \parallel R_{ij} \parallel b_j)$, using its private key. After verifying $Sign(S_i, i \parallel R_{ij} \parallel b_j)$, the center examines a registered random number in a log. If a different random number with the same ID is already registered, the center determines that there exist clones of the terminal with the ID.

(4) If the center does not find the clone with the ID i, the center calculates an encrypted session key as follows:

$$F_{ij} = ENC_S(R_{ij}, SK_j \parallel b_j) \qquad (15)$$

where $ENC_S(R_{ij}, message)$ is an encrypted message using a symmetric key cryptosystem and the shared random number R_{ij}. The center transmits the encrypted session key F_{ij} to the terminal T_i. The terminal T_i decrypts F_{ij} by the random number to get the session key of the round b_j. If the center discovers clones, the center revokes both clones and the original and does not distribute the session key. After getting the session key, the terminal should delete the random number R_{ij} in order to keep a high security.

5.5. Critical points

There are several points to examine.

5.5.1. Random number generator

The scheme uses a random number as the difference between an original terminal and a clone. If the clone can predict the random number the original generates, the clone can decrypt the data F_{ij} to get the session key without being detected. Therefore, the random number generator itself should be hard to be copied, as well as it should generate a good random number sequence, which is hard to predict. For mobile phone terminals, information on the geographic location, the level of electronic field, the remaining battery power, and the elapsed time can be useful to individualize the random number.

5.5.2. *Public key scheme*

A terminal encrypts the random number by the center's public key, so that even the clone with the same secret key as the original cannot decrypt it, if the clone eavesdrops the transmission. Thus, public key schemes are indispensable for this clone discovery scheme. Terminals including mobile phone terminals available for electronic commerce usually possess the ability to run public key schemes.

5.5.3. *Log management*

The center should keep a log of random numbers received from a terminal together with the terminal's ID. At the first phase of each round, the center clears the log. After verifying the signature in E_{ij}, the center checks if the random number R_{ij} is the same as the registered one with the same ID. If any random number has not been registered yet, the center registers the random number in the log, such as (i, R_{ij}). If the same random number has already been registered, the center considers that the terminal re-transmitted the same random number. Re-transmission may occur in mobile or wireless communication. If the different random number with the same ID has already been registered, the center decides that there exists a clone of the terminal.

5.5.4. *Access forcing*

If a terminal wishes to continue getting valuable data, in every round the terminal should generate a fresh random number and pass it to the center. According to the protocol, if a terminal does not transmit a random number until the deadline, it cannot get the session key for the next round, even if there are no clones.

5.5.5. *Round length*

Assume the scheme is applied to broadcast content distribution. The center and terminals should run the clone discovery protocol as well as the main protocol for content distribution. Thus, powerless terminals cannot operate the protocol properly if the round is too short. However, a long round increases the risk that successfully made clones may enjoy getting the content illegally without attracting the center's attention.

5.6. *Security*

A type of attack likely to occur would be the "one-shot attack," in which an attacker reads out by some means the ID, the private key and other information from the target terminal and makes a clone. After getting the above information, the attacker returns the terminal without any change to the holder of the terminal. In other words, the attacker does not use the original terminal at the very moment when it tries to fool the center.

(1) Even if such a one-shot reading out may be successfully done, the clones cannot predict correctly the random number generated by the original if the state of the random number generator is totally different from the observed one. Without the random number, the clone cannot get the session key for the next round.

(2) When a clone transmits a random number to the center before the original does, the center can detect the existence of a clone if the original terminal follows the protocol and transmits its random number.

(3) Even if the attacker eavesdrops the transactions between the original terminal and the center, an attacker cannot get the random number if the employed public key scheme is secure enough.

(4) Even if the attacker sends the data in the previous round, the center can detect the replay since both of E_{ij} and F_{ij} contain the round number.

Acknowledgments

The content of this note is based on a lecture given at IMS of National University of Singapore in September 2001. The author thanks Professor Harald Niederreiter and IMS for providing such a nice opportunity.

This work was partially supported by MEXT Grant-in-Aid for Scientific Research 13224040.

References

1. R. Anderson, *Security Engineering*, John Wiley & Sons, pp. 352–353, 2001.
2. J. Anzai, N. Matsuzaki, and T. Matsumoto, "A method for masked sharing of group keys (3)," *Technical Report of IEICE*, ISEC99-38, pp. 1–8, 1999.
3. J. Anzai, N. Matsuzaki, and T. Matsumoto, "A flexible method for masked sharing of group keys," *IEICE Trans. Fundamentals*, vol. E84-A, no. 1, pp. 239–246, 2001. Preliminary version appeared as J. Anzai, N. Matsuzaki, and T. Matsumoto, "A quick group key distribution scheme with entity

revocation," *Advances in Cryptology – ASIACRYPT '99*, LNCS vol. 1716, pp. 333–347, Springer-Verlag, 1999.

4. J. Anzai, N. Matsuzaki, and T. Matsumoto, "Clone discovery," to appear.

5. S. Berkovits, "How to broadcast a secret," *Advances in Cryptology – EURO-CRYPT '91*, LNCS vol. 547, pp. 535–541, Springer-Verlag, 1992.

6. C. Blundo, L. Mattos, and D. Stinson, "Generalized Beimel-Chor schemes for broadcast encryption and interactive key distribution," *Theoretical Computer Science*, 200(1-2), pp. 313–334, 1998.

7. C. Blundo, L. Mattos, and D. Stinson, "Trade-offs between communication and storage in unconditionally secure schemes for broadcast encryption and interactive key distribution," *Advances in Cryptology — CRYPTO '96*, LNCS vol. 1109, pp. 387–400, Springer-Verlag, 1996.

8. R. Canetti, J. Garay, G. Itkis, D. Micciancio, M. Naor, and B. Pinkas, "Multicast security: a taxonomy and efficient constructions," *Proceedings of INFOCOM '99*, vol. 2, pp. 708–716, 1999.

9. R. Canetti, T. Malkin, and K. Nissim, "Efficient communication-storage tradeoffs for multicast encryption," *Advances in Cryptology — EURO-CRYPT '99*, LNCS vol. 1592, pp. 459–474, Springer-Verlag, 1999.

10. B. Chor, A. Fiat, and M. Naor, "Tracing traitors," *Advances in Cryptology — CRYPTO '94*, LNCS vol. 839, pp. 257–270, Springer-Verlag, 1994.

11. R. Cramer, V. Shoup, "A practical public key cryptosystem provably secure against adaptive chosen ciphertext attack," *Advances in Cryptology — CRYPTO '98*, LNCS vol. 1462, pp. 13–25, Springer-Verlag, 1998.

12. Y. Desmedt, Y. Frankel, "Threshold cryptosystems," *Advances in Cryptology — CRYPTO '89*, LNCS vol. 435, pp. 307–315, Springer-Verlag, 1989.

13. A. Fiat, M. Naor, "Broadcast encryption," *Advances in Cryptology — CRYPTO '93*, LNCS vol. 773, pp. 480–491, Springer-Verlag, 1993.

14. P. Kocher, J. Jaffe, and B. Jun, "Differential power analysis," *Advances in Cryptology — CRYPTO '99*, LNCS vol. 1666, pp. 388–397, Springer-Verlag, 1999.

15. R. Kumar, S. Rajagopalan, and A. Sahai, "Coding constructions for blacklisting problems without computational assumptions," *Advances in Cryptology — CRYPTO '99*, LNCS vol. 1666, pp. 609–623, Springer-Verlag, 1999.

16. K. Kurosawa, Y. Desmedt, "Optimum traitor tracing and asymmetric scheme," *Advances in Cryptology — EUROCRYPT '98*, LNCS vol. 1403, pp. 145–157, Springer-Verlag, 1998.

17. M. Luby, J. Staddon, "Combinatorial bounds for broadcast encryption," *Advances in Cryptology — EUROCRYPT '98*, LNCS vol. 1403, pp. 512–526, Springer-Verlag, 1998.

18. M. Mambo, A. Nishikawa, S. Tsujii, and E. Okamoto, "Efficient secure broadcast communication systems," *Technical Report of IEICE*, ISEC93-34, pp. 21–31, 1993.

19. T. Matsushita, Y. Watanabe, K. Kobara, and H. Imai, "A sufficient content distribution scheme for mobile subscribers," *Proceedings of International Symposium on Information Theory and Its Applications, ISITA 2000*, pp. 497–500, 2000.

20. N. Matsuzaki, J. Anzai, "Secure group key distribution schemes with terminal revocation," *Proceedings of JWIS'98 (IEICE Technical Report*, ISEC98-52), pp. 37–44, 1998.

21. A. Menezes, P. van Oorschot, and S. Vanstone, *Handbook of Applied Cryptography*, CRC Press, pp. 113–114, 1997.

22. K. Nyberg, R.A. Rueppel, "Message recovery for signature schemes based on the discrete logarithm problem," *Advances in Cryptology — EUROCRYPT '94*, LNCS vol. 950, pp. 182–193, Springer-Verlag, 1995.

23. J.R. Rao, P. Rohatgi, H. Scherzer, and S. Tinguely, "Partitioning attacks: or how to rapidly clone some GSM cards," *IEEE Symposium on Security and Privacy*, 2002.

24. A. Shamir, "How to share a secret," *Communications of ACM*, vol. 22, no. 11, pp. 612–613, 1979.

25. U. S. Dept. of Commerce/National Institute of Standards and Technology, "Digital signature standard," *Federal Information Processing Standards Publication 186-1*, 1998.

26. D. Wallner, E. Harder, and R. Agee, "Key management for multicast: issues and architectures," RFC2627, IETF, June 1999.

27. C. Wong, M. Gouda, and S. Lam, "Secure group communications using Key graphs," *Proceedings of ACM SIGCOMM '98*, 1998. Also available as Technical Report TR 97-23, Department of Computer Science, The University of Texas at Austin, July 1997.

ALGEBRAIC FUNCTION FIELDS OVER FINITE FIELDS

Harald Niederreiter

Department of Mathematics, National University of Singapore
2 Science Drive 2, Singapore 117543, Republic of Singapore
E-mail: nied@math.nus.edu.sg

Algebraic function fields are important in several areas of information theory such as coding theory and cryptography. These tutorial lecture notes provide a quick introduction to the theory of algebraic function fields. The focus is on finite constant fields since this is the only case of interest for applications to information theory. The subject is approached from the viewpoint of valuation theory. The main topics covered in these lecture notes are valued fields, valuations of algebraic function fields, divisors, the Riemann-Roch theorem, the zeta function of an algebraic function field, and the Hasse-Weil bound.

1. Introduction

The aim of these lecture notes is to provide the reader with an easy access to the rather sophisticated theory of algebraic function fields. For information theorists, the main motivation for studying algebraic function fields is their importance in coding theory, in particular for the construction of algebraic-geometry codes. These applications to coding theory are treated in detail in the books [3, 4, 6, 7]. Recently, algebraic function fields have gained additional significance because of new applications to cryptography and the construction of low-discrepancy sequences (see [3] for these applications).

For the above applications, the only case of interest is that of algebraic function fields over a finite field. Therefore, these lecture notes will mostly concentrate on this case. After a brief review of finite fields and the theory of valuations, the special case of rational function fields over finite fields is considered in detail. Many of the typical features of algebraic function fields over finite fields can be readily understood in this special case. The aim in our discussion of general algebraic function fields over finite fields is to get as quickly as possible to the notion of genus, the zeta function, and the Hasse-Weil bound, all of which are crucial in applications.

The following is a rough definition (in the language of field theory): an *algebraic function field* (in one variable) is a finite extension of a field of rational functions (in one variable). If the rational function field is over a field k, then we speak of an algebraic function field over k. A rigorous definition will be given in Section 6.

Many basic aspects of the theory of algebraic function fields are analogous to the theory of algebraic number fields, i.e., the theory of the finite extensions of the field $\mathbf{Q}$ of rational numbers. Historically, the theory of algebraic function fields is a creation of the 19th century and the early 20th century. Fundamental contributions to this theory were made by B. Riemann, G. Roch, D. Hilbert, A. Hurwitz, E. Artin, F. K. Schmidt, H. Hasse, A. Weil, and C. Chevalley.

An algebraic function field over a finite field is also called a *global function field*. A field is called a *global field* if it is either a global function field or an algebraic number field. A classical monograph on global fields is [9].

2. Finite Fields

A *finite field* is a field with finitely many elements. The cardinality of a finite field is called its *order*. For a general treatment of finite fields we refer to [2] where also the proofs of all theorems in this section can be found.

Example 2.1: For every positive integer m we can consider the residue class ring $\mathbf{Z}_m := \mathbf{Z}/m\mathbf{Z}$ of the integers modulo m. The ring $\mathbf{Z}_m$ is a field if and only if m is a prime number. Thus, for every prime number p we have a finite field $\mathbf{Z}_p$ of order p. Note that $\mathbf{Z}_p$ is a finite prime field, i.e., it has no proper subfields. As a set, we can identify $\mathbf{Z}_p$ with $\{0, 1, \ldots, p-1\}$.

Any finite field $\mathbf{F}$ has a uniquely determined prime characteristic p (thus $p \cdot 1 = 0$ in $\mathbf{F}$) and the order of $\mathbf{F}$ is p^n, where n is the dimension of $\mathbf{F}$ as a vector space over its finite prime field (the latter field is isomorphic to $\mathbf{Z}_p$).

Conversely, for every prime power $q = p^n$ there does indeed exist a finite field of order q. Such a finite field can be explicitly constructed as follows. Take the prime number p and an irreducible polynomial $f \in \mathbf{Z}_p[x]$ of degree n; it can be shown that such an f exists (see [2, Section 3.2]). Then the residue class ring $\mathbf{Z}_p[x]/(f)$ is a finite field of order $p^n = q$.

Theorem 2.2: *Any two finite fields of the same order are isomorphic as fields.*

In this sense, it is meaningful to speak of *the* finite field $\mathbf{F}_q$ of prime-power order q. Some authors use the term *Galois field* GF(q), in recognition of the pioneering work of Galois in this area.

The finite field $\mathbf{F}_q$ of characteristic p may be viewed as the splitting field of the polynomial $x^q - x$ over $\mathbf{Z}_p$. In fact, $\mathbf{F}_q$ consists exactly of all roots of this polynomial.

Theorem 2.3: *Every subfield of* $\mathbf{F}_q$, $q = p^n$, *has order* p^d *with a positive divisor d of n. Conversely, if d is a positive divisor of n, then there is a unique subfield of* $\mathbf{F}_q$ *of order* p^d.

Example 2.4: Let $q = 2^{12}$. Then the subfields of $\mathbf{F}_q$ have the orders 2^1, 2^2, 2^3, 2^4, 2^6, 2^{12}.

The roots of irreducible polynomials over a finite field are simple and enjoy a special relationship among themselves.

Theorem 2.5: *The roots of an irreducible polynomial over* $\mathbf{F}_q$ *of degree m are given by the m distinct elements*

$$\alpha^{q^j} \in \mathbf{F}_{q^m}, \quad j = 0, 1, \ldots, m - 1,$$

where α is a fixed root.

It follows that any finite extension $\mathbf{F}_{q^m}/\mathbf{F}_q$ is a Galois extension with a cyclic Galois group generated by the *Frobenius automorphism*

$$\sigma(\beta) = \beta^q \quad \text{for all } \beta \in \mathbf{F}_{q^m}.$$

A famous classical theorem describes the multiplicative structure of $\mathbf{F}_q$.

Theorem 2.6: *The multiplicative group* $\mathbf{F}_q^*$ *of nonzero elements of* $\mathbf{F}_q$ *is cyclic.*

Definition 2.7: A generator of the cyclic group $\mathbf{F}_q^*$ is called a *primitive element* of $\mathbf{F}_q$.

Example 2.8: 2 is a primitive element of $\mathbf{F}_{11} = \mathbf{Z}_{11}$ since $2^{10} = 1$ in $\mathbf{F}_{11}$ by Fermat's theorem, but $2^1 = 2 \neq 1$, $2^2 = 4 \neq 1$, $2^5 = 32 = 10 \neq 1$ in $\mathbf{F}_{11}$.

3. Valued Fields

Example 3.1: Let $\mathbf{Q}$ be the field of rational numbers and take a prime number p. Using unique factorization in $\mathbf{Z}$, we can write every nonzero $r \in \mathbf{Q}$ in the form

$$r = p^m \frac{a}{b}$$

with a unique $m \in \mathbf{Z}$, where p does not divide the nonzero integers a and b. Then we define

$$|r|_p = p^{-m}.$$

We also set $|0|_p = 0$. The map $|\cdot|_p \colon \mathbf{Q} \to \mathbf{R}_{\geq 0}$ has the following properties:

(i) $|r|_p = 0$ if and only if $r = 0$;
(ii) $|rs|_p = |r|_p |s|_p$ for all $r, s \in \mathbf{Q}$;
(iii) $|r + s|_p \leq \max(|r|_p, |s|_p) \leq |r|_p + |s|_p$ for all $r, s \in \mathbf{Q}$.

The properties (i) and (ii) are trivial. To prove (iii), let $r, s \in \mathbf{Q}^*$ and write as above

$$r = p^m \frac{a}{b}, \quad s = p^n \frac{c}{d},$$

where without loss of generality $m \leq n$. Then

$$r + s = p^m \frac{a}{b} + p^n \frac{c}{d} = p^m \frac{ad + p^{n-m} bc}{bd},$$

and so

$$|r + s|_p \leq p^{-m} = \max(|r|_p, |s|_p).$$

The map $|\cdot|_p$ is called the (normalized) p-*adic absolute value* on $\mathbf{Q}$. There is another obvious map on $\mathbf{Q}$ satisfying (i), (ii), and the inequality (iii) in the wider sense, namely the ordinary absolute value $|\cdot|$. Any map on $\mathbf{Q}$ satisfying (i), (ii), (iii) (again in the wider sense), and having at least one value $\neq 1$ on $\mathbf{Q}^*$ is called an *absolute value* on $\mathbf{Q}$.

It is a well-known theorem that any absolute value on $\mathbf{Q}$ is equivalent (in the sense of one being a power of the other) to either $|\cdot|$ or $|\cdot|_p$ for some prime number p (see [5, Section 1.3] and [9, Section 1-4]). For any $r \in \mathbf{Q}^*$ we have

$$|r| \cdot \prod_p |r|_p = 1,$$

where the product is over all prime numbers p. This product formula is typical for global fields. To verify this product formula, note that everything on the left-hand side is multiplicative, and so it suffices to consider a nonzero integer r with canonical factorization

$$r = \pm p_1^{m_1} \cdots p_k^{m_k},$$

say. Then $|r|_{p_i} = p_i^{-m_i}$ for $1 \leq i \leq k$ and $|r|_p = 1$ for all other primes p. Therefore

$$|r| \cdot \prod_p |r|_p = p_1^{m_1} \cdots p_k^{m_k} \cdot p_1^{-m_1} \cdots p_k^{-m_k} = 1.$$

Let us now concentrate on $|\cdot|_p$. For all $r \in \mathbf{Q}^*$ we set

$$\nu_p(r) = -\log_p |r|_p = m$$

in the earlier notation. We also put $\nu_p(0) = \infty$. The map $\nu_p \colon \mathbf{Q} \to \mathbf{Z} \cup \{\infty\}$ has the following properties:

(i) $\nu_p(r) = \infty$ if and only if $r = 0$;
(ii) $\nu_p(rs) = \nu_p(r) + \nu_p(s)$ for all $r, s \in \mathbf{Q}$;
(iii) $\nu_p(r + s) \geq \min(\nu_p(r), \nu_p(s))$ for all $r, s \in \mathbf{Q}$;
(iv) $\nu_p(\mathbf{Q}^*) \neq \{0\}$.

This example gives rise to the following general concept.

Definition 3.2: Let K be an arbitrary field. A (nonarchimedean) *valuation* of K is a map $\nu \colon K \to \mathbf{R} \cup \{\infty\}$ which satisfies:

(i) $\nu(x) = \infty$ if and only if $x = 0$;
(ii) $\nu(xy) = \nu(x) + \nu(y)$ for all $x, y \in K$;
(iii) $\nu(x + y) \geq \min(\nu(x), \nu(y))$ for all $x, y \in K$;
(iv) $\nu(K^*) \neq \{0\}$.

If the image $\nu(K^*)$ is a discrete set in $\mathbf{R}$, then ν is called *discrete*. If $\nu(K^*) = \mathbf{Z}$, then ν is called *normalized*. The ordered pair (K, ν) is called a *valued field*. Property (iv) excludes what is sometimes called the trivial valuation of K, which is ∞ at $0 \in K$ and 0 on K^*. General references for valued fields are [5] and [9].

Obviously, ν_p is a (discrete) normalized valuation of $\mathbf{Q}$ for every prime number p. We call ν_p the *p-adic valuation* of $\mathbf{Q}$.

We will be interested only in discrete valuations. In the following two results, let (K, ν) be a valued field.

Lemma 3.3: (i) $\nu(1) = \nu(-1) = 0$;
(ii) $\nu(-x) = \nu(x)$ for all $x \in K$;
(iii) $\nu(xy^{-1}) = \nu(x) - \nu(y)$ for all $x \in K$, $y \in K^*$.

Proof: (i) $1^2 = 1 \Rightarrow 2\nu(1) = \nu(1) \Rightarrow \nu(1) = 0$: and $(-1)^2 = 1 \Rightarrow 2\nu(-1) = \nu(1) = 0 \Rightarrow \nu(-1) = 0$. In both cases, property (ii) in Definition 3.2 was used.

(ii) $\nu(-x) = \nu((-1) \cdot x) = \nu(-1) + \nu(x) = \nu(x)$.

(iii) $0 = \nu(1) = \nu(yy^{-1}) = \nu(y) + \nu(y^{-1}) \Rightarrow \nu(y^{-1}) = -\nu(y)$. The result follows again from property (ii) in Definition 3.2.   □

Proposition 3.4: If $x, y \in K$ with $\nu(x) \neq \nu(y)$, then

$$\nu(x + y) = \min(\nu(x), \nu(y)).$$

Proof: Suppose without loss of generality that $\nu(x) < \nu(y)$. Then $\nu(x + y) \geq \nu(x)$ by property (iii) in Definition 3.2. If we had $\nu(x + y) > \nu(x)$, then by the same property and Lemma 3.3(ii),

$$\nu(x) = \nu((x + y) - y) \geq \min(\nu(x + y), \nu(-y))$$
$$= \min(\nu(x + y), \nu(y)) > \nu(x),$$

a contradiction.   □

Example 3.5: Here is a simple illustration of Proposition 3.4. Let $K = \mathbf{Q}$ and ν the p-adic valuation of $\mathbf{Q}$. Take nonzero $x, y \in \mathbf{Z}$. Then we can write $x = p^m a$ and $y = p^n c$ with a and c not divisible by p. Suppose $\nu(x) = m < \nu(y) = n$. Then

$$x + y = p^m a + p^n c = p^m(a + p^{n-m}c)$$

with $a + p^{n-m}c \equiv a \not\equiv 0 \pmod{p}$. Therefore

$$\nu(x + y) = m = \min(\nu(x), \nu(y)).$$

4. Places and Valuation Rings

All valuations in this section are discrete.

Two valuations ν and μ of the field K are called *equivalent* if there exists a constant $c > 0$ such that

$$\nu(x) = c\mu(x) \qquad \text{for all } x \in K.$$

Obviously, this yields an equivalence relation between valuations of K. An equivalence class of valuations of K is called a *place* of K.

Since $\nu(K^*)$ is a nonzero discrete subgroup of $(\mathbf{R}, +)$, we have $\nu(K^*) = \alpha\mathbf{Z}$ for some positive $\alpha \in \mathbf{R}$. Thus, there exists a uniquely determined normalized valuation of K that is equivalent to ν. In other words, every place P of K contains a uniquely determined normalized valuation of K, which is denoted by ν_P. Thus, we can identify places of K and (discrete) normalized valuations of K.

Definition 4.1: Let P be a place of K. Then

$$O_P := \{x \in K : \nu_P(x) \geq 0\}$$

is called the *valuation ring* of P.

Using the properties of valuations in Section 3, we see that O_P is an integral domain with $1 \in O_P$.

Proposition 4.2: *The multiplicative group of units of O_P is given by*

$$U_P := \{x \in K : \nu_P(x) = 0\}.$$

Proof: For $x \in O_P$ we have: x is a unit of $O_P \Leftrightarrow x^{-1} \in O_P \Leftrightarrow \nu_P(x^{-1}) \geq 0 \Leftrightarrow \nu_P(x) \leq 0 \Leftrightarrow \nu_P(x) = 0$. □

Proposition 4.3: *O_P has a unique maximal ideal given by*

$$\mathsf{M}_P := \{x \in K : \nu_P(x) > 0\}.$$

Proof: It is trivial that M_P is an ideal of O_P. Let J with $\mathsf{M}_P \subset \mathsf{J} \subseteq O_P$ be an ideal of O_P. Take $x \in \mathsf{J} \setminus \mathsf{M}_P$, then $\nu_P(x) = 0$ and x is a unit of O_P by Proposition 4.2. Thus $1 = xx^{-1} \in \mathsf{J}$, hence $\mathsf{J} = O_P$, and so M_P is maximal. Finally, let $\mathsf{M} \subset O_P$ be an arbitrary maximal ideal of O_P. Since M cannot contain a unit of O_P, we must have $\mathsf{M} \subseteq \mathsf{M}_P$, and so $\mathsf{M} = \mathsf{M}_P$. □

M_P is in fact a principal ideal: take any $t \in O_P$ with $\nu_P(t) = 1$, then it is easy to check that $\mathsf{M}_P = t O_P$. Such a t is called a *local parameter* at P.

Definition 4.4: The field O_P/M_P is called the *residue class field* of P. The ring homomorphism

$$x \in O_P \mapsto x + \mathsf{M}_P \in O_P/\mathsf{M}_P$$

is called the *residue class map* of P.

Example 4.5: Let ν_p be the p-adic valuation of $\mathbf{Q}$ (see Section 3), which is a normalized valuation of $\mathbf{Q}$. In the following, we write all rational numbers

a/b in reduced form, i.e., $\gcd(a, b) = 1$. It is easy to see that

$$O_p = \left\{ \frac{a}{b} \in \mathbf{Q} : \gcd(b, p) = 1 \right\},$$

$$U_p = \left\{ \frac{a}{b} \in \mathbf{Q} : \gcd(a, p) = \gcd(b, p) = 1 \right\},$$

$$\mathsf{M}_p = \left\{ \frac{a}{b} \in \mathbf{Q} : p \,|\, a \right\}.$$

For any $b \in \mathbf{Z}$ we write $\bar{b}$ for the residue class of b modulo p. If $\gcd(b, p) = 1$, then $\bar{b} \in \mathbf{Z}_p = \mathbf{Z}/p\mathbf{Z}$ has a multiplicative inverse $\bar{b}^{-1} \in \mathbf{Z}_p$. The map $\psi : O_p \to \mathbf{Z}_p$ given by

$$\psi\left(\frac{a}{b}\right) = \bar{a}\bar{b}^{-1} \in \mathbf{Z}_p \qquad \text{for all } \frac{a}{b} \in O_p$$

is well defined. Clearly, ψ is a surjective ring homomorphism with kernel M_p, and so the residue class field O_p/M_p is isomorphic to the finite field $\mathbf{Z}_p$.

Remark 4.6: Valuation rings in K can be characterized purely algebraically as the maximal proper Noetherian subrings of K, or also as the maximal proper principal ideal domains contained in K (see [5, Section 2.1]). This provides an alternative approach to the concept of a place.

The proof of the following theorem can be found in [5, Section 2.1] and [9, Section 1-2].

Theorem 4.7: (Weak Approximation Theorem) *Let $P_1, \dots, P_n$ be distinct places of K. Then for any given elements $x_1, \dots, x_n \in K$ and integers $m_1, \dots, m_n$, there exists an element $x \in K$ such that*

$$\nu_{P_i}(x - x_i) \geq m_i \qquad \text{for } i = 1, \dots, n.$$

Example 4.8: Let $K = \mathbf{Q}$. Choosing n distinct places of K means choosing distinct primes $p_1, \dots, p_n$. Let $x_1, \dots, x_n \in \mathbf{Z}$ and take positive integers $m_1, \dots, m_n$. Then by the Chinese remainder theorem there exists an $x \in \mathbf{Z}$ with

$$x \equiv x_i \pmod{p_i^{m_i}} \qquad \text{for } i = 1, \dots, n.$$

This just means that

$$\nu_{p_i}(x - x_i) \geq m_i \qquad \text{for } i = 1, \dots, n.$$

Thus, the weak approximation theorem can be viewed as an analog of the Chinese remainder theorem for K.

5. Rational Function Fields

Let k be an arbitrary field and let $k(x)$ be the rational function field over k in the variable x (in rigorous algebraic terms, x is transcendental over k). In this context, k is called the *constant field* of $k(x)$. The elements of $k(x)$ can be represented as $f(x)/g(x)$ with $f(x), g(x) \in k[x]$, $g(x) \neq 0$, and $\gcd(f(x), g(x)) = 1$.

Valuations of $k(x)$ can be constructed in an analogous way as for $\mathbf{Q}$ (compare with Section 3). The role of the prime numbers is now played by irreducible polynomials. Fix a monic irreducible polynomial $p(x) \in k[x]$. Using unique factorization in $k[x]$, we can write every nonzero $r(x) \in k(x)$ in the form

$$r(x) = p(x)^m \frac{f(x)}{g(x)}$$

with a unique $m \in \mathbf{Z}$, where $p(x)$ does not divide the nonzero polynomials $f(x)$ and $g(x)$. Then we put

$$\nu_{p(x)}(r(x)) = m.$$

We also put $\nu_{p(x)}(0) = \infty$. Then it is easily checked that $\nu_{p(x)}$ is a (discrete) normalized valuation of $k(x)$.

There is another normalized valuation of $k(x)$ which is obtained from the degree map. If $r(x) = f(x)/g(x)$ is nonzero, then we set

$$\nu_\infty(r(x)) = \deg(g(x)) - \deg(f(x)).$$

We also set $\nu_\infty(0) = \infty$. Again, it is easily checked that ν_∞ is a normalized valuation of $k(x)$. Indeed, the properties (i) and (iv) in Definition 3.2 are trivial. Property (ii) follows from $\deg(gh) = \deg(g) + \deg(h)$ for all $g, h \in k[x]$. To prove (iii), we have in an obvious notation,

$$\nu_\infty\left(\frac{f}{g} + \frac{e}{h}\right) = \nu_\infty\left(\frac{fh + eg}{gh}\right) = \deg(gh) - \deg(fh + eg)$$

$$\geq \deg(gh) - \max(\deg(fh), \deg(eg))$$

$$= \min(\deg(gh) - \deg(fh), \deg(gh) - \deg(eg))$$

$$= \min(\deg(g) - \deg(f), \deg(h) - \deg(e))$$

$$= \min\left(\nu_\infty\left(\frac{f}{g}\right), \nu_\infty\left(\frac{e}{h}\right)\right).$$

The valuations $\nu_{p(x)}$, with $p(x) \in k[x]$ monic irreducible, and ν_∞ are pairwise nonequivalent since $\nu_{p(x)}(p(x)) = 1$, whereas $\nu_{q(x)}(p(x)) = 0$ for monic irreducible $q(x) \neq p(x)$ and $\nu_\infty(p(x)) < 0$. Thus, we get a set of places $\{p(x) \in k[x] : p(x) \text{ monic irreducible}\} \cup \{\infty\}$ of $k(x)$.

Remark 5.1: If k is algebraically closed, then the monic irreducible polynomials over k are exactly the linear polynomials $x - a$ with $a \in k$. Thus, the above set of places can be identified with $k \cup \{\infty\}$. If $k = \mathbf{C}$, then this yields the complex plane with the point at ∞, i.e., the Riemann sphere. Thus, the above set of places of $\mathbf{C}(x)$ is in bijective correspondence with the points (or "places") on the Riemann sphere. This explains how the term "place" originated historically.

We assume from now on that the constant field k is finite, although some results hold for arbitrary k.

Lemma 5.2: *If k is finite, then for any valuation ν of $k(x)$ we have $\nu(a) = 0$ for all $a \in k^*$.*

Proof: If $k = \mathbf{F}_q$, then $a^{q-1} = 1$ for all $a \in \mathbf{F}_q^*$, and so by Lemma 3.3(i),

$$0 = \nu(1) = \nu(a^{q-1}) = (q-1)\nu(a),$$

which yields $\nu(a) = 0$. $\qquad\qquad\square$

Theorem 5.3: *If k is finite, then the set of all places of $k(x)$ is given by $\{p(x) \in k[x] : p(x) \text{ monic irreducible}\} \cup \{\infty\}$.*

Proof: We have to show that if P is a place of $k(x)$ and $\nu \in P$, then ν is equivalent to either $\nu_{p(x)}$ or ν_∞.

Case 1: $\nu(x) \geq 0$. Then $k[x] \subseteq O_P$ by Lemma 5.2. Since $\nu(k(x)^*) \neq \{0\}$ by definition, there exists a nonzero $h(x) \in k[x]$ with $\nu(h(x)) > 0$. Thus, $J := k[x] \cap M_P$ is a nonzero ideal of $k[x]$. Furthermore, $J \neq k[x]$ since $1 \notin J$. Since M_P is a prime ideal of O_P, it follows that J is a prime ideal of $k[x]$. Consequently, there exists a monic irreducible $p(x) \in k[x]$ such that $J = (p(x))$. In particular, $c := \nu(p(x)) > 0$. If $g(x) \in k[x]$ is not divisible by $p(x)$, then $g(x) \notin M_P$, and so $\nu(g(x)) = 0$. Thus, if we write a nonzero $r(x) \in k(x)$ in the form

$$r(x) = p(x)^m \frac{f(x)}{g(x)}$$

with $f(x)$ and $g(x)$ not divisible by $p(x)$, then

$$\nu(r(x)) = m\nu(p(x)) = c\nu_{p(x)}(r(x)),$$

and so ν is equivalent to $\nu_{p(x)}$.

Case 2: $\nu(x) < 0$. Then $c := \nu(x^{-1}) > 0$ and $x^{-1} \in M_P$. Take any nonzero $f(x) \in k[x]$ of degree d, say. Then

$$f(x) = \sum_{i=0}^{d} a_i x^i = x^d \sum_{i=0}^{d} a_i x^{i-d}$$

$$= x^d \sum_{i=0}^{d} a_{d-i} x^{-i}$$

with all $a_i \in k$. Furthermore,

$$\sum_{i=0}^{d} a_{d-i} x^{-i} = a_d + \sum_{i=1}^{d} a_{d-i} x^{-i} = a_d + s(x)$$

with $s(x) \in M_P$. Since $a_d \neq 0$, we have $\nu(a_d) = 0$, and so

$$\nu\left(\sum_{i=0}^{d} a_{d-i} x^{-i}\right) = 0$$

by Proposition 3.4. It follows that

$$\nu(f(x)) = \nu(x^d) = -d\nu(x^{-1}) = c\nu_{\infty}(f(x)),$$

and so ν is equivalent to ν_{∞}. $\qquad\square$

Remark 5.4: Note that in the proof of Theorem 5.3 we have not used that ν is discrete, and so the proof shows that every valuation of $k(x)$ is automatically discrete.

The places $p(x)$ are often called the *finite places* of $k(x)$ and the place ∞ is often called the *infinite place* of $k(x)$.

Example 5.5: As in Example 4.5 one shows that the residue class field of the place $p(x)$ is isomorphic to $k[x]/(p(x))$. For the place ∞ we have

$$O_{\infty} = \left\{ \frac{f(x)}{g(x)} \in k(x) : \deg(f(x)) \leq \deg(g(x)) \right\},$$

$$U_{\infty} = \left\{ \frac{f(x)}{g(x)} \in k(x) : \deg(f(x)) = \deg(g(x)) \right\},$$

$$M_{\infty} = \left\{ \frac{f(x)}{g(x)} \in k(x) : \deg(f(x)) < \deg(g(x)) \right\}.$$

Every $r(x) \in O_\infty$ can be written in the form

$$r(x) = \frac{a_d x^d + a_{d-1} x^{d-1} + \cdots + a_0}{x^d + b_{d-1} x^{d-1} + \cdots + b_0}$$

with all a_i, b_j in k. The map $\psi \colon O_\infty \to k$ given by

$$\psi(r(x)) = a_d \in k \qquad \text{for all } r(x) \in O_\infty$$

is well defined. It is easily seen that ψ is a surjective ring homomorphism with kernel M_∞, and so the residue class field of the place ∞ is isomorphic to k.

Remark 5.6: For every nonzero $r(x) \in k(x)$ we have

$$\nu_\infty(r(x)) + \sum_{p(x)} \nu_{p(x)}(r(x)) \deg(p(x)) = 0,$$

where the sum is over all monic irreducible $p(x) \in k[x]$. Note that the sum makes sense since $\nu_{p(x)}(r(x)) = 0$ for all but finitely many $p(x)$. Because of the properties of valuations, it suffices to verify the formula for nonzero monic $f(x) \in k[x]$. If

$$f(x) = \prod_{i=1}^{n} p_i(x)^{m_i}$$

is the canonical factorization of $f(x)$ in $k[x]$, then

$$\sum_{p(x)} \nu_{p(x)}(f(x)) \deg(p(x)) = \sum_{i=1}^{n} m_i \deg(p_i(x)) = \deg(f(x)) = -\nu_\infty(f(x)).$$

This formula is an additive analog of the product formula for $\mathbf{Q}$ in Example 3.1.

6. Algebraic Function Fields and Their Valuations

We have already given a rough definition of algebraic function fields in Section 1. The rigorous definition is as follows. *We consider only algebraic function fields over a finite field k.*

Definition 6.1: A field F is an *algebraic function field* over the finite field k if there exists a transcendental element $z \in F$ over k such that F is a finite extension of the rational function field $k(z)$.

The simplest example of an algebraic function field is of course a rational function field over a finite field.

Proposition 6.2: *Every valuation of an algebraic function field is discrete.*

Proof: Let F be an algebraic function field. With the notation in Definition 6.1, put $K = k(z)$. Let ν be an arbitrary valuation of F and let μ be the restriction of ν to K. It suffices to prove that the index $[\nu(F^*) : \mu(K^*)]$ is finite. Since $\nu(F^*)$ is an infinite subgroup of $(\mathbf{R}, +)$, this shows then that $\mu(K^*) = \{0\}$ is not possible. Hence μ is a valuation of K, thus discrete by Remark 5.4, and so ν is discrete.

Let $x_1, \ldots, x_n \in F^*$ be such that $\nu(x_1), \ldots, \nu(x_n)$ are in distinct cosets modulo $\mu(K^*)$. We claim that $x_1, \ldots, x_n$ are linearly independent over K. This will then show

$$[\nu(F^*) : \mu(K^*)] \leq [F : K] < \infty.$$

So suppose we have

$$\sum_{i=1}^{n} b_i x_i = 0,$$

where without loss of generality all $b_i \in K^*$. If we had $\nu(b_i x_i) = \nu(b_j x_j)$ for some $i \neq j$, then

$$\nu(x_i) - \nu(x_j) = \nu(b_j) - \nu(b_i) = \mu(b_j b_i^{-1}) \in \mu(K^*),$$

a contradiction to the choice of $x_1, \ldots, x_n$. Thus $\nu(b_1 x_1), \ldots, \nu(b_n x_n)$ are all distinct, and so Proposition 3.4 yields

$$\infty = \nu \left(\sum_{i=1}^{n} b_i x_i \right) = \min_{1 \leq i \leq n} \nu(b_i x_i).$$

This means $b_i x_i = 0$ for $1 \leq i \leq n$, again a contradiction. $\square$

In the above proof we have shown, in particular, that the restriction of a valuation of F to $k(z)$ yields a valuation of $k(z)$. Obviously, for equivalent valuations of F the restrictions are again equivalent. Thus, a place Q of F corresponds by restriction to a unique place P of $k(z)$. We say that Q *lies over* P or that P *lies under* Q. Therefore, every place of F lies either over a place of $k(z)$ corresponding to a monic irreducible polynomial in $k[z]$ or over the infinite place of $k(z)$.

Proposition 6.3: *Let F be an algebraic function field over k. Then the residue class field of every place of F is a finite extension of (an isomorphic copy of) k.*

Proof: Let Q be a place of F that lies over the place P of $K := k(z)$ (with z as in Definition 6.1). Let $R_Q := O_Q/M_Q$ and $R_P := O_P/M_P$ be the corresponding residue class fields and note that $O_P \subseteq O_Q$. The map ρ: $R_P \to R_Q$ given by

$$\rho(b + M_P) = b + M_Q \qquad \text{for all } b \in O_P$$

is well defined since $M_P \subseteq M_Q$. It is clear that ρ is an injective ring homomorphism, and so R_Q contains the isomorphic copy $\rho(R_P)$ of R_P as a subfield.

Let $x_1, \dots, x_n \in O_Q$ be such that $x_1 + M_Q, \dots, x_n + M_Q$ are linearly independent over $\rho(R_P)$. We claim that $x_1, \dots, x_n$ are linearly independent over K. This will then show

$$[R_Q : \rho(R_P)] \leq [F : K] < \infty.$$

Since R_P is a finite extension (of an isomorphic copy) of k (see Example 5.5), this proves the proposition. So suppose we have

$$\sum_{i=1}^{n} b_i x_i = 0$$

with $b_1, \dots, b_n \in K$ not all 0. Without loss of generality

$$\nu_P(b_1) = \min_{1 \leq i \leq n} \nu_P(b_i).$$

Then $b_1 \neq 0$ and

$$x_1 + \sum_{i=2}^{n} b_i b_1^{-1} x_i = 0.$$

By the condition on $\nu_P(b_1)$ we have $b_i b_1^{-1} \in O_P$ for $2 \leq i \leq n$. Passing to the residue classes modulo M_Q, we get

$$(x_1 + M_Q) + \sum_{i=2}^{n} \rho(b_i b_1^{-1} + M_P)(x_i + M_Q) = 0 + M_Q,$$

a contradiction to the choice of $x_1, \dots, x_n$.   □

The following result (see [7, Section III.1] and [9, Section 2-4] for its proof) shows, in particular, that every valuation of a rational function field over k can be extended to a valuation of an algebraic function field over k.

Theorem 6.4: *Let F be a finite extension of the rational function field $k(z)$. Then every place of $k(z)$ lies under at least one and at most $[F : k(z)]$ places of F.*

Let F again be an algebraic function field over the finite field k. Let $\tilde{k}$ be the algebraic closure of k in F, i.e.,

$$\tilde{k} = \{x \in F : x \text{ is algebraic over } k\}.$$

Clearly, $\tilde{k}$ is a field with $k \subseteq \tilde{k} \subseteq F$. The field $\tilde{k}$ is called the *full constant field* of F. The following result shows that $\tilde{k}$ is again a finite field.

Proposition 6.5: *The field $\tilde{k}$ is a finite extension of k.*

Proof: By Theorem 6.4 there exists a place Q of F. Take any $x \in \tilde{k}^*$. By means of its minimal polynomial over k we get

$$x^d + c_{d-1}x^{d-1} + \cdots + c_0 = 0$$

with $c_0, \ldots, c_{d-1} \in k$ and $c_0 \neq 0$. If we had $\nu_Q(x) < 0$, then

$$\nu_Q(x^d + c_{d-1}x^{d-1} + \cdots + c_0) = \nu_Q(x^d) < 0$$

by Proposition 3.4, a contradiction. If we had $\nu_Q(x) > 0$, then

$$\nu_Q(x^d + c_{d-1}x^{d-1} + \cdots + c_0) = \nu_Q(c_0) = 0$$

by Proposition 3.4, again a contradiction. Thus we must have $\nu_Q(x) = 0$. Now consider the map $\psi \colon \tilde{k} \to R_Q := O_Q/\mathsf{M}_Q$ given by

$$\psi(x) = x + \mathsf{M}_Q \qquad \text{for all } x \in \tilde{k}.$$

The above considerations show that ψ is injective, hence $|\tilde{k}| \leq |R_Q| < \infty$, the latter by Proposition 6.3.   □

Note that $k(z) \subseteq \tilde{k}(z) \subseteq F$, and so F is a finite extension of $\tilde{k}(z)$. Furthermore, z is transcendental over $\tilde{k}$ by Proposition 6.5, and so F is also an algebraic function field over $\tilde{k}$. In the following, we will usually assume that k is already the full constant field of F. If we want to stress this, then we use the notation F/k for an algebraic function field with full constant field k.

In view of Proposition 6.3, the following definition makes sense.

Definition 6.6: The *degree* $\deg(P)$ of a place P of F/k is defined to be the degree of the residue class field of P over k. A place of F/k of degree 1 is also called a *rational place* of F/k.

Example 6.7: Let $F = k(x)$ be the rational function field over k. It is obvious that for any nonconstant rational function $r(x) \in F$ there exists a finite place P of F such that $\nu_P(r(x)) \neq 0$. Thus, by the proof of Proposition 6.5, the full constant field of F is k. By Example 5.5, the degree of a finite place $p(x)$ of F is equal to the degree of the polynomial $p(x)$ and the degree of the place ∞ of F is equal to 1. If $k = \mathbf{F}_q$, then F has thus exactly $q + 1$ rational places.

For an algebraic function field F/k, we denote by $\mathcal{P}_F$ the set of all places of F. Note that $\mathcal{P}_F$ is a denumerable set. The following result (see [7, Section I.6]) is a strengthening of Theorem 4.7 in the case under consideration.

Theorem 6.8: (Strong Approximation Theorem) *Let $\mathcal{S}$ be a proper nonempty subset of $\mathcal{P}_F$ and let $P_1, \dots, P_n \in \mathcal{S}$ be distinct. Then for any given elements $x_1, \dots, x_n \in F$ and integers $m_1, \dots, m_n$, there exists an element $x \in F$ such that*

$$\nu_{P_i}(x - x_i) = m_i \qquad \text{for } i = 1, \dots, n,$$
$$\nu_P(x) \geq 0 \qquad \text{for all } P \in \mathcal{S} \setminus \{P_1, \dots, P_n\}.$$

7. Divisors

Let F/k be an algebraic function field with full constant field k (k finite).

Definition 7.1: A *divisor* D of F is a formal sum

$$D = \sum_{P \in \mathcal{P}_F} m_P P$$

with all $m_P \in \mathbf{Z}$ and $m_P \neq 0$ for at most finitely many $P \in \mathcal{P}_F$.

A place $P \in \mathcal{P}_F$ is also a divisor (put $m_P = 1$, $m_Q = 0$ for all $Q \in \mathcal{P}_F$ with $Q \neq P$). In this context, a place is called a *prime divisor*.

The divisors of F form a group under the obvious addition law

$$D + E = \sum_{P \in \mathcal{P}_F} m_P P + \sum_{P \in \mathcal{P}_F} n_P P$$
$$= \sum_{P \in \mathcal{P}_F} (m_P + n_P) P.$$

The zero element is the zero divisor

$$0 := \sum_{P \in \mathcal{P}_F} m_P P \qquad \text{with all } m_P = 0.$$

The additive inverse of $D = \sum_{P \in \mathcal{P}_F} m_P P$ is

$$-D = \sum_{P \in \mathcal{P}_F} (-m_P)P.$$

The abelian group of all divisors of F is called the *divisor group* $\text{Div}(F)$ of F. It can also be described as the free abelian group generated by the places (prime divisors) of F.

Remark 7.2: Note the following simple analogy between polynomials and divisors. A polynomial is an assignment which associates with each monomial x^i, $i = 0, 1, \ldots$, a coefficient a_i such that all but finitely many $a_i = 0$. A divisor is an assignment which associates with each place $P \in \mathcal{P}_F$ an integer coefficient m_P such that all but finitely many $m_P = 0$. Addition of divisors operates in the same way as addition of polynomials.

Definition 7.3: The *support* $\text{supp}(D)$ of the divisor $D = \sum_{P \in \mathcal{P}_F} m_P P$ is given by

$$\text{supp}(D) = \{P \in \mathcal{P}_F : m_P \neq 0\}.$$

By the definition of a divisor, $\text{supp}(D)$ is a finite subset of $\mathcal{P}_F$. If $D = \sum_{P \in \mathcal{P}_F} m_P P$, it is often convenient to write $m_P = \nu_P(D)$. Thus, a divisor D can also be represented in the form

$$D = \sum_{P \in \text{supp}(D)} \nu_P(D)P.$$

Definition 7.4: If $D \in \text{Div}(F)$ is as above, then the *degree* $\deg(D)$ of D is defined by

$$\deg(D) = \sum_{P \in \text{supp}(D)} \nu_P(D) \deg(P).$$

Proposition 7.5: *The degree map* $\deg \colon \text{Div}(F) \to \mathbf{Z}$ *is a group homomorphism.*

Proof: This is a straightforward verification. $\qquad\qquad\qquad\qquad\square$

Consequently, the divisors of F of degree 0 form a subgroup $\text{Div}^0(F)$ of $\text{Div}(F)$. We can introduce a partial order on $\text{Div}(F)$ by saying that $D_1 \leq D_2$ if

$$\nu_P(D_1) \leq \nu_P(D_2) \qquad \text{for all } P \in \mathcal{P}_F.$$

A divisor $D \geq 0$ is called *positive* (or *effective*).

If F is the rational function field and $f \in F^*$, then it is obvious that $\nu_P(f) \neq 0$ for at most finitely many $P \in \mathcal{P}_F$. The same is true for an arbitrary algebraic function field F/k. Thus, the following definition makes sense.

Definition 7.6: Let F be an algebraic function field and $f \in F^*$. Then the *principal divisor* $\operatorname{div}(f)$ of f is defined by

$$\operatorname{div}(f) = \sum_{P \in \mathcal{P}_F} \nu_P(f)P.$$

Remark 7.7: If f is in the full constant field k of F and $f \neq 0$, then it follows as in the proof of Lemma 5.2 that $\nu_P(f) = 0$ for all $P \in \mathcal{P}_F$. Therefore we get $\operatorname{div}(f) = 0$. The converse holds also, since it can be shown that for any $f \in F \setminus k$, i.e., for any transcendental f over k, there exists at least one $P \in \mathcal{P}_F$ with $\nu_P(f) \neq 0$ (see [7, Corollary I.1.19]).

If F is the rational function field and $f \in F^*$, then

$$\deg(\operatorname{div}(f)) = \sum_{P \in \mathcal{P}_F} \nu_P(f) \deg(P) = 0$$

by Remark 5.6 and the information about the degree of $P \in \mathcal{P}_F$ in Example 6.7. The same formula holds also for an arbitrary algebraic function field F/k (see [7, Section I.4]).

Proposition 7.8: *The degree of every principal divisor is equal to 0.*

It is easily checked that the set $\operatorname{Princ}(F)$ of principal divisors of F forms a subgroup of $\operatorname{Div}^0(F)$; note e.g. that

$$\operatorname{div}(fg) = \operatorname{div}(f) + \operatorname{div}(g) \qquad \text{for all } f, g \in F^*.$$

The factor group

$$\operatorname{Cl}(F) := \operatorname{Div}^0(F)/\operatorname{Princ}(F)$$

is finite (see [7, Section V.1]) and its cardinality $h(F) := |\operatorname{Cl}(F)|$ is called the *divisor class number* of F.

8. The Riemann-Roch Theorem

F/k is again an algebraic function field with full constant field k (k finite).

For any divisor D of F/k we form the *Riemann-Roch space*

$$\mathcal{L}(D) := \{f \in F^* : \operatorname{div}(f) + D \geq 0\} \cup \{0\}.$$

Explicitly, this means that $\mathcal{L}(D)$ consists of all $f \in F$ with

$$\nu_P(f) \geq -\nu_P(D) \qquad \text{for all } P \in \mathcal{P}_F.$$

Proposition 8.1: $\mathcal{L}(D)$ *is a vector space over* k.

Proof: $\mathcal{L}(D)$ is nonempty since $0 \in \mathcal{L}(D)$. Let $f_1, f_2 \in \mathcal{L}(D)$ and $a \in k$. Then for any $P \in \mathcal{P}_F$,

$$\nu_P(f_1 + f_2) \geq \min(\nu_P(f_1), \nu_P(f_2)) \geq -\nu_P(D)$$

and

$$\nu_P(af_1) = \nu_P(a) + \nu_P(f_1) \geq -\nu_P(D).$$

Thus, $f_1 + f_2 \in \mathcal{L}(D)$ and $af_1 \in \mathcal{L}(D)$. $\qquad\qquad\square$

Example 8.2: For the zero divisor we have $\mathcal{L}(0) = k$. Note that for $f \in F^*$ we have $f \in \mathcal{L}(0) \Leftrightarrow \nu_P(f) \geq 0$ for all $P \in \mathcal{P}_F$. But

$$\deg(\mathrm{div}(f)) = \sum_{P \in \mathcal{P}_F} \nu_P(f) \deg(P) = 0$$

by Proposition 7.8, and so $f \in \mathcal{L}(0) \Leftrightarrow \nu_P(f) = 0$ for all $P \in \mathcal{P}_F \Leftrightarrow f \in k^*$, the latter by Remark 7.7.

Remark 8.3: If $\deg(D) < 0$, then necessarily $\mathcal{L}(D) = \{0\}$. For if we had a nonzero $f \in \mathcal{L}(D)$, then by applying the degree map to

$$\mathrm{div}(f) + D \geq 0$$

we get $0 + \deg(D) \geq 0$, a contradiction.

The vector space $\mathcal{L}(D)$ has, in fact, a finite dimension over k which is denoted by $\ell(D)$. Thus, by the above, we have $\ell(0) = 1$ and $\ell(D) = 0$ if $\deg(D) < 0$. Information on $\ell(D)$ is given in the following fundamental theorem (see [7, Sections I.4 and I.5]).

Theorem 8.4: (Riemann-Roch Theorem) *Let* F/k *be an algebraic function field. Then there exists a constant* c *such that for any divisor* D *of* F *we have*

$$\ell(D) \geq \deg(D) + 1 - c.$$

As a consequence of the Riemann-Roch theorem, we can define the number

$$g = \max_{D \in \mathrm{Div}(F)} (\deg(D) - \ell(D) + 1).$$

The integer $g = g(F)$ is called the *genus* of F and is the most important invariant of an algebraic function field. By putting $D = 0$ in the definition, we see that $g \geq 0$. Note that by definition we have

$$\ell(D) \geq \deg(D) + 1 - g \qquad \text{for all } D \in \mathrm{Div}(F).$$

By the following result (see [7, Section I.5]), we have equality if $\deg(D)$ is sufficiently large.

Theorem 8.5: (Supplement to the Riemann-Roch Theorem) *If* $\deg(D) \geq 2g - 1$, *then*

$$\ell(D) = \deg(D) + 1 - g.$$

Example 8.6: If F is a rational function field, then it is easy to verify that

$$\ell(D) \geq \deg(D) + 1 \qquad \text{for all } D \in \mathrm{Div}(F).$$

Therefore $g(F) = 0$. In fact, rational function fields over finite fields can be characterized by the property of having genus 0.

An algebraic function field of genus 1 is also called an *elliptic function field*. Elliptic function fields F/k with $k = \mathbf{F}_q$ can be characterized (see [7, Section VI.1]). In all cases, F is a quadratic extension of $K := k(x)$. If q is odd, then $F = K(y)$ for some $y \in F$ with

$$y^2 = f(x),$$

where $f \in k[x]$ is squarefree of degree 3. If q is even, then $F = K(y)$ for some $y \in F$ with either

$$y^2 + y = f(x)$$

with $f \in k[x]$ of degree 3 or

$$y^2 + y = x + \frac{1}{ax + b}$$

with $a, b \in k$ and $a \neq 0$.

There is no general explicit formula for the genus of an algebraic function field. However, for certain special families of algebraic function fields, such as in the example below, there is such a formula. In general, the computation of the genus of an algebraic function field is a nontrivial problem.

An important tool for genus computations is the Hurwitz genus formula (see [7, Section III.4]).

Example 8.7: Let $k = \mathbf{F}_q$ with q odd and let $K := k(x)$ be the rational function field. Let $F = K(y)$ be the quadratic extension defined by

$$y^2 = f(x),$$

where $f \in k[x]$ is squarefree of degree $d \geq 1$. Then

$$g(F) = \left\lfloor \frac{d-1}{2} \right\rfloor.$$

A proof can be found in [7, Section III.7].

9. Zeta Function and Hasse-Weil Bound

As usual, F/k is an algebraic function field with full constant field k (k finite). A general reference for the material in this section is [7, Chapter V].

Proposition 9.1: *An algebraic function field F/k has at most finitely many rational places.*

Proof: By Definition 6.1, there exists a $z \in F \setminus k$ such that $[F : k(z)] < \infty$. All rational places of F lie over rational places of $k(z)$ (use the first part of the proof of Proposition 6.3). By Theorem 6.4, for each rational place P of $k(z)$ there are at most $[F : k(z)]$ rational places of F lying over P. Moreover, if $k = \mathbf{F}_q$, then there are only $q + 1$ rational places of $k(z)$ by Example 6.7. Hence the number of rational places of F is at most $(q + 1)[F : k(z)]$. □

Now we can define the zeta function of an algebraic function field $F/\mathbf{F}_q$. For each integer $n \geq 1$, consider the composite field

$$F_n := \mathbf{F}_{q^n} \cdot F.$$

This is an algebraic function field over $\mathbf{F}_{q^n}$ (called a *constant field extension* of F). Let N_n denote the number of rational places of $F_n/\mathbf{F}_{q^n}$, which is a finite number by Proposition 9.1.

Definition 9.2: The *zeta function* of $F/\mathbf{F}_q$ is defined to be the formal power series

$$Z(F, t) = \exp\left(\sum_{n=1}^{\infty} \frac{N_n}{n} t^n \right) \in \mathbf{C}[[t]]$$

over the complex numbers.

Example 9.3: We compute the zeta function of the rational function field F over $\mathbf{F}_q$. We have $N_n = q^n + 1$ for all $n \geq 1$ by Example 6.7. Hence we get

$$\log Z(F,t) = \sum_{n=1}^{\infty} \frac{q^n + 1}{n} t^n = \sum_{n=1}^{\infty} \frac{(qt)^n}{n} + \sum_{n=1}^{\infty} \frac{t^n}{n}$$

$$= -\log(1 - qt) - \log(1 - t)$$

$$= \log \frac{1}{(1-t)(1-qt)},$$

that is,

$$Z(F,t) = \frac{1}{(1-t)(1-qt)}.$$

Theorem 9.4: (Weil Theorem) *Let $F/\mathbf{F}_q$ be an algebraic function field of genus g. Then:*

(i) *$Z(F,t)$ is a rational function of the form*

$$Z(F,t) = \frac{L(F,t)}{(1-t)(1-qt)},$$

where $L(F,t) \in \mathbf{Z}[t]$ is a polynomial of degree $2g$ with $L(F,0) = 1$ and leading coefficient q^g. Moreover, $L(F,1)$ is equal to the divisor class number $h(F)$ of F.

(ii) *Factor $L(F,t)$ into the form*

$$L(F,t) = \prod_{j=1}^{2g} (1 - \omega_j t) \in \mathbf{C}[t].$$

Then $|\omega_j| = q^{1/2}$ for $1 \leq j \leq 2g$.

Theorem 9.5: (Hasse-Weil Bound) *Let $F/\mathbf{F}_q$ be an algebraic function field of genus g. Then the number $N(F)$ of rational places of $F/\mathbf{F}_q$ satisfies*

$$|N(F) - (q+1)| \leq 2g q^{1/2}.$$

Proof: By the definition of $Z(F,t)$ we obtain

$$N(F) = N_1 = \frac{d(\log Z(F,t))}{dt} \bigg|_{t=0}$$

$$= Z'(F,0).$$

On the other hand, by Theorem 9.4(i) we get

$$Z'(F,0) = \left(\frac{L'(F,t)}{L(F,t)} + \frac{1}{1-t} + \frac{q}{1-qt} \right)\bigg|_{t=0}$$
$$= a_1 + 1 + q,$$

where a_1 is the coefficient of t in $L(F,t)$. Comparing coefficients in the identity in Theorem 9.4(ii), we obtain

$$a_1 = -\sum_{j=1}^{2g} \omega_j.$$

Combining the three identities above yields

$$N(F) = q + 1 - \sum_{j=1}^{2g} \omega_j.$$

Therefore

$$|N(F) - (q+1)| = |\sum_{j=1}^{2g} \omega_j| \le \sum_{j=1}^{2g} |\omega_j| = 2gq^{1/2}$$

by Theorem 9.4(ii). $\qquad\square$

Remark 9.6: A refined approach yields the *Serre bound*

$$|N(F) - (q+1)| \le g\lfloor 2q^{1/2} \rfloor.$$

We refer to [3, Section 1.6] and [7, Section V.3] for proofs of this bound.

In particular, we get an upper bound on $N(F)$ depending only on g and q. Hence the following definition makes sense.

Definition 9.7: For a fixed prime power q and an integer $g \ge 0$, let

$$N_q(g) := \max N(F),$$

where the maximum is over all algebraic function fields $F/\mathbf{F}_q$ of genus g.

It is trivial that $N_q(0) = q + 1$. From the Serre bound we get

$$N_q(g) \le q + 1 + g\lfloor 2q^{1/2} \rfloor.$$

If we put

$$A(q) = \limsup_{g \to \infty} \frac{N_q(g)}{g},$$

then we conclude that $A(q) \leq \lfloor 2q^{1/2} \rfloor$ for all q. Vlăduţ and Drinfeld [8] have improved this to

$$A(q) \leq q^{1/2} - 1 \qquad \text{for all } q,$$

and from a result of Ihara [1] it follows that

$$A(q) = q^{1/2} - 1 \qquad \text{for all squares } q.$$

Further information on $A(q)$ can be found in [3, Chapter 5]. The quantity $A(q)$ is very important for applications to algebraic-geometry codes (see [3, Section 6.2] and [7, Section VII.2]).

Acknowledgment

This research was partially supported by the grant POD0103223 with Temasek Laboratories in Singapore.

References

1. Y. Ihara, Some remarks on the number of rational points of algebraic curves over finite fields, *J. Fac. Sci. Univ. Tokyo Sect. IA Math.* **28** (1981), 721–724.
2. R. Lidl and H. Niederreiter, *Introduction to Finite Fields and Their Applications*, rev. ed., Cambridge University Press, Cambridge, 1994.
3. H. Niederreiter and C.P. Xing, *Rational Points on Curves over Finite Fields: Theory and Applications*, Cambridge University Press, Cambridge, 2001.
4. O. Pretzel, *Codes and Algebraic Curves*, Oxford University Press, Oxford, 1998.
5. P. Ribenboim, *The Theory of Classical Valuations*, Springer, New York, 1999.
6. S.A. Stepanov, *Codes on Algebraic Curves*, Kluwer, New York, 1999.
7. H. Stichtenoth, *Algebraic Function Fields and Codes*, Springer, Berlin, 1993.
8. S.G. Vlăduţ and V.G. Drinfeld, Number of points of an algebraic curve, *Funct. Anal. Appl.* **17** (1983), 53–54.
9. E. Weiss, *Algebraic Number Theory*, McGraw-Hill, New York, 1963.

AUTHENTICATION SCHEMES

Dingyi Pei

State Key Laboratory of Information Security
Graduate School of Chinese Academy of Sciences
19(A) Yu Quang Road, Beijing 100039, P. R. China
and
Institute of Information Security, Guangzhou University
Guihuagang, East 1, Guangzhou 510405, P. R. China
E-mail: gztcdpei@scut.edu.cn

These notes introduce optimal authentication schemes, both without and with arbitration. We describe how the construction of optimal authentication schemes is reduced to the construction of certain combinatorial designs. Then based on rational normal curves over finite fields, a new family of optimal authentication schemes with three participants is constructed.

Contents

1. Authentication Model   283
   1.1. Model with three participants   284
   1.2. Model with arbitration   286
2. Authentication Schemes with Three Participants   290
   2.1. Entropy   290
   2.2. Information-theoretic bound   294
   2.3. Optimal authentication schemes   296
3. Optimal Authentication Schemes Based on Rational Normal Curves   300
   3.1. SPBD based on RNC   300
   3.2. New family of optimal authentication schemes   305
   3.3. Encoding rules   306
4. Authentication Schemes with Arbitration   311
   4.1. Lower bounds   312
   4.2. Combinatorial structure of optimal schemes with arbitration   314
References   321

1. Authentication Model

Secrecy and authentication are the two fundamental aspects of information security. Secrecy provides protection for sensitive messages against eavesdropping by an unauthorized person, while authentication provides protection for messages against impersonating and tampering by an active deceiver. Secrecy and authentication are two independent concepts. It is possible that only secrecy, or only authentication, or both are concerned in an information system. We mainly study one of the authentication methods – authentication schemes – in this article and don't touch other authentication methods, like digital signature.

In this section we discuss the authentication models introduced by G.J. Simmons [9, 10].

1.1. *Model with three participants*

We consider the authentication model that involves three participants: a transmitter, a receiver and an opponent. The transmitter wants to communicate a sequence of source states to the receiver. In order to deceive the receiver, the opponent impersonates the transmitter to send a fraudulent message to the receiver, or to tamper the message sent to the receiver. The transmitter and the receiver must act with the common purpose to deal with the spoofing attack from the opponent. They are assumed to trust each other in this model. If the transmitter and the receiver may also cheat each other, it is necessary to introduce the fourth participant – an arbiter. The model with three participants usually is called an A-code and that with an arbiter is called an A^2-code in the literature. We introduce A-codes in this subsection and A^2-codes in the next subsection.

Let S denote the set of all source states which the transmitter may convey to the receiver. In order to protect against attacks from the opponent, source states are encoded using one from a set of encoding rules $\mathcal{E}$ before transmission. Let $\mathcal{M}$ denote the set of all possible encoded messages. Usually the number of encoded messages is much larger than that of source states. An encoding rule $e \in \mathcal{E}$ is a one-to-one mapping from S to $\mathcal{M}$. The range of the mapping $e(S)$ is called the set of valid messages of e, which is a subset of $\mathcal{M}$. Prior to transmission the transmitter and the receiver agree upon an encoding rule e kept secret from the opponent. The transmitter uses e to encode source states. The encoded messages are transmitted through a public insecure channel. When a message is received, the receiver checks whether it lies in the range $e(S)$. If it is the case, then the message is accepted as authentic, otherwise it is rejected. The receiver recovers the source states from the received messages by determining their (unique) preimages under the agreed encoding rule e. We assume that the opponent has a complete understanding of the system including all encoding rules. The only thing he does not know is the particular encoding rule agreed upon by the transmitter and the receiver. The opponent can be successful in his spoofing attack if and only if his fraudulent message is valid for the used encoding rule. The set of valid messages usually is different from rule to rule. In order to decrease the possibility of successful deception from the opponent, the used encoding rule must be alternated frequently.

The three sets $(\mathcal{S}, \mathcal{M}, \mathcal{E})$ form an authentication scheme.

We assume that the opponent has the ability to impersonate the transmitter to send messages to the receiver, or to tamper the messages sent by the transmitter. We speak of a spoofing attack of order r if after observation of the first r $(r \geq 0)$ messages sent by the transmitter using the same encoding rule, the opponent places a fraudulent message into the channel attempting to make the receiver accept it as authentic. It is an impersonation attack when r is zero. Let P_r denote the expected probability of successful deception for an optimum spoofing attack of order r. We are going to find an expression for P_r.

We think of source states, encoded messages and encoding rules as random variables denoted by S, M and E, respectively, i.e. there are probability distributions on the sets $\mathcal{S}, \mathcal{M}$ and $\mathcal{E}$, respectively. Let $p(S = s)$ denote the probability of the event that the variable S takes the value $s \in \mathcal{S}$. Similarly we have $p(E = e)$ and $p(M = m)$ where $e \in \mathcal{E}$ and $m \in \mathcal{M}$. For simplicity, we abbreviate by omitting the names of random variables in a probability distribution when this causes no confusion. For instance, we write $p(s), p(e)$ and $p(m)$ for the above probabilities. Similarly we write the conditional probability $p(e \mid m)$ instead of $p(E = e \mid M = m)$.

For any $m^r = (m_1, m_2, \ldots, m_r) \in \mathcal{M}^r$ and $e \in \mathcal{E}$, let $f_e(m^r) = (f_e(m_1), \ldots, f_e(m_r))$ denote the unique elements $(s_1, \ldots, s_r) \in \mathcal{S}^r$, when $m_i \in e(\mathcal{S})$ for $1 \leq i \leq r$, such that $s_i = f_e(m_i)$ is the pre-image of m_i under e.

We consider only impersonation $(r = 0)$ and plaintext substitution. The latter means that the opponent is considered to be successful only when, after observing a sequence of messages $m_1, \ldots, m_r$, he chooses a fraudulent message m' that is valid for the used encoding rule e, and $m' \neq m_i$ $(1 \leq i \leq r)$. In this case, the receiver is informed of a source state which the transmitter does not intend to convey. If the receiver gets a particular messages twice, he cannot decide whether the message was sent twice by the transmitter or was repeated by an opponent. Hence we assume that the transmitter never sends a particular source state twice under the same encoding rule. Let S^r denote the random variable of the first r source states which the transmitter intends to convey to the receiver. Accordingly we assume that

$$p(S^r = (s_1, \dots, s_r)) > 0 \tag{1}$$

if and only if the components $s_1, \dots, s_r$ are pairwise distinct. The reason for the "if" part will be explained later on.

We assume also that

$$p(e) > 0 \qquad \text{for each } e \in \mathcal{E}, \tag{2}$$

otherwise the encoding rule is never used and can be dismissed from $\mathcal{E}$.

Assume that the random variables E and S^r are independent, i.e.

$$p(e, s^r) = p(e)p(s^r) \qquad \text{for all } e \in \mathcal{E}, \ s^r \in \mathcal{S}^r.$$

Let $m^r = (m_1, \dots, m_r) \in \mathcal{M}^r, m_i \ (1 \le i \le r)$ are distinct, define

$$\mathcal{E}(m^r) = \{e \in \mathcal{E} \mid m_i \in e(S), 1 \le i \le r\}.$$

$\mathcal{E}(m^r)$ may be empty. We require that $\mathcal{E}(m)$ is not empty for each $m \in \mathcal{M}$, otherwise the message is never used and can be dismissed from $\mathcal{M}$.

Let M^r denote the random variable of the first r messages sent by the transmitter. The probability distributions on E and on S^r determine the distribution on M^r. For any $e \in \mathcal{E}$ and $m^r \in \mathcal{M}^r$, if $e \notin \mathcal{E}(m^r)$, then $p(e, m^r) = 0$; if $e \in \mathcal{E}(m^r)$, then

$$p(e, m^r) = p(e)p(f_e(m^r)). \tag{3}$$

Let $P(m \mid m^r)$ be the probability that the message m is valid given that m^r has been observed. Then

$$P(m \mid m^r) = \sum_{e \in \mathcal{E}(m^r * m)} p(e \mid m^r),$$

where $m^r * m$ denotes the message sequence $m_1, \dots, m_r, m$. Given that m^r has observed, the opponent's optimum strategy is to choose the message m' that maximizes $P(m \mid m^r)$. Thus, the unconditional probability of success in an optimum spoofing attack of order r is

$$P_r = \sum_{m^r \in \mathcal{M}^r} p(m^r) \max_{m \in \mathcal{M}} P(m \mid m^r). \tag{4}$$

1.2. *Model with arbitration*

In the authentication model with three participants discussed in the previous subsection, the transmitter and the receiver are assumed trusted, they don't cheat each other. But it is not always the case: the transmitter may deny a message that he has sent, and the receiver may attribute a fraudulent message to the transmitter. A trusted third party, called arbiter, is introduced.

Let S and M denote the set of source states and encoded messages, respectively, as before. We define encoding rules of the transmitter and decoding rules of the receiver as follows. An encoding rule is a one-to-one mapping from S to M. Let $\mathcal{E}_T$ denote the set of all encoding rules. A decoding rule is a mapping from M onto $S \cup \{\text{reject}\}$. Each message corresponds to a source state or to "reject". In the former case the message is called valid for the decoding rule, while in the latter case the message will be rejected by the receiver. The set of all decoding rules is denoted by $\mathcal{E}_R$.

Suppose $f \in \mathcal{E}_R$ is a decoding rule of the receiver and $s \in S$ is a source state, let $M(f, s)$ denote the set of all valid messages for f corresponding to s. The sets $M(f, s) \subset M$ are disjoint for different source states s.

We say that an encoding rule $e \in \mathcal{E}_T$ is valid for the decoding rule $f \in \mathcal{E}_R$ if $e(s) \in M(f, s)$ for any $s \in S$.

Prior to transmission, the receiver selects a decoding rule $f \in \mathcal{E}_R$ and secretly gives it to the arbiter. The arbiter selects one message from $M(f, s)$ for each source state $s \in S$ forming an encoding rule $e \in \mathcal{E}_T$ and secretly gives it to the transmitter to be used. In this case the encoding rule e is valid for the decoding rule f. Receiving a message, the receiver checks whether it is a valid message for f (i.e. it is in some subset $M(f, s)$). If it is, then he accepts it as authentic and recovers the corresponding source state. When the transmitter and the receiver are disputing whether one message has been sent or not sent by the transmitter, the arbiter checks whether the message under dispute is valid for the encoding rule given to the transmitter. If it is valid, then the arbiter thinks that the message was sent by the transmitter since only the transmitter knows the encoding rule. In the opposite case, the arbiter thinks that the message was not sent by the transmitter. Since the receiver does not know how the arbiter constructs the encoding rule e, so it is not easy for him to choose a fraudulent message which is valid for e, i.e. it is not easy for the receiver to attribute a fraudulent message to the transmitter. Similarly, since the transmitter does not know the decoding rule f selected by the receiver, it is hard for the transmitter to choose a fraudulent message which is valid for f but is not valid for e, i.e. it is hard for the transmitter to deny a message that he has sent. As to the opponent, he may succeed in a deception only if he can find a valid message for f.

The following three types of spoofing attacks are considered.

The spoofing attack O_r by the opponent: after observing a sequence of r distinct messages $m_1, m_2, \ldots, m_r$ using the same encoding rule, the opponent sends a message m $(m \neq m_i, 1 \leq i \leq r)$ to the receiver and succeeds if the receiver accepts the message as authentic and the message represents a different source state from those represented by m_i $(1 \leq i \leq r)$.

The spoofing attack R_r by the receiver: after receiving a sequence of r distinct messages $m_1, m_2, \ldots, m_r$, the receiver claims to have received a different message m and succeeds if the message m is valid for the encoding rule used by the transmitter.

The spoofing attack T by the transmitter: the transmitter sends a message to the receiver and then denies having sent it. The transmitter succeeds if this message is accepted by the receiver as authentic and it is not valid for the encoding rule used by the transmitter.

Let P_{O_r}, P_{R_r} and P_T denote the success probability for the optimal spoofing attack of the three kinds defined above, respectively. We are going to find their expressions.

Assume (1) still holds. For a given decoding rule $f \in \mathcal{E}_R$, define

$$\mathcal{M}(f) = \bigcup_{s \in \mathcal{S}} \mathcal{M}(f, s)$$

and

$$\mathcal{E}_T(f) = \{e \in \mathcal{E}_T | e \text{ is valid for } f\}.$$

$\mathcal{M}(f)$ is the set of all valid messages for f. $\mathcal{E}_T(f)$ is the set of all encoding rules which are valid for f.

For a given encoding rule $e \in \mathcal{E}_T$, let

$$\mathcal{M}(e) = \{e(s) \mid s \in \mathcal{S}\} \subset \mathcal{M}$$

and

$$\mathcal{E}_R(e) = \{f \in \mathcal{E}_R | e \text{ is valid for } f\}.$$

For a given $m^r = (m_1, \ldots, m_r) \in \mathcal{M}^r$, define the set

$$\mathcal{E}_R(m^r) = \{f \in \mathcal{E}_R | m_i \in \mathcal{M}(f), f(m_i) \neq f(m_j), 1 \leq i < j \leq r\}.$$

$\mathcal{E}_R(m^r)$ is the set of all decoding rules for which m^r is valid. For any given $f \in \mathcal{E}_R$ define

$$\mathcal{E}_T(f, m^r) = \{e | e \in \mathcal{E}_T(f), m_i \in \mathcal{M}(e), 1 \le i \le r\}.$$

This is the set of all decoding rules for which f and m^r are valid.

Similarly to the previous subsection, let S^r and M^r denote the random variables of the first r source states and the first r encoded messages, respectively, E_R and E_T denote the random variables of decoding rules and encoding rules, respectively. We assume also that $p(E_R = f) > 0$ for any $f \in \mathcal{E}_R$ and $p(E_T = e \mid E_R = f) > 0$ for any $e \in \mathcal{E}_T(f)$. It follows immediately that $p(E_T = e) > 0$ for any $e \in \mathcal{E}_T$ and $p(E_R = f \mid E_T = e) > 0$ for any $f \in \mathcal{E}_R(e)$.

For any message $m \in \mathcal{M}$ we assume that there exists at least one decoding rule f such that $m \in \mathcal{M}(f)$, otherwise m is never used and can be dismissed from $\mathcal{M}$. Similarly, for any message $m \in \mathcal{M}(f)$ there exists at least one encoding rule $e \in \mathcal{E}_T(f)$ such that $m \in \mathcal{M}(e)$, otherwise m can be dismissed from $\mathcal{M}(f)$.

Let $P(m|m^r)$ denote the probability of the event that the message m is acceptable by the receiver, given that the first r messages $m^r = (m_1, m_2, \ldots, m_r)$ have been observed, where $m_1, \ldots, m_r, m$ represent different source states. We have

$$P(m|m^r) = \sum_{f \in \mathcal{E}_R(m^r * m)} p(f|m^r).$$

Similarly to (4) we define

$$P_{O_r} = \sum_{m^r \in M^r} p(m^r) \max_{m \in \mathcal{M}} P(m|m^r). \tag{5}$$

Let $P(m|f, m^r)$ denote the probability of the event that the message m could be valid for the encoding rule used by the transmitter, given the decoding rule f and the first r messages $m^r = (m_1, \ldots, m_r)$. We have

$$P(m|f, m^r) = \sum_{e \in \mathcal{E}_T(f, m^r * m)} p(e|f).$$

We define

$$P_{R_r} = \max_{f \in \mathcal{E}_R} \sum_{m^r \in M^r} p(m^r|f) \max_{m \in \mathcal{M}} P(m|f, m^r)). \tag{6}$$

For a given $f \in \mathcal{E}_R$ and $e \in \mathcal{E}_T(f)$ define

$$\mathcal{M}'(e) = \mathcal{M} \backslash \mathcal{M}(e),$$

$$\mathcal{M}'_f(e) = \mathcal{M}(f) \backslash \mathcal{M}(e) \subset \mathcal{M}'(e).$$

Let $P(m'|e)$ denote the probability of the event that the message $m' \in \mathcal{M}'(e)$ is acceptable by the receiver, given the encoding rule e. We have

$$P(m'|e) = \sum_{f \in \mathcal{E}_R(e,m')} p(f|e),$$

where

$$\mathcal{E}_R(e, m') = \{f | f \in \mathcal{E}_R(e), m' \in \mathcal{M}'_f(e)\}.$$

We define

$$P_T = \max_{e \in \mathcal{E}_T} \max_{m' \in \mathcal{M}'(e)} P(m'|e). \tag{7}$$

2. Authentication Schemes with Three Participants

We study authentication schemes with three participants in this section. When we say authentication schemes, it always means the schemes with three participants in this section. The success probability P_r of optimal spoofing attack of order r was introduced in Subsection 1.1. The information-theoretic lower bound of P_r and the necessary and sufficient conditions for achieving this bound will be given in Subsection 2.2 of this section. In order to discuss the information-theoretic bound, an important concept of information theory – entropy – is introduced in Subsection 2.1 and some often used properties of entropy are proven. One important aim in constructing authentication schemes is to make P_r as small as possible. When P_r $(0 \leq r \leq t-1)$ achieve their information-theoretic bounds and the number of encoding rules achieves also its lower bound, the authentication schemes are called optimal of order t. It is proved in Subsection 2.3 that each optimal authentication scheme corresponds to a strong partially balanced design (SPBD will be defined in Subsection 2.3), and vice versa, each SPBD can be used to construct an optimal authentication scheme. Thus, to construct optimal authentication schemes is reduced to find SPBDs. We will construct a new family of SPBDs, therefore a new family of optimal authentication schemes, in Section 3.

2.1. *Entropy*

Suppose all possible values (or states) of the variable X are $x_1, \ldots, x_n$, the probability of X taking x_i is $p(x_i)$, hence

$$\sum_{i=1}^{n} p(x_i) = 1. \tag{8}$$

We call X a random variable. If $n = 1$, then $p(x_1) = 1$ and X always takes the value x_1. Therefore X is totally determined. If probabilities of X taking x_i $(1 \le i \le n)$ are all equal, i.e. $p(x_1) = p(x_2) = \cdots = p(x_n) = 1/n$, then X is most undetermined. For a general random variable, its indeterminacy is between these two cases. We have a quantity to measure the indeterminacy, it is called entropy. The entropy of a random variable X is defined as

$$H(X) = -\sum_{i=1}^{n} p(x_i) \log p(x_i).$$

Here the base of the logarithm function is 2.

Lemma 2.1: *We have $0 \le H(X) \le \log n$. If X is totally determined, then $H(X) = 0$. If $p(x_1) = \cdots = p(x_n) = 1/n$, then $H(X)$ achieves its maximum $\log n$.*

Proof: It is only necessary to prove the inequality $0 \le H(X) \le \log n$, the other conclusions are trivial. We write p_i for $p(x_i)$. Put

$$F = -\sum_{i=1}^{n} p_i \log p_i - \lambda \sum_{i=1}^{n} p_i.$$

If

$$\frac{\partial F}{\partial p_i} = -\log(e p_i) - \lambda = 0, \qquad i = 1, 2, \ldots, n,$$

(here e is the base of the natural logarithm function), then $p_1, p_2, \ldots, p_n$ are all equal with the value $1/n$. Hence $H(X)$ achieves its maximum value $\log n$ in this case. The conclusion $H(X) \ge 0$ is obvious. $\square$

Based on the above lemma, we can think that the entropy $H(X)$ measures the indeterminacy of X. If each value of X is represented by r bits and X takes any sequence $(a_1, a_2, \ldots, a_r)$ of r bits with equal probability, then $H(X) = \log 2^r = r$. We see that bit can be taken as the unit of entropy. In the above example, the probability of X taking $(a_1, a_2, \ldots, a_r)$ is $1/2^r$, the logarithm $-\log(1/2^r) = r$ can be explained as the amount of information provided when the event takes place. When an event with probability p takes place, the amount of information it provides is $\log p^{-1}$. The entropy $H(X)$ can be explained as expected amount of information provided when X takes a value.

Let X and Y be two random variables. Let $p(x, y)$ be the union probability of $X = x$ and $Y = y$ simultaneously, and $p(y \mid x)$ be the conditional probability of $Y = y$ given $X = x$. We have

$$p(x, y) = p(x)p(y \mid x) = p(y)p(x \mid y). \tag{9}$$

If $p(x \mid y) = p(x)$ for any x and y, then $p(x, y) = p(x)p(y)$. We say X and Y are independent. Define the union entropy of X and Y by

$$H(X, Y) = -\sum_{x,y} p(x, y) \log p(x, y),$$

where the summation runs through all pairs (x, y) such that $p(x, y) > 0$ (in the following all summations have this restriction, we will not mention it again). Define conditional entropy by

$$H(X \mid Y) = -\sum_{x,y} p(x, y) \log p(x \mid y)$$

and

$$H(Y \mid X) = -\sum_{x,y} p(x, y) \log p(y \mid x).$$

$H(X \mid Y)$ denotes the indeterminacy of X when Y is determined. If X and Y are independent, then

$$H(X \mid Y) = -\sum_{x,y} p(x, y) \log p(x)$$

$$= -\sum_{x} \log p(x) \sum_{y} p(x, y)$$

$$= -\sum_{x} p(x) \log p(x) = H(X),$$

similarly we have $H(Y \mid X) = H(Y)$.

Lemma 2.2:

$$H(X, Y) = H(X) + H(Y \mid X) = H(Y) + H(X \mid Y).$$

Proof: It is only necessary to prove the first equality since $H(X, Y) = H(Y, X)$. We have

$$H(X) + H(Y \mid X) = -\sum_{x} p(x) \log p(x) - \sum_{x,y} p(x, y) \log p(y \mid x)$$

$$= -\sum_{x,y} p(x, y) \log \left(p(x)p(y \mid x) \right)$$

$$= -\sum_{x,y} p(x, y) \log p(x, y) = H(X, Y).$$

Here (9) is used. □

If X and Y are independent, it follows from Lemma 2.2 that $H(X,Y) = H(X) + H(Y)$.

Lemma 2.3: (*Jensen inequality*) *Suppose* $x_i > 0, p_i > 0$ $(1 \leq i \leq n)$ *and* $p_1 + p_2 + \cdots + p_n = 1$. *Then*

$$\sum_{i=1}^{n} p_i \log x_i \leq \log \left(\sum_{i=1}^{n} p_i x_i \right).$$

The equality holds if and only if x_i $(1 \leq i \leq n)$ *are all equal.*

Proof: It is equivalent to prove that

$$x_1^{p_1} x_2^{p_2} \cdots x_n^{p_n} \leq p_1 x_1 + p_2 x_2 + \cdots p_n x_n.$$

The left side is the geometric mean value and the right side is the arithmetic mean value. This is a well known inequality. □

Lemma 2.4:

$$H(X \mid Y) \leq H(X).$$

Proof: We have

$$H(X \mid Y) - H(X) = - \sum_{x,y} p(x,y) \log p(x \mid y) + \sum_{x,y} p(x,y) \log p(x)$$

$$= \sum_{x,y} p(x,y) \log \frac{p(x)p(y)}{p(x,y)}.$$

It follows from Lemma 2.3 that

$$H(X \mid Y) - H(x) \leq \log \left(\sum_{x,y} p(x,y) \frac{p(x)p(y)}{p(x,y)} \right) = 0. \qquad \square$$

Lemma 2.4 means that the indeterminacy of X could decrease and could not increase when Y is determined. One may find some information about X from determined Y. The decrease of indeterminacy $H(X) - H(X \mid Y)$ is the lost amount of information.

It was Shannon who studied the theory of secrecy based on the information theory first. This made a milestone in the development of cryptology. Simmons [9] developed an analogous theory for authentication.

2.2. *Information-theoretic bound*

Proposition 2.5: (*Pei* [3]) *The inequality*

$$P_r \geq 2^{H(E|M^{r+1})-H(E|M^r)} \tag{10}$$

holds for any integer $r \geq 0$. *The equality holds iff for any* $m^r * m \in \mathcal{M}^{r+1}$ *with* $\mathcal{E}(m^r * m) \neq \emptyset$, *the ratio*

$$\frac{p(e|m^r)}{p(e|m^r * m)} \tag{11}$$

is independent of m^r, m *and* $e \in \mathcal{E}(m^r * m)$. *When this equality holds, the probability* P_r *equals the above ratio.*

Remark. If $e \in \mathcal{E}(m^r * m)$, then by the assumptions (1), (2) and (3),

$$p(m^r)p(e \mid m^r) = p(e, m^r) = p(e)p(f_e(m^r)) > 0,$$

the numerator and the denominator of the fraction (11) are nonzero. This is the reason of the "if" part of (1).

Proof: Let M_{r+1} denote the random variable of the $(r + 1)$-th message. For a given $m^r = (m_1, \ldots, m_r) \in \mathcal{M}^r$, let

$$\text{supp}(M_{r+1}, E \mid m^r) = \{(m, e) | e \in \mathcal{E}(m^r * m), \ m \neq m_i \ (1 \leq i \leq r)\}$$

denote the support of the conditional probability distribution of the random variable pair (M_{r+1}, E) conditional on $M^r = m^r$. Then underbounding a maximum by an average gives

$$\max_{m \in \mathcal{M}} P(m|m^r) \geq \sum_{m \in \mathcal{M}} p(M_{r+1} = m \mid m^r)P(m|m^r)$$

$$= \sum_{(m,e) \in \text{supp}(M_{r+1}, E \mid m^r)} p(M_{r+1} = m \mid m^r)p(e|m^r)$$

$$= \widetilde{E}\left(\frac{p(M_{r+1} = m \mid m^r)p(e|m^r)}{p(M_{r+1} = m, e \mid m^r)}\right), \tag{12}$$

where $\widetilde{E}$ is the expectation on $\text{supp}(M_{r+1}, E \mid m^r)$. By use of Jensen's inequality, we obtain

$$\log \max_{m \in \mathcal{M}} P(m|m^r) \geq \widetilde{E}\left(\log \frac{p(M_{r+1} = m \mid m^r)p(e|m^r)}{p(r+1 = m, e \mid m^r)}\right)$$

$$= H(M_{r+1}, E|M^r = m^r) - H(M_{r+1}|M^r = m^r) - H(E|M^r = m^r), \tag{13}$$

where

$$H(M_{r+1}, E|M^r = m^r)$$

$$= \sum_{(m,e)\in\text{supp}(M_{r+1},E|m^r)} p(M_{r+1} = m, e|m^r) \log p(M_{r+1} = m, e|m^r),$$

$$H(M_{r+1}|M^r = m^r)$$

$$= - \sum_{m:p(M_{r+1}=m|m^r)>0} p(M_{r+1} = m|m^r) \log p(M_{r+1} = m|m^r),$$

$$H(E|M^r = m^r)$$

$$= - \sum_{e\in\mathcal{E}(m^r)} p(e|m^r) \log p(e|m^r).$$

Finally we make another use of Jensen's inequality to obtain

$$\log P_r = \log \sum_{m^r \in \mathcal{M}^r} p(m^r) \max_{m\in\mathcal{M}} P(m|m^r)$$

$$\geq \sum_{m^r \in \mathcal{M}^r} p(m^r) \log \max_{m\in\mathcal{M}} P(m|m^r)$$

$$\geq H(M_{r+1}, E|M^r) - H(M_{r+1}|M^r) - H(E|M^r)$$

$$= H(E|M^{r+1}) - H(M^r). \tag{14}$$

Lemma 2.2 is used in the last equality.

From the above derivation, we see that equality holds in this bound if and only if the following two conditions are satisfied:

(i) $P(m|m^r)$ is independent of those m and m^r with $p(M_{r+1} = m|m^r) > 0$, so that its average and maximum value coincide in (12).

(ii) In order to get equality in (13), for every $m^r * m \in \mathcal{M}^{r+1}$ with $\mathcal{E}(m^r * m) \neq \emptyset$, the ratio

$$\frac{p(M_{r|1} = m|m^r)p(e|m^r)}{p(M_{r+1} = m, e|m^r)} = \frac{p(e|m^r)}{p(e|m^r * m)}$$

is independent of m, m^r and $e \in \mathcal{E}(m^r * m)$.

Condition (i) can be deduced from condition (ii) since, if $p(M_{r+1} = m|m^r) > 0$, then $\mathcal{E}(m^r * m) \neq \emptyset$ and

$$P(m|m^r) = \sum_{e \in \mathcal{E}(m^r * m)} p(e|m^r)$$

$$= \frac{p(e|m^r)}{p(e|m^r * m)} \sum_{e \in \mathcal{E}(m^r * m)} p(e|m^r * m)$$

$$= \frac{p(e|m^r)}{p(e|m^r * m)}.$$

This completes the proof of Proposition 2.5.   □

The lower bound in Proposition 2.5 is called the information-theoretic bound. Simmons proved it in the case of $r = 0$. The author gave its proof for the general case and found the necessary and sufficient condition for achieving the bound in the form stated in Proposition 2.5. The result was first announced at Asiacrypt'91 (see [3, 8]).

2.3. *Optimal authentication schemes*

By Proposition 2.5 we know that

$$P_0 P_1 \cdots P_{r-1} \geq 2^{H(E|M^r) - H(E)} \geq 2^{-H(E)}. \tag{15}$$

Hence by Lemma 2.1 we have

$$|\mathcal{E}| \geq 2^{H(E)} \geq (P_0 P_1 \cdots P_{r-1})^{-1}. \tag{16}$$

Definition 2.6: An authentication scheme is called *optimal of order t* if its P_r ($0 \leq r \leq t - 1$) achieve their information-theoretic bounds and the number of its encoding rules also achieves its lower bound $(P_0 P_1 \cdots P_{t-1})^{-1}$.

We study the characterization of the optimal authentication schemes and find their construction method in this subsection.

For any positive integer r define

$$\overline{\mathcal{M}^r} = \{m^r \in \mathcal{M}^r \mid \mathcal{E}(m^r) \neq \emptyset\}.$$

We assume that $\overline{\mathcal{M}^1} = \mathcal{M}$, otherwise some messages are never used and can be dismissed. For any $m^r \in \overline{\mathcal{M}^r}$ define

$$\mathcal{M}(m^r) = \{m \in \mathcal{M} \mid \mathcal{E}(m^r * m) \neq \emptyset\}.$$

Corollary 2.7: *Suppose that* $|\mathcal{S}| = k, |\mathcal{M}| = v$, *the positive integer* $t \leq k$, *and*

$$P_r = 2^{H(E|M^{r+1}) - H(E|M^r)}, \quad 0 \leq r \leq t - 1.$$

Then

(i) *For any $m^r \in \overline{\mathcal{M}^r}$ $(1 \le r \le t)$, the probability $p(f_e(m^r))$ is independent of $e \in \mathcal{E}(m^r)$.*

(ii) *For any $m^r * m \in \overline{\mathcal{M}^r}, 1 \le r \le t - 1$, we have*

$$P_0 = \sum_{e \in \mathcal{E}(m)} p(e), \quad P_r = \frac{\sum_{e \in \mathcal{E}(m^r * m)} p(e)}{\sum_{e \in \mathcal{E}(m^r)} p(e)}.$$

Therefore

$$\sum_{e \in \mathcal{E}(m^i)} p(e) = P_0 P_1 \cdots P_{i-1}$$

for any $m^i \in \overline{\mathcal{M}^i}$ $(1 \le i \le t)$.

(iii) $P_0 = k/v$ *and* $|\mathcal{M}(m^r)| = (k - r)P_r^{-1}$ *for any $m^r \in \overline{\mathcal{M}^r}$.*

(iv) $|\overline{\mathcal{M}^i}| = C_k^i (P_0 P_1 \cdots P_{i-1})^{-1}$, $2 \le i \le t$. *(It is trivial that $|\overline{\mathcal{M}^1}| = v$.)*

Proof: (i) If $e \in \mathcal{E}(m^r * m)$, we have (Proposition 2.5)

$$P_r = \frac{p(e \mid m^r)}{p(e \mid m^r * m)}$$

$$= \frac{p(e, m^r)p(m^r * m)}{p(e, m^r * m)p(m^r)}$$

$$= \frac{p(f_e(m^r)) \sum_{e' \in \mathcal{E}(m^r * m)} p(e')p(f_{e'}(m^r * m))}{p(f_e(m^r * m)) \sum_{e' \in \mathcal{E}(m^r)} p(e')p(f_{e'}(m^r))}. \tag{17}$$

Taking $r = 0$ in (17) we obtain

$$P_0 = \frac{\sum_{e' \in \mathcal{E}(m)} p(e')p(f_{e'}(m))}{p(f_e(m))}.$$

It follows that $p(f_e(m))$ does not dependent on $e \in \mathcal{E}(m)$ and properties (i) and (ii) hold for $r = 1$. Then (i) and (ii) can be proved by induction on r.
We have

$$vP_0 = \sum_{m \in \mathcal{M}} \sum_{e \in \mathcal{E}(m)} p(e) = k \sum_{e \in \mathcal{E}} p(e) = k,$$

$$\sum_{m \in \mathcal{M}(m^r)} \sum_{e \in \mathcal{E}(m^r * m)} p(e) = (k - r) \sum_{e \in \mathcal{E}(m^r)} p(e).$$

Thus, (iii) follows from (ii).
Finally, it is trivial that $\overline{\mathcal{M}^1} = v$. By (iii), when $r \ge 2$, we have

$$|\overline{\mathcal{M}^r}| = \frac{1}{r} \sum_{m^{r-1} \in \overline{\mathcal{M}^{r-1}}} |\mathcal{M}(m^{r-1})| = \frac{(k - r - 1)|\overline{\mathcal{M}^{r-1}}|}{rP_{r-1}}.$$

Thus, (iv) follows by induction on r. □

We write $\mathcal{M}(e)$ instead of $e(\mathcal{S})$ for each $e \in \mathcal{E}$ in the following, it is a subset of $\mathcal{M}$ with k elements. Thus, we have a family of k-subsets

$$\{\mathcal{M}(e) \subset \mathcal{M} \mid e \in \mathcal{E}\}. \tag{18}$$

A subset is called a block in the combinatorial design theory also. We will see that the success probability of spoofing attack P_r is determined by the distribution of those blocks of (18) in $\mathcal{M}$ at great extent. For an optimal authentication scheme, this family of blocks must has some special properties. We have to introduce some concepts of combinatorial designs first.

Definition 2.8: Let v, b, k, λ, t be positive integers with $t \leq k$. A *partially balanced design* (PBD) t-$(v, b, k; \lambda, 0)$ is a pair (M, E) where M is a set of v points and E is a family of b subsets of M, each of cardinality k (called blocks) such that any t-subset of M either occurs together in exactly λ blocks or does not occur in any block.

Definition 2.9: If a partially balanced t-design t-$(v, b, k; \lambda_t, 0)$ is also a partially balanced r-design r-$(v, b, k; \lambda_r, 0)$ for $0 \leq r < t$ as well, then it is called a *strong partially balanced t-design* (SPBD) and is denoted by t-$(v, b, k; \lambda_1, \lambda_2, \dots, \lambda_t, 0)$.

If every t-subset of M always occurs in exactly λ blocks, then a t-$(v, b, k; \lambda, 0)$ design is just a t-design t-(v, b, k, λ) which has already been extensively studied in the theory of block designs. The concept of partially balanced t-designs is a generalization of the concept of t-designs. A t-design is always strong.

Usually we assume that any element of $\mathcal{M}$ appears at least in one block $\mathcal{M}(e)$, otherwise the element can be dismissed from $\mathcal{M}$. So we only consider the SPBD which is also a 1-design.

Theorem 2.10: *(Pei [3]) An authentication scheme* $(\mathcal{S}, \mathcal{M}, \mathcal{E})$ *is optimal of order t if and only if the pair*

$$(\mathcal{M}, \{\mathcal{M}(e) \mid e \in \mathcal{E}\}) \tag{19}$$

is an SPBD t-$(v, b, k; \lambda_1, \lambda_2, \dots, \lambda_t, 0)$, *the variable E has uniform distribution and the distributions of S^r $(1 \leq r \leq t)$ satisfy the condition: the probability $p(S^r = f_e(m^r))$ is independent of $e \in \mathcal{E}(m^r)$ for any $m^r \in \overline{\mathcal{M}^r}$. Here*

$$v = |\mathcal{M}|, \quad b = |\mathcal{E}|, \quad k = |\mathcal{S}|,$$
$$\lambda_r = (P_r P_{r+1} \cdots P_{t-1})^{-1} \ (1 \leq r \leq t-1), \quad \lambda_t = 1.$$

Proof: Assume that the authentication scheme $(\mathcal{S}, \mathcal{M}, \mathcal{E})$ is optimal of order t. Since the number of encoding rules achieves its lower bound $(P_0 P_1 \cdots, P_{t-1})^{-1}$, it follows from (15) and (16) that $|\mathcal{E}| = 2^{H(E)}$ and $H(E \mid M^t) = 0$. The former equality means that E has uniform distribution and the latter means that for any $m^t \in \mathcal{M}^t$, $|\mathcal{E}(m^t)| = 0$ or 1. Hence we know that $\lambda_t = 1$.

Since E is uniform, it follows from (ii) of Corollary 2.7 that

$$P_r = \frac{|\mathcal{E}(m^r * m)|}{|\mathcal{E}(m^r)|}, \quad 1 \le r \le t - 1, \tag{20}$$

for any $m^r * m \in \overline{\mathcal{M}^{r+1}}$.

For any $m^r \in \mathcal{M}^r, 1 \le r \le t - 1$, we have $|\mathcal{E}(m^r)| = 0$ or $m^r \in \overline{\mathcal{M}^r}$. In the latter case there exists m^t such that $m^r \subset m^t$. It follows from (20) that

$$|\mathcal{E}(m^r)| = (P_r P_{r+1} \cdots P_{t-1})^{-1}(= \lambda_r), \quad 1 \le r \le t - 1.$$

So far we have proved that the pair (19) is an SPBD with the given parameters. The condition satisfied by S^r $(1 \le r \le t)$ is nothing but the condition (i) of Corollary 2.7. Thus, the conditions given in the theorem are necessary.

Now we show that the conditions are also sufficient. Assume the conditions hold. For any $m^r * m \in \overline{\mathcal{M}^{r+1}}, 0 \le r \le t - 1$, and $e \in \mathcal{E}(m^r * m)$, we have

$$\frac{p(e \mid m^r)}{p(e \mid m^r * m)} = \frac{p(e, m^r)p(m^r * m)}{p(e, m^r * m)p(m^r)}$$

$$= \frac{p(f_e(m^r)) \sum_{e' \in \mathcal{E}(m^r * m)} p(e')p(f_{e'}(m^r * m))}{p(f_e(m^r * m)) \sum_{e' \in \mathcal{E}(m^r)} p(e')p(f_{e'}(m^r))}$$

$$= \frac{|\mathcal{E}(m^r * m)|}{|\mathcal{E}(m^r)|} = \frac{\lambda_{r+1}}{\lambda_r},$$

where $\lambda_0 = b$. The above ratios are constant, hence P_r $(0 \le r \le t - 1)$ achieve their information-theoretic bounds (Proposition 2.5) and

$$P_r = \frac{\lambda_{r+1}}{\lambda_r} \quad 0 \le r \le t - 1. \tag{21}$$

It is obvious that

$$b = \lambda_0 = (P_0 P_1 \cdots P_{t-1})^{-1}.$$

The theorem is proved now. $\qquad\square$

3. Optimal Authentication Schemes Based on Rational Normal Curves

We have shown in the previous section that to construct optimal authentication schemes is reduced to construct SPBDs. We know from the literature that a special kind of combinatorial design, called orthogonal array ([1]), is also an SPBD (with t the strength of the array). Orthogonal arrays can be used to construct optimal Cartesian authentication schemes [12] (Cartesian authentication scheme is a special kind of authentication scheme, we do not explain it here). Some special family of partially balanced incomplete block designs (PBIB [7]) can provide SPBDs with $t = 2$. In this section we construct a new family of SPBDs by means of rational normal curves (RNC) over finite fields. Then we discuss the authentication schemes constructed from the new SPBDs.

3.1. *SPBD based on RNC*

Let F_q be the finite field with q elements and $n \geq 2$ be a positive integer. Let $PG(n, F_q)$ be the projective space of dimension n over F_q. A point of $PG(n, F_q)$ is denoted by $(x_0, \dots, x_n)$, where x_i ($0 \leq i \leq n$) are elements of F_q and x_i are not all zero. If λ is a non-zero element of F_q, then $\lambda(x_0, \dots, x_n)$ and $(x_0, \dots, x_n)$ denote the same point of $PG(n, F_q)$. Let T be a non-singular matrix over F_q of order $n + 1$. The one-to-one transformation of points in $PG(n, F_q)$ defined by

$$PG(n, F_q) \longrightarrow PG(n, F_q)$$
$$(x_0, \dots, x_n) \longmapsto (x_0, \dots, x_n)T$$

is called a projective transformation of $PG(n, F_q)$. The group of all projective transformations of $PG(n, F_q)$ is denoted by $PGL_{n+1}(F_q)$. It is the factor group of the linear group $GL_{n+1}(F_q)$ of order $n+1$ over its subgroup $\{\lambda I_{n+1} \mid \lambda \neq 0\}$, where I_{n+1} is the unit matrix.

We define a curve C in $PG(n, F_q)$ to be the image of the map

$$PG(1, F_q) \longrightarrow PG(n, F_q)$$
$$(x_0, x_1) \longmapsto (x_0^n, x_0^{n-1}x_1, \dots, x_1^n).$$

$PG(1, F_q)$ consists of the following $q + 1$ points:

$$(1, \alpha), \quad \alpha \in F_q; \quad (0, 1).$$

Therefore the curve C consists of the following $q + 1$ points:

$$(1, \alpha, \alpha^2, \dots, \alpha^n), \quad \alpha \in F_q,$$
$$(0, 0, 0, \dots, 0, 1). \tag{22}$$

It is easy to see that the curve C is the common zero locus of the following polynomials in $X_0, \dots, X_n$:

$$X_i^2 - X_{i-1}X_{i+1}, \ (1 \leq i \leq n-1) \tag{23}$$
$$X_1 X_{n-1} - X_0 X_n.$$

We call the image of the curve C under any projective transformation a rational normal curve (RNC). In other words, an RNC is the image of a map

$$PG(1, F_q) \longrightarrow PG(n, F_q)$$
$$(x_0, x_1) \longmapsto (A_0(x_0, x_1), \dots, A_n(x_0, x_1))$$

where $A_0(X_0, X_1), \dots, A_n(X_0, X_1)$ is an arbitrary basis of the space of homogeneous polynomials of degree n in X_0, X_1.

There are $q + 1$ points on each RNC of $PG(n, F_q)$.

Points $p_i = (x_{i0}, \dots, x_{in})$ $(i = 1, \dots, m)$ are called linearly independent (also called in general position) if the rank of the matrix (x_{ij}) $(1 \leq i \leq m, 0 \leq j \leq n)$ is m.

Lemma 3.1: *Any $n + 1$ $(n \leq q)$ points on an RNC in $PG(n, F_q)$ are linearly independent.*

Proof: It is only necessary to prove the lemma for the curve C. By using the Vandermonde determinant

$$\begin{vmatrix} 1 & \alpha_1 & \alpha_1^2 & \cdots & \alpha_1^n \\ 1 & \alpha_2 & \alpha_2^2 & \cdots & \alpha_2^n \\ \cdots & \cdots & \cdots & \cdots & \cdots \\ 1 & \alpha_{n+1} & \alpha_{n+1}^2 & \cdots & \alpha_{n+1}^n \end{vmatrix} = (-1)^{n(n+1)/2} \prod_{1 \leq i \leq j \leq n+1} (\alpha_i - \alpha_j)$$

and

$$\begin{vmatrix} 1 & \alpha_1 & \alpha_1^2 & \cdots & \alpha_1^n \\ 1 & \alpha_2 & \alpha_2^2 & \cdots & \alpha_2^n \\ \cdots & \cdots & \cdots & \cdots & \cdots \\ 0 & 0 & 0 & \cdots & 1 \end{vmatrix} = (-1)^{n(n+1)/2} \prod_{1 \leq i \leq j \leq n} (\alpha_i - \alpha_j),$$

the lemma can be proved easily. $\square$

Lemma 3.2: *Suppose that $q \geq n + 2$. Through any $n + 3$ points in $PG(n, F_q)$, any $n + 1$ points of which are linearly independent, passes a unique RNC.*

Proof: By taking a projective transformation, we may assume that the $n + 3$ given points have the following form:

$$
\begin{aligned}
p_1 &= (1, 0, \ldots, 0), \\
p_2 &= (0, 1, \ldots, 0), \\
&\cdots \\
p_{n+1} &= (0, 0, \ldots, 1), \\
p_{n+2} &= (1, 1, \ldots, 1), \\
p_{n+3} &= (\nu_0, \nu_1, \ldots, \nu_n).
\end{aligned}
\tag{24}
$$

Since any $n + 1$ points among them are linearly independent, we have

$$
\nu_i \quad (0 \leq i \leq n)
$$

are distinct non-zero elements in F_q.

Define a polynomial

$$
G(X_0, X_1) = \prod_{i=0}^{n} (X_0 - \nu_i^{-1} X_1)
$$

in X_0, X_1. The polynomials

$$
H_i(X_0, X_1) = G(X_0, X_1)/(X_0 - \nu_i^{-1} X_1) \quad (0 \leq i \leq n)
$$

form a basis for the space of homogeneous polynomials of degree n, because if there were a linear relation

$$
\sum_{i=0}^{n} a_i H_i(X_0, X_1) = 0, \quad a_i \in F_q,
$$

then substituting $(X_0, X_1) = (1, \nu_i)$ we could deduce that $a_i = 0$ $(0 \leq i \leq n)$. Thus the RNC defined by the map

$$
\sigma: \quad (x_0, x_1) \longmapsto (H_0(x_0, x_1), \ldots, H_n(x_0, x_1)) \tag{25}
$$

passes through the $n + 3$ points p_i $(1 \leq i \leq n + 3)$, since $\sigma(\mu_i, \nu_i) = p_i$ $(1 \leq i \leq n + 1)$, $\sigma(1, 0) = p_{n+2}$ and $\sigma(0, 1) = p_{n+3}$. Conversely, any RNC passing through the points p_i $(1 \leq i \leq n + 3)$ can be written in the form (25). This completes the proof of the lemma. $\square$

Lemma 3.3: *Let $n \geq 2$ be an integer and $q \geq n+2$ be a prime power. The number of RNCs in $PG(n, F_q)$ is*

$$q^{n(n+1)/2-1} \prod_{i=3}^{n+1} (q^i - 1).$$

Proof: We consider all sets, each of which consists of such $n + 3$ points that any $n + 1$ points among them are linearly independent. Let N denote the number of such sets. Suppose $\{p_1, p_2, \ldots, p_{n+3}\}$ is such a set. There are totally $(q^{n+1} - 1)/(q - 1)$ points in $PG(n, F_q)$. The point p_1 can be any point of $PG(n, F_q)$. So there are $(q^{n+1} - 1)/(q - 1)$ ways to choose p_1. After the points $p_1, \ldots, p_i$ $(1 \leq i \leq n)$ have been chosen, the point p_{i+1} may have

$$\frac{q^{n+1} - 1}{q - 1} - \frac{q^i - 1}{q - 1} = \frac{q^i(q^{n+1-i} - 1)}{q - 1}$$

choices, since there are $(q^i - 1)/(q - 1)$ points which are linearly dependent on $p_1, \ldots, p_i$. Now we suppose that the points $p_1, \ldots, p_{n+1}$ have been chosen. By taking a projective transformation, we can assume that the points $p_1, \ldots, p_{n+1}$ are of the form (24). Hence the point p_{n+2} is of the form $(a_0, \ldots, a_n)$ with $a_i \neq 0$ $(0 \leq i \leq n)$, it has

$$\frac{(q - 1)^{n+1}}{q - 1} = (q - 1)^n$$

choices. Finally, if the points $p_1, \ldots, p_{n+2}$ have been chosen, we can also assume that $p_1, \ldots, p_{n+1}$ are of the form in (24). Hence the point p_{n+3} is of the form $(b_0, \ldots, b_n)$ with $b_i \neq 0$, $b_i \neq b_j$ $(0 \leq i < j \leq n)$, it has

$$(q - 2)(q - 3) \cdots (q - n - 1)$$

choices. Therefore we have proved that

$$N = \prod_{i=0}^{n} \frac{q^i(q^{n+1-i} - 1)}{q - 1} \cdot (q - 1)^n \prod_{i=2}^{n+1} (q - i) \cdot ((n + 3)!)^{-1}$$

$$= q^{n(n+1)/2}((n + 3)!)^{-1} \prod_{i=2}^{n+1} (q^i - 1)(q - i).$$

By Lemma 3.2 and the fact that there are $q + 1$ points on every RNC, it follows that the number of RNCs in $PG(n, F_q)$ is

$$\frac{N}{C_{q+1}^{n+3}} = q^{n(n+1)/2-1} \prod_{i=3}^{n+1} (q^i - 1),$$

where C_{q+1}^{n+3} is the binomial coefficient. This completes the proof. $\qquad\square$

Let $\mathcal{M}$ denote the set of all points in $PG(n, F_q)$ and $\mathcal{E}$ the set of all RNCs in $PG(n, F_q)$. The following theorem will prove that the pair $(\mathcal{M}, \mathcal{E})$ is an SPBD.

Theorem 3.4: *(Pei [4]) Let $t \geq 5$ be an integer. For any prime power $q \geq t - 1$ there exists an SPBD t-$(v, b, k; \lambda_1, \ldots, \lambda_t, 0)$, where*

$$v = (q^{t-2} - 1)/(q - 1), \quad k = q + 1,$$

$$b = q^{(t-2)(t-3)/2-1} \prod_{i=3}^{t-2} (q^i - 1),$$

and

$$\lambda_1 = q^{(t-2)(t-3)/2-1} \prod_{i=2}^{t-3} (q^i - 1),$$

$$\lambda_r = q^{(r+t-3)(t-2-r)/2} (q - 1)^{r-2} \prod_{i=1}^{t-2-r} (q^i - 1) \prod_{i=1}^{r-2} (q - i),$$

$$(2 \leq r \leq t - 3)$$

$$\lambda_{t-2} = (q - 1)^{t-4} \prod_{i=1}^{t-4} (q - i),$$

$$\lambda_{t-1} = \prod_{i=2}^{t-3} (q - i).$$

(Here we use the convention that $\prod_{i=1}^{r-2}(q - i) = 1$ if $r = 2$.)

Proof: Take $PG(t-3, F_q)$ as the set of points and all RNCs in $PG(t-3, F_q)$ as blocks. For any given t points of $PG(t-3, F_q)$, if any $t - 2$ points among them are linearly independent, then there exists a unique RNC passing through them by Lemma 3.2, otherwise there is no RNC passing through them by Lemma 3.1. Thus $\lambda_t = 1$.

Given any r $(1 \leq r \leq t - 2)$ points, if they are linearly dependent, then there is no RNC passing through them, and if they are linearly independent, then by the proof of Lemma 3.3 there are λ_r RNCs passing through them, where

$$\lambda_1 = \prod_{i=1}^{t-3} \frac{q^{t-2} - q^i}{q - 1} \cdot (q - 1)^{t-3} \prod_{i=2}^{t-2} (q - i) \prod_{i=0}^{t-2} (q - i)^{-1}$$

$$= q^{(t-2)(t-3)/2-1} \prod_{i=2}^{t-3} (q^i - 1),$$

$$\lambda_r = \prod_{i=r}^{t-3} \frac{q^{t-2} - q^i}{q-1} \cdot (q-1)^{t-3} \prod_{i=2}^{t-2}(q-i) \prod_{i=r-1}^{t-2}(q-i)^{-1}$$

$$= q^{(r+t-3)(t-2-r)/2}(q-1)^{r-2} \prod_{i=1}^{t-2-r}(q^i - 1) \prod_{i=1}^{r-2}(q-i), \quad (2 \le r \le t-3)$$

and

$$\lambda_{t-2} = (q-1)^{t-3} \prod_{i=2}^{t-2}(q-i) \prod_{i=t-3}^{t-2}(q-i)^{-1}$$

$$= (q-1)^{t-4} \prod_{i=1}^{t-4}(q-i).$$

Finally, for any given $t-1$ points, if there exist $t-2$ points among them which are linearly dependent, then there is no RNC passing through them, otherwise the number of RNCs passing through them is λ_{t-1}, where

$$\lambda_{t-1} = \prod_{i=2}^{t-2}(q-i)(q-t+2)^{-1} = \prod_{i=2}^{t-3}(q-i).$$

This completes the proof. □

3.2. *New family of optimal authentication schemes*

Based on the SPBDs constructed in the previous subsection, we can construct a new family of optimal authentication schemes. Let the set of all points on the curve C be the set S of source states and the set of all points in $PG(n, F_q)$ be the set M of encoded messages. The set S can be also understood as the union of all elements of F_q with an extra special element denoted by ∞. Let C^* be any RNC in $PG(n, F_q)$, take one fixed projective transformation T^* which carries C to C^* and define an encoding rule σ_{T^*} which acts on S in the following way:

$$\sigma_{T^*}(\alpha) = (1, \alpha, \dots, \alpha^n)T^*, \quad \alpha \in F_q,$$

$$\sigma_{T^*}(\infty) = (0, \dots, 0, 1)T^*.$$

Thus, we obtain the set of encoding rules $\mathcal{E}$, the number of encoding rules is equal to the number of RNCs in $PG(n, F_q)$, i.e.

$$|\mathcal{E}| = q^{n(n+1)/2-1} \prod_{i=3}^{n+1}(q^i - 1).$$

Prior to transmission the transmitter and receiver randomly choose a matrix T from $\mathcal{E}$. The receiver knows the matrix T and its inverse T^{-1}, therefore he can recover the source state s by calculating

$$\sigma_T(s)T^{-1}. \tag{26}$$

Only when the vector in (26) belongs to the set (22) the receiver accepts $\sigma_T(s)$ as authentic, otherwise he refuses it.

Assume that the random variables E and S^r $(1 \le r \le n+3)$ satisfy the conditions given in Theorem 2.10, then the probabilities P_r $(0 \le r \le n+2)$ of the authentication scheme achieve their lower bounds. Furthermore we have

$$P_0 = \frac{q^2 - 1}{q^{n+1} - 1},$$

$$P_1 = \frac{q - 1}{q^n - 1},$$

$$P_r = \frac{(q-1)(q-r+1)}{q^{n+1} - q^r}, \quad (2 \le r \le n-1)$$

$$P_n = \frac{q - n + 1}{q^n}, \tag{27}$$

$$P_{n+1} = \frac{q - n}{(q-1)^n},$$

$$P_{n+2} = \prod_{i=2}^{n} (q - i)^{-1}.$$

3.3. *Encoding rules*

We consider encoding rules of the authentication schemes constructed n the previous subsection. Let $PO_{n+1}(F_q)$ denote the subgroup of $PGL_{n+1}(F_q)$ consisting of projective transformations which carry the curve C into itself. Two projective transformations T_1 and T_2 carry the curve C into the same RNC if and only if they belong to the same coset of $PO_{n+1}(F_q)$ in $PGL_{n+1}(F_q)$. Thus, we can take any representation system for cosets of $PO_{n+1}(F_q)$ in $PGL_{n+1}(F_q)$ as the set of encoding rules. To find such a representation system is not easy. We will give such a representation system for the case of $t = 5$ and q is odd later. Instead of choosing such a representation system we can also use the following way, which needs a certain amount of calculation, to generate encoding rules. As mentioned above, the transmitter and receiver can determine an encoding rule by randomly choosing a non-singular matrix T of order $n+1$ over F_q. Suppose that the matrices T_1 and T_2 have been randomly chosen successively, it is necessary to check whether T_1 and T_2 belong to the same coset of $PO_{n+1}(F_q)$, i.e., to check whether $T_1 T_2^{-1}$ is an element of $PO_{n+1}(F_q)$. This can be done by checking whether the point set (22) is carried into itself by the projective transformation $T_1 T_2^{-1}$. Particularly, it is easy to see whether the point $(0,\ldots,0,1)T_1 T_2^{-1}$ belongs to the set (22). If it is the case, we can try another matrix T_2 immediately. The probability of the event that $(0,\ldots,0,1)T_1 T_2^{-1}$ belongs to the set (22) is small (see below). In this way we can successively choose several matrices T such that any two of them belong to the different cosets of $PO_{n+1}(F_q)$. This work can be done in advance of the transmission.

Since

$$|PGL_{n+1}(F_q)| = q^{n(n+1)/2} \prod_{i=2}^{n+1}(q^i - 1),$$

we have, by Lemma 3.3, that

$$|PO_{n+1}(F_q)| = \frac{PGL_{n+1}(F_q)}{|E|} = q(q^2 - 1).$$

Therefore the probability of the event that two successively chosen matrices belong to the same coset of $PO_{n+1}(F_q)$ is

$$\frac{|PO_{n+1}(F_q)| - 1}{|PGL_{n+1}(F_q)| - 1} \sim q^{-(n-1)(n+3)}.$$

The probability of the event that the $(n+1)$-th row of $T_1 T_2^{-1}$ is a point of the curve C is

$$\frac{(q+1)(q-1)}{q^{n+1} - 1} \sim q^{-(n-1)}.$$

Let us come back to the problem: to find a representation system for cosets of $PO_{n+1}(F_q)$ in $PGL_{n+1}(F_q)$. At first we give the structure of the group $PO_{n+1}(F_q)$. For $0 \le i, k \le n$ and $t_1, t_2 \in F_q$, define

$$b_{i,k} = \begin{cases} C_k^i, & i \le k, \\ 0, & i > k, \end{cases}$$

and

$$c_{i,k} = \sum_{j=\max(0,i-k)}^{\min(i,n-k)} C_{n-k}^j C_k^{i-j} t_1^{k-(i-j)} t_2^{i-j},$$

where C_k^i is the binomial coefficient. For any elements $a \in F_q^*$ (the set of non-zero elements of F_q) and $t \in F_q$, define two matrices:

$$T_{n+1}(a,t) = (b_{i,k}a^i t^{k-i})_{(0 \leq i,k \leq n)},$$

$$Q_{n+1}(a,t) = (b_{(n-i),k}a^{n-i}t^{k-n+i})_{(0 \leq i,k \leq n)}.$$

For any elements $a \in F_q^*$ and $t_1, t_2 \in F_q$ with $t_1 \neq t_2$, define the matrix

$$R_{n+1}(a,t_1,t_2) = (c_{i,k}a^i)_{(0 \leq i,k \leq n)}.$$

Theorem 3.5: *The group $PO_{n+1}(F_q)$ consists of the following $q(q^2 - 1)$ matrices:*

$$T_{n+1}(a,t), \quad Q_{n+1}(a,t), \quad R_{n+1}(a,t_1,t_2), \tag{28}$$

where $a \in F_q^*$, $t, t_1, t_2 \in F_q$ with $t_1 \neq t_2$. The number of elements for these three types is $q(q-1)$, $q(q-1)$, $q(q-1)^2$, respectively.

Proof: Since

$$|PO_{n+1}(F_q)| = \frac{|PGL_{n+1}(F_q)|}{|E|} = q(q^2 - 1),$$

it is only necessary to show that the set of points (22) is carried into itself by any matrix in (28). The last conclusion of the theorem is obvious.

We have

$$\sum_{i=0}^{n} b_{i,k}a^i t^{k-i} \alpha^i = \sum_{i=0}^{k} C_k^i (a\alpha)^i t^{k-i} = (t + a\alpha)^k,$$

hence

$$(1, \alpha, \ldots, \alpha^n)T_{n+1}(a,t) = (1, (t + a\alpha), \ldots, (t + a\alpha)^n).$$

Also we have

$$(0, \ldots, 0, 1)T_{n+1}(a,t) = (0, \ldots, 0, a^n).$$

This shows that $T_{n+1}(a,t) \in PO_{n+1}(F_q)$.

Similarly we have

$$(1, \alpha, \ldots, \alpha^n)Q_{n+1}(a,t) = \begin{cases} \alpha^n(1, t + a/\alpha, \ldots, (t + a/\alpha)^n), & \alpha \neq 0, \\ (0, \ldots, 0, a^n), & \alpha = 0, \end{cases}$$

and

$$(0, \ldots, 0, 1)Q_{n+1}(a,t) = (1, t, \ldots, t^n).$$

Therefore $Q_{n+1}(a,t) \in PO_{n+1}(F_q)$.

Finally, since

$$\sum_{i=0}^{n} c_{i,k} a^i \alpha^i = \sum_{i=0}^{n} \sum_{j=max(0,i-k)}^{min(i,n-k)} C_{n-k}^{j} C_{k}^{i-j} t_1^{k-(i-j)} t_2^{i-j} (a\alpha)^i$$

$$= \sum_{j=0}^{n-k} C_{n-k}^{j} (a\alpha)^j \sum_{s=0}^{k} C_{k}^{s} t_1^{k-s} (t_2 a\alpha)^s$$

$$= (1 + a\alpha)^{n-k} (t_1 + t_2 a\alpha)^k,$$

then

$$(1, \alpha, \dots, \alpha^n) R_{n+1}(a, t_1, t_2)$$

$$= \begin{cases} (1 + a\alpha)^n (1, \frac{t_1 + t_2 a\alpha}{1+a\alpha}, \cdots, \left(\frac{t_1 + t_2 a\alpha}{1+a\alpha}\right)^n), & 1 + a\alpha \neq 0, \\ (0, \dots, 0, (t_1 - t_2)^n), & 1 + a\alpha = 0. \end{cases}$$

Also

$$(0, \dots, 0, 1) R_{n+1}(a, t_1, t_2) = a^n (1, t_2, \dots, t_2^n).$$

Therefore $R_{n+1}(a, t_1, t_2) \in PO_{n+1}(F_q)$. Thus, the proof of the theorem is now complete. $\square$

The elements of the form $T_{n+1}(a, t)$ form a subgroup of $PO_{n+1}(F_q)$ of order $q(q+1)$. We denote this subgroup by T_{n+1}. A representation system for cosets of T_{n+1} in $PGL_{n+1}(F_q)$ will be given in Theorem 3.6.

Now we assume that $(q, n) = 1$.

Theorem 3.6: *All elements of $PGL_{n+1}(F_q)$ whose last two rows have the form*

$$\begin{array}{l} 0 \cdots 0\ 1\ x_{n-1,v+1} \cdots x_{n-1,u-1}\ 0\ x_{n-1,u+1} \cdots x_{n-1,n} \\ 0 \cdots \qquad\qquad\qquad\qquad 0\quad 1\ x_{n,u+1}\ \cdots\ x_{n,n} \end{array} \quad (v < u)$$

or

$$\begin{array}{l} 0 \cdots 0\ 0 \qquad\quad\cdots\qquad 0\ 1\ x_{n-1,v+1} \cdots x_{n-1,n} \\ 0 \cdots 0\ 1\ x_{n,u+1} \cdots \qquad\qquad\qquad\qquad x_{n,n} \end{array} \quad (u < v)$$

form a representation system for cosets of T_{n+1} in $PGL_{n+1}(F_q)$.

Proof: Let $Y = (y_{i,j})$ be an element of $PGL_{n+1}(F_q)$. The last two rows of the product $T_{n+1}(a, t) \cdot Y$ have the form

$$\begin{array}{ccc} a^{n-1}(y_{n-1,0} + nt y_{n,0}) & \cdots & a^{n-1}(y_{n-1,n} + nt y_{n,n}) \\ a^n y_{n,0} & \cdots & a^n y_{n,n} \end{array}$$

Therefore any two elements, given in the theorem, corresponding to different values of u belong to different cosets. Now suppose that Y and $T_{n+1}(a,t) \cdot Y$ are two elements given in the theorem, corresponding to the same value of u, then it follows that $a = 1$, $t = 0$ (notice that $(n, q) = 1$). Hence $T_{n+1}(a,t) = I$. This shows that any two elements given in the theorem belong to different cosets.

The number of the elements given in the theorem is

$$\frac{q^{n+1} - 1}{q - 1} \cdot \frac{q^n - 1}{q - 1} \cdot \prod_{i=2}^{n} (q^{n+1} - q^i)$$

$$= q^{n(n+1)/2 - 1}(q + 1) \prod_{i=3}^{n+1} (q^i - 1),$$

which is equal to

$$\frac{|PGL_{n+1}(Fq)|}{|T_{n+1}|} = \frac{q^{n(n+1)/2} \prod_{i=2}^{n+1}(q^i - 1)}{q(q - 1)}.$$

This completes the proof of the theorem .   □

The following theorem gives a representation system for $PO_3(F_q)$ in $PGL_3(F_q)$. We omit its proof here.

Theorem 3.7: *The following* $q^2(q^3 - 1)$ *elements form a representation system for* $PO_3(F_q)$ *in* $PGL_3(F_q)$:

(i)
$$\begin{pmatrix} x_{00} & x_{01} & x_{02} \\ 0 & 1 & 0 \\ 0 & 0 & 1 \end{pmatrix}$$
where $x_{00} \neq 0$,

(ii)
$$\begin{pmatrix} x_{00} & x_{01} & x_{02} \\ 1 & x_{11} & 0 \\ 0 & 0 & 1 \end{pmatrix}$$
where $x_{00}x_{11} - x_{01} \neq 0$,

(iii)
$$\begin{pmatrix} x_{00} & x_{01} & x_{02} \\ 0 & 0 & 1 \\ 0 & 1 & x_{22} \end{pmatrix}$$
where $x_{00} \neq 0$,

(iv)
$$\begin{pmatrix} x_{00} & -x_{00}^2 + k & x_{02} \\ 1 & 0 & x_{12} \\ 0 & 1 & x_{22} \end{pmatrix}$$

where k is a nonsquare element in F_q^*, $x_{02} - x_{00}x_{12} + (x_{00}^2 - k)x_{22} \neq 0$, and only one matrix is taken from the following two matrices:

$$\begin{pmatrix} x_{00} & -x_{00}^2 + k & x_{02} \\ 1 & 0 & x_{12} \\ 0 & 1 & x_{22} \end{pmatrix}, \quad \begin{pmatrix} x_{00} & -x_{00}^2 + k & x'_{02} \\ 1 & 0 & x'_{12} \\ 0 & 1 & x'_{22} \end{pmatrix}$$

with the relation

$$\begin{pmatrix} x'_{02} \\ x'_{12} \\ x'_{22} \end{pmatrix} = \begin{pmatrix} x_{00}^2/k & x_{00}(1 - x_{00}^2/k) & k(1 - x_{00}^2/k)^2 \\ 2x_{00}/k & 1 - 2x_{00}^2/k & -2x_{00}(1 - x_{00}^2/k) \\ 1/k & -x_{00}/k & x_{00}^2/k \end{pmatrix} \begin{pmatrix} x_{02} \\ x_{12} \\ x_{22} \end{pmatrix}$$

(v) $\begin{pmatrix} x_{00} & x_{01} & x_{02} \\ 0 & 0 & 1 \\ 1 & x_{21} & x_{22} \end{pmatrix}$ *where* $x_{00} \neq 0$, $x_{01} - x_{00}x_{21} \neq 0$

and only one matrix is taken from the following two matrices:

$$\begin{pmatrix} x_{00} & x_{01} & x_{02} \\ 0 & 0 & 1 \\ 1 & x_{21} & x_{22} \end{pmatrix}, \quad \begin{pmatrix} x_{00} & x_{00}x_{01} & x_{00}x_{02} \\ 0 & 0 & 1 \\ 1 & x_{01}/x_{00} & x_{02}/x_{00} \end{pmatrix}.$$

(vi) $\begin{pmatrix} x_{00} & x_{01} & x_{02} \\ 0 & 1 & x_{12} \\ 1 & x_{21} & x_{22} \end{pmatrix}$ *where* x_{00} *is a nonsquare element in* F_q^*,

there exist no elements r and $s \neq 0$ such that $x_{00} = r^2 - s^2$ and $x_{01} = x_{00}x_{21} - r$, only $q^3(q-1)^2/4$ matrices among them are taken, any two of which belong to different cosets of $PO_3(F_q)$.

The number of matrices in (i), (ii), (iii), (iv), (v) and (vi) is $q^2(q-1)$, $q^3(q-1)$, $q^3(q-1)$, $q^3(q-1)^2/4$, $q^3(q-1)^2/2$ and $q^3(q-1)^2/4$, respectively.

We have not been able to find a representation system for the $q^3(q-1)^2/4$ cosets in (vi). When we use the authentication code, to generate an encoding rule is to choose randomly a matrix among the representation system given in Theorem 3.7. If two successively chosen matrices belong to (vi), we must check whether they belong to the same coset of $PO_3(F_q)$.

4. Authentication Schemes with Arbitration

We have discussed several properties of the authentication schemes with three participants: the information-theoretic bound for the success probability of spoofing attack, the lower bound for the number of encoding rules, the characterization of optimal schemes in Section 2. In this section we will discuss similar problems for the authentication schemes with arbitration.

4.1. *Lower bounds*

By the same method of Proposition 2.5 we can prove the information-theoretic bound for P_{O_r}, P_{R_r} and P_T, respectively, and find the necessary and sufficient conditions for achieving these bounds.

Define the set

$$\mathcal{M}_R^r = \{m^r \in \mathcal{M}^r \mid \mathcal{E}_R(m^r) \neq \emptyset\}.$$

Proposition 4.1: *The inequality*

$$P_{O_r} \geq 2^{H(E_R|M^{r+1})-H(E_R|M^r)} \tag{29}$$

holds for any integer $r \geq 0$. The equality holds iff for any $m^r * m \in \mathcal{M}_R^{r+1}$, the ratio

$$\frac{p(f|m^r)}{p(f|m^r * m)}$$

is independent of m^r, m and $f \in \mathcal{E}_R(m^r * m)$. When this equality holds, the probability P_{O_r} equals $P(m|m^r)$, and the above ratio also.

Proposition 4.2: *The inequality*

$$P_{R_r} \geq 2^{H(E_T|E_R,M^{r+1})-H(E_T|E_R,M^r)} \tag{30}$$

holds for any integer $r \geq 0$. The equality holds iff for any $m^r * m \in \mathcal{M}_R^{r+1}$ and $f \in \mathcal{E}_R(m^r * m)$ with $\mathcal{E}_T(f, m^r * m) \neq \emptyset$ the ratio

$$\frac{p(e|f, m^r)}{p(e|f, m^r * m)}$$

is independent of m^r, m, $f \in \mathcal{E}_R(m^r * m)$ and $e \in \mathcal{E}_T(f, m^r * m)$. When this equality holds, the probability P_{R_r} equals $P(m|f, m^r)$ and the above ratio also.

Proposition 4.3: *The inequality*

$$P_T \geq 2^{H(E_R|E_T,M')-H(E_R|E_T)} \tag{31}$$

holds. The equality in it holds iff for any $e \in \mathcal{E}_T, m' \in \mathcal{M}'(e)$ with $\mathcal{E}_R(e, m') \neq \emptyset$ the ratio

$$\frac{p(f|e)}{p(f|e, m')}$$

is independent of e, m' and $f \in \mathcal{E}_R(e, m')$. When this equality holds, the probability P_T equals $P(m'|e)$ and the above ratio also.

The following two propositions give the lower bounds of $|\mathcal{E}_R|$ and $|\mathcal{E}_T|$, respectively.

Proposition 4.4: *The number of decoding rules of the receiver has the lower bound*

$$|\mathcal{E}_R| \geq (P_{O_0} P_{O_1} \cdots P_{O_{t-1}} P_T)^{-1} \qquad (32)$$

(t is any positive integer). Suppose that P_T and P_{O_r} $(0 \leq r \leq t-1)$ achieve their lower bounds in (31) and (29), then the equality in (32) holds iff

$$H(E_R|E_T, M') = 0, \quad H(E_R|M^t) = H(E_R|E_T)$$

and E_R has a uniform probability distribution.

Proof: According to Propositions 4.1 and 4.3, we have

$$P_{O_0} P_{O_1} \cdots P_{O_{t-1}} P_T \geq 2^{H(E_R|M^t) - H(E_R) + H(E_R|E_T, M') - H(E_R|E_T)}. \qquad (33)$$

Since

$$H(E_R|E_T, M') \geq 0, \qquad (34)$$

$$H(E_R|M^t) \geq H(E_R|E_T) \qquad (35)$$

and

$$|\mathcal{E}_R| \geq 2^{H(E_R)}, \qquad (36)$$

it follows that

$$|\mathcal{E}_R| \geq 2^{H(E_R)} \geq (P_{O_0} P_{O_1} \cdots P_{O_{t-1}} P_T)^{-1}. \qquad (37)$$

When P_{O_r} $(1 \leq r \leq t-1)$ and P_T achieve their lower bounds in (29) and (31) as assumed, the equality in (33) holds. Therefore the equality in (32) holds iff the equalities in (34), (35) and (36) hold. The equality in (36) means that E_R has a uniform probability distribution. $\square$

Proposition 4.5: *The number of encoding rules of the transmitter has the lower bound*

$$|\mathcal{E}_T| \geq (P_{O_0} P_{O_1} \cdots P_{O_{t-1}} P_{R_0} \cdots P_{R_{t-1}})^{-1}. \qquad (38)$$

Suppose $P_{O_r}, P_{R_r} (0 \leq r \leq t-1)$ achieve their lower bounds in (29) and (30), then $|\mathcal{E}_T|$ achieves its lower bound in (38) iff

$$H(E_T|E_R, M^t) = 0, \quad H(E_R|M^t) = H(E_R|E_T)$$

and E_T has a uniform distribution.

The proof of Proposition 4.5 is similar to that of Proposition 4.4.

Remark. Since

$$H(E_T|E_R, M^t) = H(E_R, E_T|M^t) - H(E_R|M^t)$$
$$= H(E_T|M^T) + H(E_R|E_T) - H(E_R|M^t),$$

the condition that $H(E_T|E_R, M^t) = 0$ and $H(E_R|M^t) = H(E_R|E_T)$ is equivalent to that of $H(E_T|M^T) = 0$.

4.2. *Combinatorial structure of optimal schemes with arbitration*

Let us denote an authentication scheme with arbitration by $(\mathcal{S}, \mathcal{M}, \mathcal{E}_R, \mathcal{E}_T)$.

Definition 4.6: If an authentication scheme $(\mathcal{S}, \mathcal{M}, \mathcal{E}_R, \mathcal{E}_T)$ satisfies the following conditions:

(i) $P_{O_r} = 2^{H(E_R|M^{r+1}) - H(E_R|M^r)} \quad 0 \le r \le t - 1$;
(ii) $P_{R_r} = 2^{H(E_T|E_R, M^{r+1}) - H(E_T|E_R, M^r)} \quad 0 \le r \le t - 1$;
(iii) $P_T = 2^{H(E_R|E_T, M') - H(E_R|E_T)}$;
(iv) $|\mathcal{E}_R| = (P_{O_0} P_{O_1} \cdots P_{O_{t-1}} P_T)^{-1}$;
(v) $\mathcal{E}_T| = (P_{O_0} P_{O_1} \cdots P_{O_{t-1}} P_{R_0} \cdots P_{R_{t-1}})^{-1}$;

then the scheme is called *optimal of order t*.

We consider the combinatorial structure of the optimal schemes with arbitration. The following Corollaries 4.7, 4.8 and 4.9 will provide a bridge between the information-theoretic lower bounds of P_{O_r}, P_{R_r} and P_T and the combinatorial structure of optimal schemes.

Corollary 4.7: *Suppose that E_R has a uniform probability distribution and*

$$P_{O_r} = 2^{H(E_R|M^{r+1}) - H(E_R|M^r)}, \quad 0 \le r \le t - 1. \tag{39}$$

Then for any $m^r * m \in \mathcal{M}_R^{r+1}$, we have

$$P_{O_r} = |\mathcal{E}_R(m^r * m)|/|\mathcal{E}_R(m^r)|, \quad 0 \le r \le t - 1.$$

Here $\mathcal{E}_R(m^0) = \mathcal{E}_R$.

Proof: If $\mathcal{R}(m^r * m) \neq \emptyset$, it follows from Proposition 4.1 and the assumption (1) that for any $f \in \mathcal{E}_R(m^r * m)$,

$$P_{O_r} = p(f|m^r)/p(f|m^r * m)$$

$$= p(f, m^r)p(m^r * m)/p(f, m^r * m)p(m^r)$$

$$= p(f) \sum_{e \in \mathcal{E}_T(f, m^r)} p(e|f)p(f(m^r))$$

$$\times \left(p(f) \sum_{e \in \mathcal{E}_T(f, m^r * m)} p(e|f)p(f(m^r)) \right)^{-1}$$

$$\times \sum_{f' \in \mathcal{E}_R(m^r * m)} p(f') \sum_{e' \in \mathcal{E}_T(f', m^r * m)} p(e'|f')p(f'(m^r * m))$$

$$\times \left(\sum_{f' \in \mathcal{E}_R(m^r)} p(f') \sum_{e' \in \mathcal{E}_T(f', m^r)} p(e'|f')p(f'(m^r)) \right)^{-1}. \qquad (40)$$

Now we prove by induction on r that the summation

$$\sum_{e \in \mathcal{E}_T(f, m^r * m)} p(e|f)p(f(m^r * m)), \quad 0 \leq r \leq t - 1, \qquad (41)$$

does not depend on $f \in \mathcal{E}_R(m^r * m)$.

When $r = 0$, we have from (40) that

$$P_{O_0} = p(f)p(m)/p(f, m)$$

$$= \sum_{f' \in \mathcal{E}_R(m)} p(f') \sum_{e \in \mathcal{E}_T(f', m)} p(e|f')p(f'(m)) \Big/ \sum_{e \in \mathcal{E}_T(f, m)} p(e|f)p(f(m)),$$

which does not depend on $f \in \mathcal{E}_R(m)$, therefore the summation in its denominator does not depend on f too. Now we suppose that the summation in (41) does not depend on $f \in \mathcal{E}_R(m^u)$ for $r = u < t - 1$, then it follows also from (40) that

$$P_{O_u} = \sum_{f' \in \mathcal{E}_R(m^u * m)} p(f') \sum_{e \in \mathcal{E}_T(f', m^u * m)} p(e|f')p(f'(m^u * m))$$

$$\times \left(\sum_{e \in \mathcal{E}_T(f, m^u * m)} p(e|f)p(f(m^u * m)) \sum_{f' \in \mathcal{E}_R(m^u)} p(f') \right)^{-1},$$

which does not depend on $f \in \mathcal{E}_R(m^u * m)$. Hence the summation in (41) for $r = u + 1$ does not depend on $f \in \mathcal{E}_R(m^u * m)$.

It follows from (40) that

$$P_{O_r} = \sum_{f \in \mathcal{E}_R(m^r \star m)} p(f) \Big/ \sum_{f \in \mathcal{E}_R(m^r)} p(f) = |\mathcal{E}_R(m^r \star m)|/|\mathcal{E}_R(m^r)|,$$

where the second equality uses the assumption that E_R has a uniform distribution. □

For a given $f \in \mathcal{E}_R$, let $E_T(f)$ denote the random variable of valid encoding rules for f, it takes its values from $\mathcal{E}_T(f)$. Similarly, for a given $e \in \mathcal{E}_T$, let $E_R(e)$ denote the random variable of decoding rules for which e is valid, it takes its values from $\mathcal{E}_R(e)$.

Corollary 4.8: *For any given $f \in \mathcal{E}_R$, suppose that $E_T(f)$ has a uniform probability distribution and*

$$P_{R_r} = 2^{H(E_T|E_R,M^{r+1})-H(E_T|E_R,M^r)}, \ 0 \le r \le t-1. \tag{42}$$

Then for any $m^r \star m \in \mathcal{M}_R^{r+1}$, $f \in \mathcal{E}_R(m^r \star m)$ with $\mathcal{E}_T(f, m^r \star m) \ne \emptyset$,

$$P_{R_r} = |\mathcal{E}_T(f, m^r \star m)|/|\mathcal{E}_T(f, m^r)|.$$

Proof: If $\mathcal{E}_T(f, m^r \star m) \ne \emptyset$, it follows from Proposition 4.2 and the assumption (1) that for any $e \in \mathcal{E}_T(f, m^r \star m)$,

$$\begin{aligned}
P_{R_r} &= p(e|f, m^r)/p(e|f, m^r \star m) \\
&= p(e, f, m^r)p(f, m^r \star m)/p(e, f, m^r \star m)p(f, m^r) \\
&= p(f(m^r)) \sum_{e' \in \mathcal{E}_T(f, m^r \star m)} p(e'|f)p(f(m^r \star m)) \\
&\quad \times \left(p(f(m^r \star m)) \sum_{e' \in \mathcal{E}_T(f, m^r)} p(e'|f)p(f(m^r)) \right)^{-1} \\
&= \sum_{e' \in \mathcal{E}_T(f, m^r \star m)} p(e'|f) \Big/ \sum_{e' \in \mathcal{E}_T(f, m^r)} p(e'|f) \\
&= |\mathcal{E}_T(f, m^r \star m)|/|\mathcal{E}_T(f, m^r)|. \quad \square
\end{aligned}$$

Corollary 4.9: *For any given $e \in \mathcal{E}_T$, suppose that $E_R(e)$ has a uniform probability distribution and*

$$P_T = 2^{H(E_R|E_T,M')-H(E_R|E_T)}. \tag{43}$$

Then for any $e \in \mathcal{E}_T$, $m' \in \mathcal{M}'(e)$ with $\mathcal{E}_R(e, m') \neq \emptyset$,

$$P_T = |\mathcal{E}_R(e, m')|/|\mathcal{E}_R(e)|.$$

Proof: It follows from Proposition 4.3 that for any $e \in \mathcal{E}_T$, $m' \in \mathcal{M}/\mathcal{M}(e)$ with $\mathcal{E}_R(e, m') \neq \emptyset$,

$$P_T = P(m'|e) = \sum_{f \in \mathcal{E}_R(e, m')} p(f|e) = |\mathcal{E}_R(e, m')|/|\mathcal{E}_R(e)|. \qquad \square$$

The optimal schemes have a requirement for the probability distributions of $\mathcal{S}^r$.

Corollary 4.10: *Suppose (39) and (42) hold. Then for any $m^r \in \mathcal{M}_R^r$, the probability $p(f(m^r))$ does not depend on $f \in \mathcal{E}_R(m^r)$ $(1 \leq r \leq t)$.*

Proof: In the proofs of Corollaries 4.7 and 4.8 it is shown that

$$\sum_{e \in \mathcal{E}_T(f, m^r)} p(e|f) p(f(m^r))$$

does not depend on $f \in \mathcal{E}_R(m^r)$ and

$$\sum_{e \in \mathcal{E}_T(f, m^r)} p(e|f) = P_{R_0} P_{R_1} \cdots P_{R_{r-1}}.$$

The conclusion follows immediately. $\qquad \square$

Now we show that for an optimal scheme the number $|\mathcal{M}(f, s)|$ is a constant for any $f \in \mathcal{E}_R$ and $s \in S$.

Corollary 4.11: *If $P_{R_0} = 2^{H(E_T|E_R, M) - H(E_T|E_R)}$, then $|\mathcal{M}(f, s)| = c$ is a constant for any $f \in \mathcal{E}_R$ and $s \in S$. Furthermore $c = P_{R_0}^{-1}$.*

Proof: Given a decoding rule f, we have assumed that for any $m \in \mathcal{M}(f)$ there exists at least one encoding rule e such that $m \in \mathcal{M}(e)$. Hence $P(m|f) > 0$ for any $m \in \mathcal{M}(f)$. If P_{R_0} achieves its information–theoretic lower bound, P_{R_0} equals $P(m|f)$ for any $f \in \mathcal{E}_R$ and $m \in \mathcal{M}(f)$ by Proposition 4.2. Since

$$\sum_{m \in \mathcal{M}(f, s)} P(m|f) = 1$$

for any $s \in S$, it follows that $|\mathcal{M}(f, s)| = P_{R_0}^{-1}$. $\qquad \square$

Definition 4.12: Let v, b, k, c, λ, t be positive integers. A PBD t-design $(\mathcal{M}, \mathcal{E})$ is called *restricted* (RPBD t-design) if $|\mathcal{M}| = v, \mathcal{E} = (\mathcal{E}_1, \cdots, \mathcal{E}_b)$, each block $\mathcal{E}_i$ is divided into k groups, and each group has c points of $\mathcal{M}$. Any t-subset of $\mathcal{M}$ either occurs in exactly λ blocks in such a way that each point of the t-subset occupies one group, or there do not exist such blocks at all.

Similarly we can introduce RSPBD t-$(v, b, k, c; \lambda_1, \lambda_2, \dots, \lambda_t, 0)$. Our main result is the following theorem.

Theorem 4.13: (*Pei and Li* [5, 6]) *The necessary and sufficient conditions for a scheme* $(\mathcal{S}, \mathcal{M}, \mathcal{E}_R, \mathcal{E}_T)$ *being optimal are as follows.*

(i) E_T, E_R, $E_T(f)$ $(\forall f \in \mathcal{E}_R)$ *and* $E_R(e)$ $(\forall e \in \mathcal{E}_T)$ *have uniform probability distribution.*

(ii) *For any given* $m^r \in \mathcal{M}_R^r$, *the probability* $p(\mathcal{S}^r = f(m^r))$ *is constant for all* $f \in \mathcal{E}_R(m^r)$.

(iii) *For any given* $e \in \mathcal{E}_T$ *the pair*

$$(\mathcal{M}'(e), \{\mathcal{M}'_f(e) | f \in \mathcal{E}_R(e)\})$$

is a 1-$(v - k, P_T^{-1}, k(c - 1); 1, 0)$ *design, where* $v = |\mathcal{M}|$, $k = |\mathcal{S}|$, c *is a positive integer (in fact,* $c = |\mathcal{M}(f, s)|$ *for all* $f \in \mathcal{E}_R, s \in \mathcal{S}$*).*

(iv) *For any given* $f \in \mathcal{E}_R$, *the pair*

$$(\mathcal{M}(f), \{\mathcal{M}(e) | e \in \mathcal{E}_T(f)\})$$

is an SPBD t-$(kc, (P_{R_0} P_{R_1} \cdots P_{R_{t-1}})^{-1}, k; \lambda_1, \dots, \lambda_t, 0)$, *where* $\lambda_t = 1, \lambda_r = (P_{R_r} \cdots P_{R_{t-1}})^{-1}, 1 \leq r \leq t - 1$.

(v) *The pair*

$$(\mathcal{M}, \{\mathcal{M}(f) | f \in \mathcal{E}_R\})$$

is an RSPBD t-$(v, (P_{O_0} \cdots P_{O_{t-1}} P_T)^{-1}, k, c; \mu_1, \dots, \mu_t, 0)$, *where*

$$\mu_t = P_T^{-1}, \mu_r = (P_{O_r} \cdots P_{O_{t-1}} P_T)^{-1}, 1 \leq r \leq t - 1.$$

Proof: Suppose the scheme $(\mathcal{S}, \mathcal{M}, \mathcal{E}_R, \mathcal{E}_T)$ is optimal first. We prove the conditions (i)–(v) are satisfied.

According to Propositions 4.4 and 4.5, E_T and E_R have uniform probability distribution. It will be proved that $E_T(f)$ and $E_R(e)$ also have uniform probability distribution later on. The condition (ii) is nothing but Corollary 4.10.

For any given $e \in \mathcal{E}_T$ we have $|\mathcal{M}'(e)| = v - k$ and $|\mathcal{M}'_f(e)| = k(c-1)$ for any $f \in \mathcal{E}_R(e)$ by Corollary 4.11. Since

$$H(E_R|E_T, M') = -\sum p(f, e, m') \log p(f|e, m') = 0$$

by Proposition 4.4, it follows that $\mathcal{E}_R(e, m') = \emptyset$ or $|\mathcal{E}_R(e.m')| = 1$ for any m'. The number of messages m' such that $|\mathcal{E}_R(e, m')| = 1$ is $k(c-1)|\mathcal{E}_R(e)|$. Thus,

$$k(c-1)|\mathcal{E}_R(e)|P_T = \sum_{m':\mathcal{E}_R(e,m')|=1} \sum_{f \in \mathcal{E}_R(e,m')} p(f|e)$$

$$= \sum_{f \in \mathcal{E}_R(e)} p(f|e) \sum_{m' \in \mathcal{M}'_f(e)} 1 = k(c-1).$$

Hence $|\mathcal{E}_R(e)| = P_T^{-1}$. This proves (iii).

The above argument also shows that $p(f|e) = P_T$ for any $f \in \mathcal{E}_R(e)$, i.e. $E_R(e)$ has also a uniform probability distribution. Furthermore, since

$$p(e|f) = \frac{p(e)p(f|e)}{p(f)},$$

$E_T(f)$ has also a uniform probability distribution. Thus, the condition (i) is satisfied.

For any given $f \in \mathcal{E}_R$, we have $|\mathcal{M}(f)| = kc$ and $|\mathcal{M}(e)| = k$ for any $e \in \mathcal{E}_T(f)$. Take any r messages $m^r = (m_1, \ldots, m_r)$ from $\mathcal{M}(f)$, then $m_i \in \mathcal{M}(e)$ $(1 \le i \le r)$ if and only if $e \in \mathcal{E}_T(f, m^r)$. We have $|H(E_T|E_R, M^t)| = 0$ by Proposition 4.5. It follows that $|\mathcal{E}_T(f, m^t)| = 0$ or 1 as above. This means that $\lambda_t = 1$. Suppose that $|\mathcal{E}_T(f, m^t)| = 1$ for some m^t, it follows that

$$|\mathcal{E}_T(f)| = (P_{R_0} \cdots P_{R_{t-1}})^{-1}$$

by Corollary 4.8 (notice that $E_R(f)$ has a uniform probability distribution). If $\mathcal{E}_T(f, m^r) \ne \emptyset$ for some m^r $(1 \le r \le t-1)$, we can also obtain

$$\lambda_r = |\mathcal{E}_T(f, m^r)| = (P_{R_r} \cdots P_{R_{t-1}})^{-1}$$

by Proposition 4.2. This proves (iv).

Take any $m^r = (m_1, \ldots, m_r)$, then $m_i \in \mathcal{M}(f)$ $(1 \le i \le r)$ and each m_i occupies one different $\mathcal{M}(f, s)$ if and only if $f \in \mathcal{E}_R(m^r)$. Since E_R has a uniform probability distribution, it follows from Corollary 4.7 that

$$|\mathcal{E}_R(m^t)| = |\mathcal{E}_R|P_{O_0} \cdots P_{O_{t-1}} = P_T^{-1}$$

when $\mathcal{E}_R(m^t) \ne \emptyset$ for some m^t. This means $\mu_t = P_T^{-1}$. If $E_R(m^r) \ne \emptyset$ for some m^r $(1 \le r \le t-1)$, it also can be shown that

$$\mu_r = |\mathcal{E}_R(m^r)| = (P_{O_r} \cdots P_{O_{t-1}})^{-1}$$

by Corollary 4.7. This proves (v).

Conversely, we suppose the conditions (i)–(v) hold now. We prove the scheme $(\mathcal{S}, \mathcal{M}, \mathcal{E}_R, \mathcal{E}_T)$ is optimal.

Given any $m^r * m \in \mathcal{M}_R^{r+1}$ and $f \in \mathcal{E}_R(m^r * m)$ $(0 \le r \le t-1)$. Using the conditions (i), (ii) and (v), it follows from (40) that

$$\frac{p(f)}{p(f|m)} = \frac{|\mathcal{E}_R(m)|}{|\mathcal{E}_R|} = \frac{\mu_1}{|\mathcal{E}_R|}$$

and

$$\frac{p(f|m^r)}{p(f|m^r * m)} = \frac{|\mathcal{E}_R(m^r * m)|}{|\mathcal{E}_R(m^r)|} = \frac{\mu_{r+1}}{\mu_r} \quad (1 \le r \le t-1).$$

This shows P_{O_r} $(0 \le r \le t-1)$ achieve their lower bounds by Proposition 4.1.

Similarly, we can calculate

$$\frac{p(e|f)}{p(e|f,m)} = \frac{|\mathcal{E}_T(f,m)|}{|\mathcal{E}_T(f)|} = P_{R_0},$$

$$\frac{p(e|f,m^r)}{p(e|f,m^r \star m)} = \frac{|\mathcal{E}_T(f,m^r \star m)|}{|\mathcal{E}_T(f,m^r)|} = \frac{\lambda_{r+1}}{\lambda_r} \quad (1 \le r \le t-1)$$

for any $m^r \star m \in \mathcal{M}_R^{r+1}$, $f \in \mathcal{E}_R(m^r \star m)$ and $e \in \mathcal{E}_T(f,m^r \star m)$, and

$$\sum_{f \in \mathcal{E}_R(e,m')} p(f|e) = \frac{|\mathcal{E}_R(e,m')|}{|\mathcal{E}_R(e)|} = P_T$$

for any $e \in \mathcal{E}_T$, $m' \in \mathcal{M}'(e)$ and $f \in \mathcal{E}_R(e,m')$. This shows that P_{R_r} $(0 \le r \le t-1)$ and P_T achieve their lower bounds by Propositions 4.2 and 4.3.

It follows from (v) that $|\mathcal{E}_R| = (P_{O_0} \cdots P_{O_{t-1}} P_T)^{-1}$, which means $|\mathcal{E}_R|$ achieves its lower bound. We see that $|\mathcal{E}_T(f)| = (P_{R_0} \cdots P_{R_{t-1}})^{-1}$ for any f and $|\mathcal{E}_R(e)| = P_T^{-1}$ for any e by (iv) and (iii), respectively. Hence

$$|\mathcal{E}_T| = \frac{|\mathcal{E}_R||\mathcal{E}_T(f)|}{|\mathcal{E}_R(e)|} = (P_{O_0} \cdots P_{O_{t-1}} P_{R_0} \cdots P_{R_{t-1}})^{-1}.$$

This completes the proof of the theorem. $\qquad\qquad\qquad\qquad\square$

Acknowledgment

These notes were written for tutorial lectures given at the Institute for Mathematical Sciences, National University of Singapore, when the author visited there as a senior member. The author is grateful to the institute for its hospitality.

References

1. T. Beth, D. Jungnickel and H. Lenz. *Design Theory.* Bibliographisches Institut, Mannheim, 1986.
2. J. L. Massey. *Cryptography - a selective survey.* Alta Frequenza, LV(1) (1986), 4–11.
3. D. Pei. *Information-theoretic bounds for authentication codes and block designs.* J. of Cryptology, 8 (1995), 177–188.
4. D. Pei. *A problem of combinatorial designs related to authentication codes.* J. of Combinatorial Designs, 6 (1998), 417–429.
5. D. Pei, Y. Li, Y. Wang and R. Safavi-Naini. *Characterization of authentication codes with arbitration.* Lecture Notes in Computer Science, Vol. 1587, Springer-Verlag, Berlin, 1999, 303–313.
6. D. Pei and Y. Li. *Optimal authentication codes with arbitration.* Preprint.
7. D. K. Ray-Chaudhuri. *Application of the geometry of quadrics for constructing PBIB design.* Annals of Mathematical Statistics, 33 (1962), 1175–1186.
8. R. Safavi-Naini and L. Tombak. *Optimal authentication system.* Lecture Notes in Computer Science, Vol. 765, Springer-Verlag, Berlin, 1994, 12–27.
9. G. J. Simmons. *Authentication theory/coding theory.* Advances in Cryptology – Crypto'84, Lecture Notes in Computer Science, Vol. 196, Springer-Verlag, Berlin, 1985, 411–431.
10. G. J. Simmons. *Message authentication with arbitration of transmitter/receiver disputes.* Advances in Cryptology-Eurocrypt'87, Lecture Notes in Computer Science, Vol. 304, Springer-Verlag, Berlin, 1988, 151–165.
11. G. J. Simmons. *A Cartesian product construction for unconditionally secure authentication codes that permit arbitration.* J. of Cryptology, 2 (1990), 77–104.
12. D. R. Stinson. *The combinatorics of authentication and secrecy codes.* J. of Cryptology, 2 (1990), 23–49.

EXPONENTIAL SUMS IN CODING THEORY, CRYPTOLOGY AND ALGORITHMS

Igor E. Shparlinski

Department of Computing, Macquarie University
Sydney, NSW 2109, Australia
E-mail: igor@ics.mq.edu.au

1. Introduction

In these lecture notes we will try to exhibit, in a very informal way, some useful and sometimes surprising relations between exponential sums, which are a celebrated tool in analytical number theory, and several important problems of such applied areas as coding theory, cryptology and algorithms.

One can certainly ask two natural questions:

- *Why Exponential Sums?*
  This is because:
  - they are beautiful and I like them;
  - exponential sums allow us to show the existence of objects with some special properties.
- *Why Coding Theory, Cryptology and Algorithms?*
  This is because:
  - they are beautiful and I like them as well;
  - to design/analyze some codes and cryptographic schemes we need to find objects with some special properties:
    * *"good"* for designs;
    * *"bad"* for attacks.

The main goal of this work is to show that exponential sums are very useful, yet user-friendly objects, provided you know how to approach them.

I will also provide a necessary background for everybody who would like to learn about this powerful tool and to be able to use it in her and his own work. I do not pretend to give a systematic introduction to the subject, but rather I intend to help to get started in making exponential sums an active working tool, at least in the situation where their application does not require any sophisticated technique or advanced analytical methods. I hope that this brief introduction to the theory of exponential sums and their applications should help to develop some feeling of the kinds of questions where exponential sums can be useful, and if you see that the actual application is beyond your level of expertise you can always seek advice from one of the numerous experts in number theory (who probably otherwise would never know about your problem).

It is well known that for many years number theory was the main area of applications of exponential sums. Such applications include (but are not limited to):

- Uniform distribution (H. Weyl);
- Additive problems such as the Goldbach and Waring problems (G. H. Hardy, J. E. Littlewood, R. Vaughan, I. M. Vinogradov);
- Riemann zeta function and distribution of prime numbers (J. E. Littlewood, N. M. Korobov, Yu. V. Linnik, E. C. Titchmarsh, I. M. Vinogradov).

However, it has turned out that exponential sums provide a valuable tool for a variety of problems of theoretical computer science, coding theory and cryptography, see [88, 89].

I will try to explain:

- What we call exponential sums.
- How we estimate exponential sums (and why we need this at all).
- What is the current state of affairs.
- What kind of questions can be answered with exponential sums.
- How various cryptographic and coding theory problems lead to questions about exponential sums.

Unfortunately, there is no systematic textbook on exponential sums. However, one can find a variety of results and applications of exponential sums in [44, 62, 52, 88, 100].

Although many sophisticated (and not so) methods and applications of exponential sums are not even mentioned in this work, I still hope that it can prepare the reader to start independent explorations of this beautiful area and maybe even try some open problems, new or old, as well as to look for new applications. In particular, a little set of tutorial problems at the end of the notes (a few of them contain some hints) may help in a smooth transition from learning to pursuing independent research.

As a rule, the choice of examples to demonstrate various methods of estimation and applications of exponential sums has been limited to ones admitting a straightforward approach, exhibiting main ideas without gory technical details. The only opposite example is the result on BCH codes in Section 7.2. It has been included to show that even with exponential sums "life is not always easy" (other examples can somehow lead to this false conclusion) and also to show one very useful trick which is discussed in Section 7.2.4.

We remark that there is one more important area of application of exponential sums which unfortunately is not considered in these notes. Namely, we do not discuss applications to pseudo-random number generators; this topic is too extensive and requires a separate treatment. We recommend however to consult [75, 76, 77] to get some impression how the area has been developing.

Acknowledgment

I would like to thank Harald Niederreiter for the very careful reading of the manuscript and the numerous helpful suggestions. Also, without his constant help and encouragement these lecture notes would have never appeared in their present form and would just remain to be merely a set of slides. I am certainly thankful to San Ling, Chaoping Xing and other colleagues involved in the organisation of this workshop, for their invitation and for the opportunity to give these lectures. I am also thankful to Arnaldo Garcia and Alev Topuzoglu who invited me to repeat a slightly extended version of the original lectures at IMPA (Rio de Janeiro) and Sabanci University (Istanbul). Last but not least, I would like to express my deepest gratitude to the great audience of these lectures, whose active participation and curiosity, asking "simple" and "hard" questions, made it a very enjoyable experience for me.

2. Exponential Sums — Basic Notions

2.1. *Getting started*

2.1.1. *Exponential sums — What are they?*

Exponential sums are objects of the form

$$S(\mathcal{X}, F) = \sum_{x \in \mathcal{X}} \mathbf{e}(F(x))$$

where
$$\mathbf{e}(z) = \exp(2\pi i z),$$
$\mathcal{X}$ is an arbitrary finite set, F is a real-valued function on $\mathcal{X}$.

In fact $\mathcal{X}$ could be a set of vectors, in this case we talk about **multiple sums**.

2.1.2. *Exponential sums — What do we want from them?*

Certainly it would be very good to have a closed form expression for the sums $S(\mathcal{X}, F)$. Unfortunately, there are very few examples when we have such formulas. On the other hand, for main applications of exponential sums we do not need to know $S(\mathcal{X}, F)$ exactly. It is quite enough to have an **upper bound** on $S(\mathcal{X}, F)$, which is the main task of this area.

First of all we remark that because $|\mathbf{e}(z)| = 1$ for every real z,
$$|S(\mathcal{X}, F)| \leq \#\mathcal{X}.$$
This is the **trivial bound**.

We are interested in getting stronger bounds. Of course, to be able to prove such a bound we need some conditions on $\mathcal{X}$ and F. For example, if F is an integer-valued function, then $\mathbf{e}(F(x)) = 1$ and $S(\mathcal{X}, F) = \#\mathcal{X}$.

2.1.3. *Exponential sums — How do we classify them?*

There are exponentially many different types of exponential sums.

If $\mathcal{X}$ is a set of vectors, we talk about **multiple sums**. In particular, in the two-dimensional case we talk about **double sums**. Double sum technique provides an invaluable tool in estimating one-dimensional sums.

Very often the set $\mathcal{X}$ is a subset of the elements of a finite field $\mathbb{F}_q$ of q elements or a residue ring $\mathbb{Z}_m$ modulo m.

Accordingly, a very important class of exponential sums consists of **rational sums**. These are the sums with functions F of the form $F(x) = f(x)/m$ where $f : \mathcal{X} \to \mathbb{Z}$ is an integer-valued function on $\mathcal{X}$. The positive integer m is called the **denominator** of the exponential sum $S(\mathcal{X}, F)$.

It is convenient to introduce one more notation
$$\mathbf{e}_m(z) = \exp(2\pi i z/m)$$
(thus $\mathbf{e}_1(z) = \mathbf{e}(z)$). Therefore we have
$$S(\mathcal{X}, F) = \sum_{x \in \mathcal{X}} \mathbf{e}_m(f(x)).$$

2.2. Timeline

Exponential sums are almost 200 years old. It is a long history of triumphs and disappointments. Below I try to outline some most important events of this dramatic history. It is certainly impossible to give a complete account of all achievements and contributors within the framework of a few lectures, so I do apologise for all omissions of many distinguished events and researchers.

2.2.1. Johann Carl Friedrich Gauss, 1801

Exponential sums were introduced to number theory by **Gauss** in [29]. The sums he introduced and studied

$$G(a, m) = \sum_{x=0}^{m-1} e_m(ax^2)$$

are called "Gaussian sums" in his honor. Sometimes this name is extended to more general sums

$$G_n(a, m) = \sum_{x=0}^{m-1} e_m(ax^n)$$

as well. Gaussian sums $G(a, m)$ is one of very few examples when one can actually evaluate exponential sums explicitly. It should be noticed that the way Gauss used these sums is very different from modern applications of exponential sums.

2.2.2. Hermann Klaus Hugo Weyl, 1916

Hermann Weyl was probably the first mathematician who understood the great power and potential of this method. Besides creating the first general method of bounding exponential sums [105], he also found very important connections with uniform distribution of sequences which underlie many further applications of this method.

2.2.3. Godfrey Harold Hardy and John Edensor Littlewood, 1920

Godfrey Hardy and John Littlewood [33] found new applications of exponential sums to some very important number-theoretic problems and invented their "circle method" which is now routinely used for a large number of applications [100]. John Littlewood [63] also introduced exponential sums in studying the Riemann zeta function.

2.2.4. *Louis Joel Mordell, 1932*

Louis Mordell [68] created a new method of estimating rational exponential sums with polynomials with prime denominator. Although the method is obsolete and superseded by the André Weil method [104], it exhibited some very important principles and it has not lost its value as a teaching tool in the theory of exponential sums.

2.2.5. *Ivan Matveevich Vinogradov, 1935*

Ivan Vinogradov developed a principally new method of estimating general exponential sums with polynomials with irrational coefficients [102] (much stronger than H. Weyl's method) and also the method of bounding exponential sums where the set $\mathcal{X}$ consists of prime numbers of a certain interval [103]. He obtained extremely strong results for such classical problems as the *Waring problem* and the *Goldbach problem* and the bounds for the zeros of the Riemann zeta function. Even now, 65 years later we do not have anything essentially stronger.

2.2.6. *Loo-Keng Hua, 1940*

Loo-Keng Hua [43] created a new method of estimating rational exponential sums with arbitrary denominator. The method is based on the Chinese Remainder Theorem to reduce the general case to the case of prime power denominator, and then using a kind of Hensel lifting to reduce the case of prime power denominator to the case of prime denominator. Almost all works on exponential sums with arbitrary denominator follow this pattern.

2.2.7. *André Weil, 1948*

André Weil [104] invented an algebraic-geometry method of estimating "rational" exponential sums with prime denominator. In many cases the results are close to best possible. It still remains the most powerful tool in this area.

2.2.8. *Pierre Deligne, 1974*

Pierre Deligne [22] has obtained a very important extension of the algebraic-geometry method to bounds on multiple sums with polynomials and rational functions with prime denominator.

2.2.9. *You, ????*

There also have been many other exceptional researchers and outstanding results and methods, but no " breakthroughs". An excellent outline of older results is given by Loo-Keng Hua [44]. Maybe its **your** turn now! The area deserves your attention.

2.3. *Some terminology*

2.3.1. *Rational exponential sums*

We concentrate on the simplest, yet most useful, well-studied and attractive class of **rational exponential sums**. That is, the function $F(x) = f(x)/m$ takes rational values with integer denominator $m > 1$.

In fact, very often we concentrate only on the case of prime denominators. Sometimes it is convenient to think that $f(x)$ is defined on elements of the finite field $\mathbb{F}_p$ of p elements.

Examples:

- $F(x) = f(x)/p$ where f is a polynomial with integer coefficients (alternatively one can think that f is a polynomial with coefficients from $\mathbb{F}_p$);
- $F(x) = g^x/p$ where $g > 1$ is an integer (alternatively one can think that $g \in \mathbb{F}_p$).

2.3.2. *Complete and incomplete exponential sums*

Very often the function $f(x)$ in $F(x) = f(x)/m$ is purely periodic modulo m with period T. Then the sum

$$S(f) = \sum_{x=1}^{T} \mathbf{e}_m(f(x))$$

is called a **complete sum**.

A shorter sum

$$S(f, N) = \sum_{x=1}^{N} \mathbf{e}_m(f(x))$$

with $1 \leq N \leq T$ is called an **incomplete sum**.

Examples:

- If $f(x)$ is a polynomial with integer coefficients, then it is periodic modulo p with period p;

- $f(x) = g^x$ where $g > 1$ is an integer with $\gcd(g, p) = 1$, then it is periodic modulo p with period t where t is the multiplicative order of g modulo p.

Typically, incomplete sums (especially when N is relatively small compared to T) are much harder to estimate.

3. Simplest Bounds and Applications

3.1. *The basic case — Linear sums*

Certainly the simplest (and easiest) exponential sums one can think of are **linear exponential sums**, that is, exponential sums with

$$F(x) = ax/m.$$

The following simple result gives a complete description of such sums (a very unusual situation ...). It provides a very good warming up exercise.

Theorem 3.1:

$$\sum_{x=0}^{m-1} \mathbf{e}_m(ax) = \begin{cases} 0, & \text{if } a \not\equiv 0 \pmod{m}, \\ m, & \text{if } a \equiv 0 \pmod{m}. \end{cases}$$

Proof: The case $a \equiv 0 \pmod{m}$ is obvious because each term is equal to 1.

The case $a \not\equiv 0 \pmod{m}$ ... is obvious as well, because it is a sum of a geometric progression with quotient $q = \mathbf{e}_m(a) \neq 1$, thus

$$\sum_{x=0}^{m-1} \mathbf{e}_m(ax) = \sum_{x=0}^{m-1} q^x = \frac{q^m - 1}{q - 1} = \frac{\mathbf{e}_m(ma) - 1}{\mathbf{e}_m(a) - 1} = \frac{1 - 1}{\mathbf{e}_m(a) - 1} = 0. \qquad \square$$

Although this result is very simple, it has proved to be an invaluable tool for many applications of exponential sums. In particular, we repeatedly use it in these notes. In fact, this fact is not so surprising if one thinks of the exponential sum of Theorem 3.1 as the characteristic function of the numbers which are divisible by m.

3.2. *Nice result almost for free*

The following statement is a very instructive example showing the great power of the exponential sum method. The result is a rather **nontrivial** statement which follows immediately from **trivial** Theorem 3.1. In fact, I am not aware of any alternative proof of this statement whose formulation has nothing to do with exponential sums.

Let $\mathcal{X}$ be **any** finite subset of $\mathbb{Z}$ and let f be a function $f : \mathcal{X} \to \mathbb{F}_p$. Let $N_k(a)$ be the number of solutions of

$$f(x_1) + \ldots + f(x_k) \equiv f(x_{k+1}) + \ldots + f(x_{2k}) + a \pmod{p}.$$

where $x_1, \ldots, x_{2k} \in \mathcal{X}$ and a is an integer.

Theorem 3.2: $N_k(a) \le N_k(0)$.

Proof: By Theorem 3.1

$$N_k(a) = \sum_{x_1, \ldots, x_{2k} \in \mathcal{X}} \frac{1}{p} \sum_{c=0}^{p-1} \mathbf{e}_p \Big(c \big(f(x_1) + \cdots + f(x_k) \\ - f(x_{k+1}) - \ldots - f(x_{2k}) - a \big) \Big).$$

Rearranging,

$$N_k(a) = \frac{1}{p} \sum_{c=0}^{p-1} \mathbf{e}_p(-ca) \left(\sum_{x \in \mathcal{X}} \mathbf{e}_p \left(cf(x) \right) \right)^k \left(\sum_{x \in \mathcal{X}} \mathbf{e}_p \left(-cf(x) \right) \right)^k.$$

Because for any real u,

$$\mathbf{e}_p(-u) = \overline{\mathbf{e}_p(u)}$$

and for any complex z,

$$z\bar{z} = |z|^2,$$

we obtain

$$N_k(a) = \frac{1}{p} \sum_{c=0}^{p-1} \mathbf{e}_p(-ca) \left| \sum_{x \in \mathcal{X}} \mathbf{e}_p \left(cf(x) \right) \right|^{2k}$$

$$\le \frac{1}{p} \sum_{c=0}^{p-1} \left| \sum_{x \in \mathcal{X}} \mathbf{e}_p \left(cf(x) \right) \right|^{2k} = N_k(0). \qquad \square$$

It is obvious that

$$\sum_{a=0}^{p-1} N_k(a) = \#\mathcal{X}^{2k}.$$

Indeed, any $2k$-tuple $(x_1, \ldots, x_{2k}) \in \mathcal{X}^{2k}$ corresponds to one and only one congruence and will be counted exactly once.

Using Theorem 3.2 and the previous observation, we immediately obtain the following inequality:

$$N_k(0) \geq \frac{1}{p}\sum_{a=0}^{p-1} N_k(a) = \frac{\#\mathcal{X}^{2k}}{p}.$$

As we have seen, Theorem 3.2 follows from the explicit expression of $N_k(a)$ via exponential sums. It also gives a lower bound on $N_k(0)$. Now we show that having some extra information about exponential sums involved in this expression, one can show that all values of $N_k(a)$ are close to their expected value $\#\mathcal{X}^{2k}/p$.

In the formula

$$N_k(a) = \frac{1}{p}\sum_{c=0}^{p-1} \mathbf{e}_p(-ca) \left| \sum_{x\in\mathcal{X}} \mathbf{e}_p\left(cf(x)\right)\right|^{2k}$$

the term corresponding to $c = 0$ is $\#\mathcal{X}^{2k}/p$. Assume that we know a **nontrivial** upper bound

$$\max_{1\leq c\leq p-1} \left|\sum_{x\in\mathcal{X}} \mathbf{e}_p\left(cf(x)\right)\right| \leq \#\mathcal{X}\Delta$$

with some $0 \leq \Delta < 1$. Then each of the other $p-1$ terms is at most $\#\mathcal{X}^{2k}\Delta^{2k}$. Therefore

$$\left|N_k(a) - \frac{\#\mathcal{X}^{2k}}{p}\right| \leq \#\mathcal{X}^{2k}\Delta^{2k}.$$

For some k we get $\Delta^{2k} < p^{-1}$ and we have an asymptotic formula.

The smaller the value of Δ, the smaller the value of k that is needed. If $\Delta = p^{-\delta}$, one can take $k = \lfloor 1/2\delta \rfloor + 1$.

　　Moral:

(1) The **expected value** of $N_k(a)$ is given by the term corresponding to $c = 0$.
(2) The **error term** depends on the quality of our bound on exponential sums.

3.3. *Gaussian sums*

Here we show that the absolute value of Gaussian sums can be explicitly evaluated. We consider only the case of prime denominators, but the arguments can easily be carried over to arbitrary denominators (although the final formula needs some adjustments). So our purpose is to evaluate the absolute value of

$$G(a, p) = \sum_{x=1}^{p} \mathbf{e}_p(ax^2)$$

where p is prime

Theorem 3.3: *For any prime $p \geq 3$ and any integer a with $\gcd(a, p) = 1$,*

$$|G(a, p)| = p^{1/2}.$$

Proof: We have

$$\begin{aligned}
|G(a)|^2 &= \sum_{x,y=1}^{p} \mathbf{e}_p\left(a\left(x^2 - y^2\right)\right) \\
&= \sum_{y=1}^{p}\sum_{x=1}^{p} \mathbf{e}_p\left(a\left((x+y)^2 - y^2\right)\right) \\
&= \sum_{y=1}^{p}\sum_{x=1}^{p} \mathbf{e}_p\left(a\left(x^2 + 2xy\right)\right) \\
&= \sum_{x=1}^{p} \mathbf{e}_p\left(ax^2\right) \sum_{y=1}^{p} \mathbf{e}_p\left(2axy\right).
\end{aligned}$$

Because $p \geq 3$ and $\gcd(a, p) = 1$, from Theorem 3.1 we see that the last sum vanishes unless $x = p$ in which case it is equal to p and $\mathbf{e}_p\left(ax^2\right) = \mathbf{e}_p\left(ap^2\right) = 1$. $\square$

Let us make a very important observation that for any polynomial $f(x)$ of degree n, squaring the sum with $\mathbf{e}_p(f(x))$ leads to a sum with $\mathbf{e}_p(f(x+y) - f(y))$ which, for every x, is a polynomial in y of degree $n-1$. The procedure can be iterated until we arrived at linear sums. This is essentially the method of H. Weyl [105].

3.4. *Linear sums once again*

In Theorem 3.1 the argument x runs through the whole least residue system $\mathbb{Z}_m$ modulo m. A natural question to ask is: what if we take shorter sums

$$T_a(h) = \sum_{x=0}^{h-1} \mathbf{e}_m(ax)$$

with $1 \leq h \leq m - 1$?

If $1 \leq a \leq m - 1$, it is still the sum of a geometric progression with quotient $q = \mathbf{e}_m(a) \neq 1$, thus

$$|T_a(h)| = \left| \frac{q^h - 1}{q - 1} \right| \leq \frac{2}{|q - 1|}.$$

We have

$$|q - 1| = |\mathbf{e}_m(a) - 1| = |\exp(\pi i a/m) - \exp(-\pi i a/m)|$$
$$= 2|\sin(\pi a/m)|.$$

Put $b = \min\{a, m - a\}$. Then

$$|\sin(\pi a/m)| = |\sin(\pi b/m)| \geq \frac{2b}{m}$$

because $\sin(\alpha) \geq 2\alpha/\pi$ for $0 \leq \alpha \leq \pi/2$.

Therefore

$$|T_a(h)| \leq \frac{m}{2 \min\{a, m - a\}} \tag{1}$$

for $1 \leq a \leq m - 1$.

This immediately implies:

Theorem 3.4:

$$\sum_{a=1}^{m-1} \left| \sum_{x=k}^{k+h-1} \mathbf{e}_m(ax) \right| = O(m \log m).$$

Proof: We have

$$\left| \sum_{x=k}^{k+h-1} \mathbf{e}_m(ax) \right| = \left| \mathbf{e}_m(ak) \sum_{x=0}^{h-1} \mathbf{e}_m(ax) \right| \leq \frac{m}{2 \min\{a, m - a\}}.$$

Therefore

$$\sum_{a=1}^{m-1} \left| \sum_{x=k}^{k+h-1} \mathbf{e}_m(ax) \right| \leq m \sum_{a=1}^{m-1} \frac{1}{2 \min\{a, m - a\}} \leq 2m \sum_{1 \leq a \leq m/2} \frac{1}{2a}$$

and the result follows. $\qquad\square$

3.5. *Distribution of functions modulo p*

Here we obtain the first general results illustrating how exponential sums can be used to gain some information about the distribution of functions modulo p.

Another interpretation of this result is a statement about the uniformity of distribution of the fractional parts

$$\left\{ \frac{f(x)}{p} \right\}, \quad x \in \mathcal{X},$$

in the unit interval $[0, 1]$.

Let k and $1 \le h \le p$ be integers. Denote

$$N_f(k, h) = \# \left\{ x \in \mathcal{X} \, : \, f(x) \equiv v \pmod{p}, \; v \in [k, k+h-1] \right\}.$$

Theorem 3.5: *If*

$$\max_{1 \le c < p} \left| \sum_{x \in \mathcal{X}} \mathbf{e}_p\left(cf(x)\right) \right| \le \#\mathcal{X}\Delta,$$

then

$$\max_{k} \max_{1 \le h \le p} \left| N_f(k, h) - \frac{\#\mathcal{X}h}{p} \right| = O(\#\mathcal{X}\Delta \log p).$$

Proof: We have

$$N_f(k, h) = \sum_{x \in \mathcal{X}} \sum_{v=k}^{k+h-1} \frac{1}{p} \sum_{c=0}^{p-1} \mathbf{e}_p\left(c(f(x) - v)\right)$$

$$= \frac{1}{p} \sum_{c=0}^{p-1} \sum_{x \in \mathcal{X}} \sum_{v=k}^{k+h-1} \mathbf{e}_p\left(-cv\right) \mathbf{e}_p\left(cf(x)\right)$$

$$= \frac{\#\mathcal{X}h}{p} + \frac{1}{p} \sum_{c=1}^{p-1} \sum_{v=k}^{k+h-1} \mathbf{e}_p\left(-cv\right) \sum_{x \in \mathcal{X}} \mathbf{e}_p\left(cf(x)\right).$$

Therefore

$$\left| N_f(k, h) - \frac{\#\mathcal{X}h}{p} \right| \le \frac{1}{p} \sum_{c=1}^{p-1} \left| \sum_{v=k}^{k+h-1} \mathbf{e}_p\left(-cv\right) \right| \left| \sum_{x \in \mathcal{X}} \mathbf{e}_p\left(cf(x)\right) \right|$$

$$= O\left(\#\mathcal{X}\Delta p^{-1} \sum_{c=1}^{p-1} \left| \sum_{v=k}^{k+h-1} \mathbf{e}_p\left(cv\right) \right| \right)$$

$$= O\left(\#\mathcal{X}\Delta \log p\right). \qquad \square$$

4. More Sophisticated Methods

4.1. *Extend and conquer*

Here we show that sometimes it is profitable to **extend** our sum over a small set of arbitrary structure to a bigger set (just potentially increasing the size of the sum) with a nice well-studied structure. Certainly we cannot do this with the original sum because the terms are complex numbers, but this idea can be combined with some tricks. Very often it is used together with the Cauchy inequality in the form

$$\left(\sum_{j=1}^{m} s_j\right)^2 \leq m \sum_{j=1}^{m} s_j^2$$

which holds for any non-negative $s_1, \dots, s_m$.

We demonstrate this principle on the following very important example. Let $\mathcal{X}$ and $\mathcal{Y}$ be arbitrary subsets of $\mathbb{F}_p$.

Define

$$W_c = \sum_{x \in \mathcal{X}} \sum_{y \in \mathcal{Y}} \mathbf{e}_p(cxy)$$

Trivially $|W_c| \leq \#\mathcal{X}\#\mathcal{Y}$. We show that very simple arguments allow us to obtain a bound which is better than trivial for $\#\mathcal{X}\#\mathcal{Y} \geq p$. Thus this bound improves the trivial bound for *very sparse* sets of arbitrary structure!

Theorem 4.1: *For any sets $\mathcal{X}, \mathcal{Y} \subseteq \mathbb{F}_p$ and $1 \leq c < p$,*

$$|W_c| \leq (\#\mathcal{X}\#\mathcal{Y}p)^{1/2}.$$

Proof: We have

$$|W_c| = \left|\sum_{x \in \mathcal{X}} \sum_{y \in \mathcal{Y}} \mathbf{e}_p(cxy)\right| \leq \sum_{x \in \mathcal{X}} \left|\sum_{y \in \mathcal{Y}} \mathbf{e}_p(cxy)\right|.$$

From the Cauchy inequality,

$$|W_c|^2 \leq \#\mathcal{X} \sum_{x \in \mathcal{X}} \left|\sum_{y \in \mathcal{Y}} \mathbf{e}_p(cxy)\right|^2.$$

We **extend** the sums over x to all $x \in \mathbb{F}_p$:

$$|W_c|^2 \leq \#\mathcal{X} \sum_{x \in \mathbb{F}_p} \left|\sum_{y \in \mathcal{Y}} \mathbf{e}_p(cxy)\right|^2.$$

This is a very important step! We add many more terms to our sums (which we can do because each term is nonnegative). Of course we lose here, but our gain is that the sum over x (taken from some mysterious set we have no information about) is now extended to a very nice set.

Now we *Conquer*:

$$\sum_{x\in\mathbb{F}_p}\left|\sum_{y\in\mathcal{Y}}\mathbf{e}_p(cxy)\right|^2 = \sum_{x\in\mathbb{F}_p}\sum_{y_1,y_2\in\mathcal{Y}}\mathbf{e}_p\left(cx\left(y_1-y_2\right)\right)$$

$$= \sum_{y_1,y_2\in\mathcal{Y}}\sum_{x\in\mathbb{F}_p}\mathbf{e}_p\left(cx\left(y_1-y_2\right)\right)$$

$$= p\sum_{\substack{y_1,y_2\in\mathcal{Y}\\y_1=y_2}}1 = \#\mathcal{Y}p.$$

$\square$

Without any assumptions on $\mathcal{X}$ and $\mathcal{Y}$ this bound remains the best possible.

4.2. *Clone, extend and conquer*

The previous principle works for double sums. Here we show how we can create multiple **clones** of our sum and thus reduce it to a double sum.

As in the previous section we use a very important example to exhibit this principle.

Let g, $\gcd(g,p)=1$, be of multiplicative order t modulo p, that is,

$$g^k \equiv 1 \pmod{p} \quad \Longrightarrow \quad k \equiv 0 \pmod{t}.$$

Define

$$S(a,b) = \sum_{x=1}^{t}\mathbf{e}_p\left(ag^x\right)\mathbf{e}_t(bx).$$

The term $\mathbf{e}_t(bx)$ is rather unattractive (and unnatural), but we will see soon why it is needed for some applications, see Theorem 4.3.

Trivially, $|S(a,b)| \leq t$.

Theorem 4.2: *For any a,b with $\gcd(a,p)=1$,*

$$|S(a,b)| \leq p^{1/2}.$$

Proof: The function $e_p(ag^x)e_t(bx)$ is periodic with period t. Thus, for $y = 1, \ldots, t$,

$$S(a, b) = \sum_{x=1}^{t} e_p\left(ag^{x+y}\right) e_t(b(x + y))$$

$$= e_t(by) \sum_{x=1}^{t} e_p\left(ag^y g^x\right) e_t(bx)$$

$$= e_t(by)S(ag^y, b).$$

Therefore, we can *clone*:

$$|S(a, b)| = |S(ag^y, b)|.$$

Now we *extend*:

$$t|S(a, b)|^2 = \sum_{y=1}^{t} |S(ag^y, b)|^2 \leq \sum_{c=0}^{p-1} |S(c, b)|^2.$$

Finally, we *conquer*:

$$t|S(a, b)|^2 \leq \sum_{c=0}^{p-1} |S(c, b)|^2$$

$$= \sum_{x_1, x_2 = 1}^{t} e_t(b(x_1 - x_2)) \sum_{c=0}^{p-1} e_p\left(c\left(g^{x_1} - g^{x_2}\right)\right)$$

$$= tp$$

because

$$g^{x_1} - g^{x_2} \equiv 0 \pmod{p}$$

if and only if

$$x_1 \equiv x_2 \pmod{t}. \qquad \square$$

For some values of t this bound remains the best possible, see also Theorem 5.2.

4.3. *Mordell's bound*

We are now ready to prove something more complicated and less straightforward than our previous estimates.

For a polynomial $f \in \mathbb{F}_p[X]$ of degree $\deg f = n$ we define

$$S(f) = \sum_{x=0}^{p-1} e_p(f(x)).$$

Without loss of generality we can assume that $f(0) = 0$.
Mordell's method follows the following 3 main stages.

Stage I. Cloning: For $\lambda \in \mathbb{F}_p^*$, $\mu \in \mathbb{F}_p$, define

$$f_{\lambda,\mu}(x) = f(\lambda x + \mu) - f(\mu).$$

Obviously $S(f) = S(f_{\lambda,\mu})$ (because $x \to \lambda x + \mu$ is a permutation on $\mathbb{F}_p$).

Stage II. Extending: The leading coefficient of $f_{\lambda,\mu}$ is $A\lambda^n$ where $A \neq 0$ is the leading coefficient of f. There are at least $p(p-1)/n$ distinct polynomials $f_{\lambda,\mu}$:

$$\frac{p(p-1)}{n}|S(f)|^{2n} \leq \sum_{\substack{\deg g \leq n \\ g(0)=0}} |S(g)|^{2n}.$$

Stage III. Conquering: Finally we obtain

$$\sum_{\substack{\deg g \leq n \\ g(0)=0}} |S(g)|^{2n} = \sum_{\substack{\deg g \leq n \\ g(0)=0}} S(g)^n \overline{S(g)}^n = \sum_{\substack{\deg g \leq n \\ g(0)=0}} S(g)^n S(-g)^n$$

$$= \sum_{\substack{\deg g \leq n \\ g(0)=0}} \sum_{x_1,\ldots,x_{2n}=0}^{p-1} e_p\left(\sum_{\nu=1}^{n} g(x_\nu) - \sum_{\nu=n+1}^{2n} g(x_\nu)\right)$$

$$= \sum_{x_1,\ldots,x_{2n}=0}^{p-1} \sum_{a_1,\ldots,a_n=0}^{p-1} e_p\left(\sum_{j=1}^{n} a_j \left(\sum_{\nu=1}^{n} x_\nu^j - \sum_{\nu=n+1}^{2n} x_\nu^j\right)\right)$$

$$= \sum_{x_1,\ldots,x_{2n}=0}^{p-1} \prod_{j=1}^{n} \sum_{a_j=0}^{p-1} e_p\left(a_j\left(\sum_{\nu=1}^{n} x_\nu^j - \sum_{\nu=n+1}^{2n} x_\nu^j\right)\right)$$

$$= p^n T,$$

where T is the number of solutions of

$$\sum_{\nu=1}^{n} x_\nu^j \equiv \sum_{\nu=n+1}^{2n} x_\nu^j \pmod{p}, \qquad j = 1,\ldots,n,$$

where $0 \leq x_1,\ldots,x_{2n} \leq p-1$.

The first n symmetric functions of $x_1, \dots, x_n$ and $x_{n+1}, \dots, x_{2n}$ are the same. Recalling the Newton formulas we see that they are roots of the same polynomial of degree n. Therefore they are permutations of each other.

There are p^n values for $x_1, \dots, x_n$ and for each fixed values of $x_1, \dots, x_n$ there are at most $n!$ values for the other n variables $x_{n+1}, \dots, x_{2n}$. Therefore

$$T \le n! p^n.$$

This yields

$$|S(f)| \le c(n) p^{1-1/n}$$

where $c(n) = (n\, n!)^{1/2n} \approx (n/e)^{1/2}$.

4.4. Shorter sums ... but large bound

Here we show a general principle how the problem of bounding incomplete sums reduces to the problem of bounding *almost the same* complete sums. Unfortunately, we lose a little bit, the bound becomes bigger by a logarithmic factor.

For g, $\gcd(g,p) = 1$, of multiplicative order t modulo p, define **incomplete** sums

$$T(a; N) = \sum_{x=1}^{N} \mathbf{e}_p\left(ag^x\right).$$

Theorem 4.3: *For any a with $\gcd(a,p) = 1$ and $1 \le N \le t$,*

$$|T(a; N)| = O(p^{1/2} \log p).$$

Proof: We have

$$|T(a; N)| = \left| \sum_{x=1}^{t} \mathbf{e}_p\left(ag^x\right) \frac{1}{t} \sum_{b=0}^{t-1} \sum_{y=1}^{N} \mathbf{e}_t(b(x - y)) \right|$$

$$= \frac{1}{t} \left| \sum_{b=0}^{t-1} S(a,b) \sum_{y=1}^{N} \mathbf{e}_t(-by) \right|$$

$$\le \frac{1}{t} \sum_{b=0}^{t-1} |S(a,b)| \left| \sum_{y=1}^{N} \mathbf{e}_t(-by) \right|$$

$$\le \frac{p^{1/2}}{t} \sum_{b=0}^{t-1} \left| \sum_{y=1}^{N} \mathbf{e}_t(-by) \right| = O(p^{1/2} \log p)$$

by Theorem 4.2 and Lemma 3.4.   □

5. Some Strongest Known Results

5.1. *Weil's kingdom*

Using algebraic geometry tools due to André Weil [104] (an upper bound for the number of solutions of equations $F(x, y) = 0$ in finite fields), one can prove much stronger bounds for various sums with

- polynomials;
- rational functions;
- algebraic functions.

Here we present only one of such bounds in the following form given by C. Moreno and O. Moreno.

Theorem 5.1: *For any polynomials $g(X), h(X) \in \mathbb{F}_p[X]$ such that the rational function $f(X) = h(X)/g(X)$ is not constant on $\mathbb{F}_p$, the bound*

$$\left| \sum_{\substack{x \in \mathbb{F}_p \\ g(x) \neq 0}} \mathbf{e}_p\left(f(x)\right) \right| \leq (\max\{\deg g, \deg h\} + r - 2)\, p^{1/2} + \delta$$

holds, where

$$(r, \delta) = \begin{cases} (v, 1), & \text{if } \deg h \leq \deg g, \\ (v + 1, 0), & \text{if } \deg h > \deg g, \end{cases}$$

and v is the number of distinct zeros of $g(X)$ in the algebraic closure of $\mathbb{F}_p$.

In the special case when $f(X)$ is a nonconstant polynomial of degree $\deg f = n$, the bound takes its well-known form

$$\left| \sum_{x \in \mathbb{F}_p} \mathbf{e}_p\left(f(x)\right) \right| \leq (n - 1)\, p^{1/2}. \tag{2}$$

Nowadays we have a purely elementary alternative to the algebraic geometry method which is due to S. A. Stepanov, N. M. Korobov, H. Stark, W. Schmidt and to several other researchers.

Surprisingly enough, in some special cases elementary methods give much stronger results. Such improvements are due to A. Garcia and F. Voloch, D. Mit'kin, R. Heath-Brown and S. V. Konyagin, for more details see [34].

It is important to remember that

$$\boxed{\text{``elementary''} \neq \text{``simple''}}$$

"Elementary" merely means that there is no explicit use of any algebraic geometry notions and tools.

For multivariate polynomials an analogue of (2) is due to P. Deligne [22], but it requires some special conditions on the polynomial in the exponent which are not so easy to verify. This limits the range of applications of that bound, while the Weil bound (2) is very easy to apply.

5.2. *Exponential functions*

Exponential functions form another natural family of functions which arise in many applications. The problem of estimating exponential sums with exponential functions has a long history, we refer to [52, 53, 54, 62, 75, 76, 88] for more details.

Using some improvements of the Weil bound due to R. Heath-Brown and S. V. Konyagin [34], one can improve Theorem 4.2. Namely, the following result has been obtained by S. V. Konyagin and I. E. Shparlinski [52], Theorem 3.4.

Theorem 5.2: *For any a, b with $\gcd(a, p) = 1$,*

$$|S(a,b)| \leq \begin{cases} p^{1/2}, & \text{if } t \geq p^{2/3}; \\ p^{1/4}t^{3/8}, & \text{if } p^{2/3} > t \geq p^{1/2}; \\ p^{1/8}t^{5/8}, & \text{if } p^{1/2} > t \geq p^{1/3}; \end{cases}$$

holds.

The main challenge is to obtain nontrivial bounds for as small values of t as possible. Theorem 5.2 works only for $t \geq p^{1/3+\varepsilon}$. For almost all primes Theorem 5.5 of [52] provides a nontrivial bound for $t \geq p^{\varepsilon}$. We present it in the form given in [70].

Theorem 5.3: *Let Q be a sufficiently large integer. For any $\varepsilon > 0$ there exists $\delta > 0$ such that for all primes $p \in [Q, 2Q]$, except at most $Q^{5/6+\varepsilon}$ of them, and any element $g_{p,T} \in \mathbb{F}_p$ of multiplicative order $T \geq p^{\varepsilon}$ the bound*

$$\max_{\gcd(c,p)=1} \left| \sum_{x=0}^{T-1} \mathbf{e}_p\left(cg_{p,T}^x\right) \right| \leq T^{1-\delta}$$

holds.

5.3. *More applications*

Combining the Weil bound (2) and Theorem 3.5, we obtain that for any polynomial f of degree n,

$$\max_k \max_{0 \le h \le p-1} |N_f(k,h) - h| = O(np^{1/2} \log p). \tag{3}$$

We recall that a number $a \not\equiv 0 \pmod{p}$ is called a *quadratic residue* if the congruence $a \equiv x^2 \pmod{p}$ has a solution and it is called a *quadratic non-residue* otherwise. Numbers a with $a \equiv 0 \pmod{p}$ do not belong to either of these two classes.

Using (3) for the quadratic polynomial $f(x) = x^2$, we see that in any interval $[k, k+h-1]$ the imbalance between the number of quadratic residues modulo p and non-residues is at most $O(p^{1/2} \log p)$. This is the famous *Polya–Vinogradov* inequality.

More precisely, let us denote by $Q_+(k,h)$ and $Q_-(k,h)$ the numbers of quadratic residues and non-residues, respectively, in the interval $[k, k+h-1]$.

Theorem 5.4: *The bound*

$$\max_k \max_{0 \le h \le p-1} \left| Q_\pm(k,h) - \frac{h}{2} \right| = O(p^{1/2} \log p)$$

holds.

Proof: Because the residue ring modulo p is a field, we see that if $a \not\equiv 0 \pmod{p}$ and the congruence $a \equiv x^2 \pmod{p}$ has a solution, then it has two distinct solutions. Taking into account that an interval $[k, k+h-1]$ with $0 \le h \le p-1$ contains at most one zero, we obtain the inequalities

$$\frac{1}{2}N_f(k,h) - 1 \le Q_+(k,h) \le \frac{1}{2}N_f(k,h)$$

and

$$h - 1 \le Q_+(k,h) + Q_-(k,h) \le h.$$

Using (3) we obtain the desired result. $\qquad\square$

In fact, our proof of Theorem 5.4 does not really need the Weil bound; it is quite enough to use Theorem 3.3.

Similarly, Theorem 5.2 and Theorem 5.3 can be used to study the distribution of the values of g^x in short intervals, see [52, 88, 89] for numerous applications of this type of result to cryptography, coding theory and computer science.

5.4. *What else can we estimate?*

There are several other classes of exponential sums which have attracted much of attention of experts in analytical number theory. Here we present a short outline of such classes.

- Exponential sums with composite denominator

$$S(f) = \sum_{x=0}^{q-1} \mathbf{e}_q(f(x)),$$

  where $q \geq 1$ is an integer, $f \in \mathbb{Z}[X]$. These sums are very well studied, thanks to works of Hua Loo Keng, Vasili Nechaev, Sergei Stečkin, see [43, 44, 97]. Basically, these results use that both *Chinese Remainder Theorem* and *Hensel Lifting* have analogues for exponential sums. Thus, after some transformations and manipulations, sums with composite denominators can be reduced to sums with prime denominators. Unfortunately, the final result loses a lot of power on its way from the Weil bound (2).

- Exponential sums with recurring sequences. For linear recurring sequences such estimates are due to N. M. Korobov and H. Niederreiter, see [62, 54, 75, 76, 88]. For nonlinear recurring sequences such estimates are due to H. Niederreiter and I. E. Shparlinski, see [77].

- *H. Weyl, J. van der Corput, I. M Vinogradov, N. M. Korobov:* sums with polynomials with irrational coefficients ... *not much progress since 1947.* Generally, the theory is somewhat similar to that of rational sums. For example, an analogue of Theorem 3.1 is the identity

$$\int_0^1 \mathbf{e}(a\alpha)d\alpha = \begin{cases} 0, & \text{if } a \neq 0, \\ 1, & \text{if } a = 0. \end{cases}$$

An analogue of (1) is the inequality

$$\left| \sum_{x=0}^{h-1} \mathbf{e}(\alpha x) \right| \leq \frac{1}{2\|\alpha\|}$$

which holds for any non-integer α, where $\|\alpha\|$ is the distance from α to the closest integer. Many other properties of rational sums are preserved as well. However, having said that, let us stress that typically such sums are much harder to estimate, in particular due to many technical complications.

- It is easy to see that $\mathbf{e}_p(\cdot)$ is an additive character of $\mathbb{F}_p$. Similar results are known for additive and multiplicative characters of arbitrary finite

fields and residue rings. Although usually for sums of multiplicative characters the theory follows the same path as for exponential sums, there are some exceptions. For example, there is no analogue of Theorem 3.4 for multiplicative character sums. On the other hand, the celebrated Burgess bound [12] has no analogue for exponential sums.
- Thousands of less general results for various interesting (and not so) special cases.

6. Twin Brothers of Exponential Sums — Character Sums

6.1. *Definitions*

A multiplicative character χ of $\mathbb{F}_q^*$ is a function

$$\chi : \mathbb{F}_q^* \to \{z \in \mathbb{C} : |z| = 1\}$$

with

$$\chi(ab) = \chi(a)\chi(b) \qquad \forall a, b \in \mathbb{F}_q^*.$$

The trivial character χ_0 is the character with $\chi_0(a) = 1$, $a \in \mathbb{F}_q^*$.
It is convenient to put $\chi(0) = 0$ for all characters χ (including χ_0).

Characters can be described in terms of the *index* or the *discrete logarithm* with respect to some fixed primitive root of $\mathbb{F}_q$.

The most "famous" character is the quadratic character or **Legendre symbol** modulo an odd prime p, which for $a \not\equiv 0 \pmod p$ is defined by

$$\left(\frac{a}{p}\right) = \begin{cases} 1, & \text{if } a \equiv x^2 \pmod p \text{ is solvable,} \\ -1, & \text{otherwise,} \end{cases}$$

or

$$\left(\frac{a}{p}\right) = \begin{cases} 1, & \text{if } a \text{ is a quadratic residue,} \\ -1, & \text{otherwise.} \end{cases}$$

Characters can be extended to residue rings.

Jacobi symbol is the residue ring analogue of the Legendre symbol.

Warning For Jacobi symbol modulo a composite m it is **not true** that

$$\left(\frac{a}{m}\right) = \begin{cases} 1, & \text{if } a \text{ is a quadratic residue,} \\ -1, & \text{otherwise.} \end{cases}$$

The theory of character sums

$$T(\chi, \mathcal{X}) = \sum_{x \in \mathcal{X}} \chi(x)$$

is similar to the theory of exponential sums ... but not quite.

6.2. *Polya–Vinogradov bound again*

Despite what we have just said about great similarities between exponential sums and character sums, one of the first results of the theory demonstrates that actually there are some important distinctions as well. Namely, the Polya–Vinogradov inequality is sometimes formulated as a bound on linear character sums, which, as this inequality shows, behave very differently compared with linear exponential sums.

Theorem 6.1: *For any integer N, $1 \le N \le p$,*

$$\sum_{x=1}^{N} \left(\frac{x}{p}\right) = O(p^{1/2} \log p).$$

Proof: Following the standard principle, let us estimate the sums

$$S(a) = \sum_{x=1}^{p} \left(\frac{x}{p}\right) e_p(ax).$$

If $a \equiv 0 \pmod{p}$ then

$$S(0) = \sum_{x=1}^{p} \left(\frac{x}{p}\right) = 0$$

because for any quadratic non-residue b

$$-S(0) = \left(\frac{b}{p}\right) S(0) = \sum_{x=1}^{p} \left(\frac{bx}{p}\right) = \sum_{x=1}^{p} \left(\frac{x}{p}\right) = S(0).$$

If $\gcd(a, p) = 1$ then

$$S(a) = \sum_{x=1}^{p} \left(\frac{x}{p}\right) e_p(ax) + \sum_{x=1}^{p} e_p(ax)$$

$$= \sum_{x=1}^{p} \left(1 + \left(\frac{x}{p}\right)\right) e_p(ax)$$

$$= \sum_{x=1}^{p-1} \left(1 + \left(\frac{x}{p}\right)\right) e_p(ax) + 1$$

$$= 2 \sum_{x \ quadr.\ res.} e_p(ax) + 1$$

$$= \sum_{y=1}^{p-1} e_p(ay^2) + 1 = G(a, p).$$

By Theorem 3.3 we have

$$|S(a)| = p^{1/2}, \quad \gcd(a, p) = 1.$$

Now

$$\left| \sum_{x=1}^{N} \left(\frac{x}{p} \right) \right| = \left| \sum_{x=1}^{p} \left(\frac{x}{p} \right) \frac{1}{p} \sum_{a=0}^{p-1} \sum_{y=1}^{N} e_p(a(x-y)) \right|$$

$$= \frac{1}{p} \left| \sum_{a=0}^{p-1} S(a) \sum_{y=1}^{N} e_p(-ay) \right|$$

$$\le \frac{1}{d} p \sum_{a=0}^{p-1} |S(a)| \left| \sum_{y=1}^{N} e_p(-ay) \right|$$

$$\le \frac{p^{1/2}}{p} \sum_{a=0}^{p-1} \left| \sum_{y=1}^{N} e_p(ay) \right| = O(p^{1/2} \log p)$$

by Theorem 3.4. □

Corollary 6.2: *For* $1 \le N \le p$, *the interval* $[1, N]$ *contains* $N/2 + O(p^{1/2} \log p)$ *quadratic residues and non-residues.*

Analysing when the above expression becomes positive we derive:

Corollary 6.3: *The smallest positive quadratic non-residue is* $N_0 = O(p^{1/2} \log p)$.

6.3. *Let's push it down!* —*Other methods are helpful as well*

The following nice trick is due to Vinogradov. It shows that if we have a non-trivial bound for character sums of length M, than we can say something interesting for much smaller intervals!

Let us fix some $M > N_0$ and count the number T of quadratic non-residues in the interval $[1, M]$.

Because each quadratic non-residue must have a prime divisor $q \ge N_0$ we obtain

$$T \le \sum_{M \ge q \ge N_0} (\lfloor M/q \rfloor + 1) \le M \sum_{M \ge q \ge N_0} 1/q + \pi(M).$$

We have $\pi(M) = O(M/\log M)$ and

$$\sum_{M \ge q \ge N_0} 1/q = \ln \ln M - \ln \ln N_0 + o(1).$$

Let $M = p^{1/2} \log^2 p$. Then $T = M/2 + o(M)$. Therefore

$$\ln \ln M - \ln \ln N_0 \geq 1/2 + o(1)$$

or

$$\frac{\ln M}{\ln N_0} \geq e^{1/2} + o(1)$$

or

$$N_0 = M^{1/e^{1/2} + o(1)} \leq p^{1/2e^{1/2} + o(1)}.$$

7. Applications to Coding Theory

7.1. *Direct applications*

Many coding theory questions can immediately be formulated as questions about bound of exponential sums:

- correlation and autocorrelation, see [3, 4, 5, 21, 24, 35, 36, 37, 38, 39, 40];
- Minimal distance of BCH codes [64];
- Size of Varshamov–Mazur codes for asymmetric channels [52, 65, 88].

   Surprisingly enough, it works the other way as well. Some coding theory lower bounds can be applied to obtain very tight lower bounds for exponential sums [6, 58, 78, 81, 98, 99]. One can certainly argue about the importance of lower bounds because all known applications are based on upper bounds. Nevertheless, they certainly improve our understanding of the area and are an intrinsic part of the theory of exponential sums.

   Several other interrelations between exponential sums and coding theory, which enrich both areas, can be found in [88].

7.2. *Less obvious applications: Dimension of BCH codes*

7.2.1. *Definitions*

Let q be a prime power and let n be an integer with $\gcd(n, q) = 1$.

   Denote by t the multiplicative order of q modulo n; and fix an element $\alpha \in \mathbb{F}_{q^t}^*$ of multiplicative order n (it exists because $n \mid q^t - 1$);.

   Let l be an integer. To construct a BCH code with *designed distance* Δ we consider the polynomial g over $\mathbb{F}_q$ of the smallest degree such that

$$g\left(\alpha^{l+y}\right) = 0, \qquad y = 1, \ldots, \Delta - 1,$$

and consider the cyclic code of length n with g as the generator polynomial. That is, the linear space of dimension $k = n - \deg g$ of n-dimensional vectors $(a_0, \ldots, a_{n-1}) \in \mathbb{F}_q^n$ such that

$$a_0 + a_1 Z + \ldots + a_{n-1} Z^{n-1} \equiv 0 \pmod{g(Z)}.$$

Generally for every linear code there are three parameters of interest: the length, the minimal distance and the dimension. For a BCH code the length n is given, the minimal distance d is at least the designed distance Δ (and this bound is known to be tight in many cases [64]). The question about the dimension is more interesting. Of course,

$$t \leq \deg g \leq (\Delta - 1)t,$$

thus the dimension

$$n - t \geq k \geq n - (\Delta - 1)t.$$

To get something stronger one should study the structure of the roots of g in more detail.

We make the following observations:

- all roots of g are powers of α because

$$g(Z) \mid \prod_{y=1}^{\Delta-1} \prod_{x=1}^{t} \left(Z - \alpha^{(l+y)q^x} \right);$$

- α^j is a root of g if and only if

$$jq^x \equiv l + y \pmod{n},$$

for some $x = 1, \ldots, t$ and $y = 1, \ldots, \Delta - 1$.

Let us denote by $J(q, n, \Delta)$ the largest possible dimension of q-ary generalized BCH codes of length n and of designed distance Δ taken over all $l = 0, \ldots, n-1$.

From the above discussion we conclude that $J(q, n, \Delta)$ is the number of $j = 0, 1, \ldots, n-1$ for which the congruence

$$jq^x \equiv l + y \pmod{n}, \qquad 1 \leq x \leq t, \ 1 \leq y \leq \Delta - 1, \qquad (4)$$

is not solvable.

Thus, the original question has been reduced to a question about the distribution of values of an exponential function to which our technique can be applied.

7.2.2. Preparations

For a divisor d of n denote by t_d the multiplicative order q modulo d (thus $t = t_n$).

Lemma 7.1: *For any $d \mid n$, the bound $t_{n/d} \geq t/d$ holds.*

Lemma 7.2: *For any integers a, b, the congruence*

$$aq^x \equiv bq^y \pmod{n}, \qquad 1 \leq x, y \leq t,$$

is solvable only when $\gcd(a, n) = \gcd(b, n) = d$, and in this case for the number of solutions $N(a, b)$ the bound

$$N(a, b) \leq td$$

holds.

Proof: As $\gcd(q, n) = 1$, the condition on a and b is evident. Also it is evident that for any fixed x there are at most $t/t_{n/d}$ possible values for y, hence $N(a, b) \leq t^2/t_{n/d} \leq td$ because of Lemma 7.1. $\qquad\square$

We define the sums

$$T(a, h) = \sum_{u=1}^{h} \mathbf{e}_n(au), \quad W_d(h) = \sum_{\gcd(a,n)=d} |T(a, h)|^2,$$

where d is a divisor of n, $d \mid n$.

Lemma 7.3: *For any $d \mid n$ with $d < n$, the bound*

$$W_d(h) \leq nh/d$$

holds.

Proof: Denote $m = n/d$. We have

$$W_d(h) \leq \sum_{a=0}^{n/d-1} |T(ad, h)|^2 - h^2 = mM - h^2,$$

where M is the number of solutions of the congruence

$$u \equiv v \pmod{m}, \qquad 1 \leq u, v \leq h.$$

Write $h = km + r$ with $0 \leq r \leq m - 1$, then $M = r(k+1)^2 + (m-r)k^2 = k^2m + 2kr + r$. Therefore

$$\begin{aligned}
W_d(h) &\leq mM - h^2 = k^2m^2 + 2kmr + mr - h^2 \\
&= (h - r)^2 + 2r(h - r) + mr - h^2 = r(m - r) \\
&\leq rm \leq hm = nh/d.
\end{aligned}$$

$\qquad\square$

7.2.3. *Main result*

Theorem 7.4: *The bound*

$$J(q, n, \Delta) \leq \frac{4n^3}{(\Delta - 1)^2 t}$$

holds.

Proof: Let $h = \lfloor \Delta/2 \rfloor$ and let N_j denote the number of solutions of the congruence

$$jq^x \equiv l + h + u - v \pmod{n}, \tag{5}$$

where

$$x = 1, \ldots, t, \quad u, v = 1, \ldots, h.$$

Then $J(q, n, \Delta) \leq |\mathfrak{I}(q, n, \Delta)|$, where $\mathfrak{I}(q, n, \Delta)$ is the set of $j = 0, 1, \ldots, n - 1$ for which this congruence is unsolvable, that is, $N_j = 0$.

Set

$$S(a) = \sum_{x=1}^{t} \mathbf{e}(aq^x/n).$$

Then $N_j = th^2/n + R_j/n$, where

$$R_j = \sum_{a=1}^{n-1} S(aj)|T(a, h)|^2 \mathbf{e}_n(-a(l + h)).$$

Let us consider

$$R = \sum_{j=0}^{n-1} R_j^2.$$

We have

$$R = \sum_{j=0}^{n-1} \sum_{a,b=1}^{n-1} S(aj)S(bj)|T(a, h)|^2 |T(b, h)|^2 \mathbf{e}_n \left(-(a + b)(l + h) \right)$$

$$= \sum_{a,b=1}^{n-1} |T(a, h)|^2 |T(b, h)|^2 \mathbf{e}_n \left(-(a + b)(l + h) \right) \sum_{j=0}^{n} S(aj)S(bj).$$

Then,

$$\sum_{j=0}^{n-1} S(aj)S(bj) = \sum_{x,y=1}^{t} \sum_{j=0}^{n-1} \mathbf{e}_n \left(j(aq^x + bq^y) \right)$$

$$= nN(a, -b).$$

For all divisors $d \mid n$ we gather together all terms corresponding to a and b with

$$\gcd(a, n) = \gcd(b, n) = d.$$

Applying Lemma 7.2, we obtain

$$R = n \sum_{d \mid n, d < n} \sum_{\gcd(a,n)=\gcd(b,n)=d} |T(a, h)|^2$$

$$\times |T(b, h)|^2 N(a, -b) \mathbf{e}_n \left((a + b)(l + h)\right)$$

$$\leq nt \sum_{d \mid n, d < n} dW_d(h)^2$$

$$\leq nt \max_{d \mid n, d < n} dW_d(h) \sum_{d \mid n, d < n} W_d(h).$$

From Lemma 7.3 and the identity

$$\sum_{\substack{d \mid n \\ d < n}} W_d(h) = \sum_{a=1}^{n-1} |T(a, h)|^2 = nh - h^2 \tag{6}$$

we derive

$$R \leq n^3 h^2 t.$$

Since $R_j = -h^2 t$ for $j \in \mathfrak{I}(q, n, \Delta)$, then

$$|\mathfrak{I}(q, n, \Delta)| h^4 t^2 = \sum_{j=0}^{n-1} R_j^2 \leq n^3 h^2 t.$$

Taking into account that $h \geq (\Delta - 1)/2$, we obtain the result. $\qquad \square$

It is useful to keep in mind that *exponential sums do not always win*. For certain values of parameters the following elementary statement provides a sharper bound.

Theorem 7.5: *The bound*

$$J(q, n, \Delta) \leq 2e^{1/2} n^{1-\alpha_q(\delta)}$$

holds, where $\delta = (\Delta - 1)/n$ and

$$\alpha_q(\delta) = \frac{\delta}{2 \ln(3q/\delta)}.$$

Thus, this can be taken as an encouragement to study other number-theoretic techniques.

7.2.4. *Discussion: Some lessons to learn*

It is easy to see that the proof of Theorem 7.4 is much more technically involved than reasonably straightforward proofs of other results presented here. In fact, one of the reasons for presenting here Theorem 7.4 has been the fact that it provides quite an instructive example of several potential difficulties which can arise and some technical tricks which can be used to get around these difficulties.

First of all, one of the reasons for the proof to be so painful has been the fact that we work with congruences modulo a *composite* number. As a result, sometimes the denominator of the exponential sums involved becomes n/d rather than n, for a divisor $d|n$.

The other reason is more subtle. It may look strange that instead of studying the congruence (4), directly associated with $J(q, n, \Delta)$, we have studied the less attractive and strangely looking congruence (5). Certainly we could easily study the congruence (4) as well, getting a simpler expression of the form $M_j = t(\Delta - 1)/n + Q_j/n$, where

$$Q_j = \sum_{a=1}^{n-1} S(aj)T(a, \Delta - 1)e_n(-al)$$

for the number of solutions of this congruence. The rest would go along the same lines, except that instead of an explicit formula (6) involving squares of $|T(a, h)|$ we would use Theorem 3.4 to estimate various sums of the first powers of $|T(a, \Delta - 1)|$, thus getting an extra $\log n$ in our estimates (I leave this as an exercise to fill all missing details and obtain an upper bound for $J(q, n, \Delta)$ along these lines). However, our saving compared to the trivial bound is only of order t. Although typically t is much greater than $\log n$, sometimes, namely when $n = q^t - 1$, t is exactly of order $\log n$. Thus, for such small values of t even such small losses as $\log n$ turn out to be fatal for the method. On the other hand, the "symmetrisation" trick with adding one more variable in the congruence we need to study helps to avoid the appearance of extra logarithms! It is important to remember, however, that there is no direct explicit relation between M_j and N_j, so this approach does not apply when we want to estimate M_j. However, for our purposes here we only need to count how often $M_j = 0$ and thus we can use the obvious property that if $M_j = 0$ then $N_j = 0$.

8. Applications to Cryptography

8.1. *Distribution of some cryptographic primitives*

8.1.1. *Security of exponentiation with precomputation*

Let g be an element of order t modulo a prime number p, that is,

$$g^T \equiv 1 \pmod{p} \quad \Leftrightarrow \quad t|T.$$

Let r be the bit length of t, $2^{r-1} \le t \le 2^r - 1$.

Many signature schemes use exponentiation $g^x \pmod{p}$ for a "random" x.

Using repeated squaring this takes about $1.5r$ multiplications on average and about $2r$ operations in the worst case.

One of the possible ways to speed-up exponentiation is to precompute the values $g^{2^j} \pmod{p}$, $j = 0, \dots, r$. Then computing $g^x \pmod{p}$ takes $0.5r$ multiplications on average, r multiplications in the worst case.

Main Problem: How can we generate *secure* pairs (x, g^x) faster (for some special x)?

Secure: Finding x from g^x for the values of x generated by this method should be as hard as for a random $x \in [0, M-1]$.

In 1998, V. Boyko, M. Peinado and R. Venkatesan [10] proposed the following algorithm.

Given $n \ge k \ge 1$:

Preprocessing Step: Generate n random integers $\alpha_1, \dots, \alpha_n \in \mathbb{Z}_M$. Compute $\beta_j \equiv g^{\alpha_j} \pmod{p}$ and store the values of α_j and β_j in a table, $j = 1, \dots, n$.

Pair Generation: Generate a random set $S \subseteq \{1, \dots, n\}$ of cardinality $\#S = k$. Compute

$$x \equiv \sum_{j \in S} \alpha_j \pmod{M}, \quad X \equiv \prod_{j \in S} \beta_j \equiv g^b \pmod{p}.$$

Cost: $k - 1$ modular multiplications.

It is easy to see that

$$X \equiv g^x \pmod{p}.$$

In 1999, P. Q. Nguyen, I. E. Shparlinski and J. Stern [72] proposed the following generalization of this scheme which involved one more integer parameter h.

Given $n \ge k \ge 1$ and $h \ge 2$:

Preprocessing Step: Generate n random integers $\alpha_1, \ldots, \alpha_n \in \mathbb{Z}_M$. Compute $\beta_j \equiv g^{\alpha_j} \pmod{p}$ and store the values of α_j and β_j in a table, $j = 1, \ldots, n$.

Extended Pair Generation: Generate a random set $S \subseteq \{1, \ldots, n\}$, $|S| = k$ and for each $j \in S$ select a random integer $x_j \in \{0, \ldots, h-1\}$. Compute

$$x \equiv \sum_{j \in S} \alpha_j x_j \pmod{M}, \quad X \equiv \prod_{j \in S} \beta_j^{x_j} \pmod{p}.$$

Cost: $k + h - 3$ modular multiplications.

One verifies that the congruence

$$X \equiv g^x \pmod{p}$$

holds. The cost estimate (which is better than naive $O(k \log h)$) follows from a result of [11].

Finally, using some bounds of exponential sums and to establish some results about the uniformity of distribution of sums

$$\sum_{j \in S} \alpha_j x_j \pmod{M}, \quad x_j \in \{1, \ldots, h-1\},$$

the security of this scheme was proved in [72]. We present this result in an informal way and refer to [72] for exact formulations (which formalises the notion of security).

Theorem (informally). *Let $n = \gamma r$ with some $\gamma > 0$. There are values of k and h with*

$$k + h = O(r / \log r)$$

and such that the scheme is as secure as the generator $x \to g^x \pmod{p}$ for arbitrary x.

The most important characteristics of this scheme are

Table size: linear in r (say $n = r$)
Speed-up: $\log r \to \infty$.

8.1.2. *Diffie–Hellman triples and RSA pairs*

Let g be a primitive element modulo p.

The following assumption is known as the *Diffie–Hellman Indistinguishability Assumption*: It is feasible to distinguish between *Diffie–Hellman triples* (g^x, g^y, g^{xy}) with random x and y and random triples $(u, v, w) \in \mathbb{F}_p^3$?

One of the possible (very naive approaches) to disprove this assumption would be to find some statistically "visible" singularities in differences in the distribution of the triples

$$(g^x, g^y, g^{xy}), \qquad x, y = 1, \ldots, p-1.$$

However, R. Canetti, J. B. Friedlander and I. E. Shparlinski [14] in 1997, and in a stronger form, R. Canetti, J. B. Friedlander, S. V. Konyagin, M. Larsen, D. Lieman and I. E. Shparlinski [13] in 1999, proved that Diffie-Hellman triples are uniformly distributed.

Recently, J. B. Friedlander and I. E. Shparlinski [25] obtained a similar statement for Diffie-Hellman triples with "sparse" x and y (one can use such x and y in order to speed-up computation). It is shown in [25] that Diffie-Hellman triples with x and y having at most $0.35 \log p$ nonzero digits are uniformly distributed.

Surprisingly enough, several results of [13] play a central role in studying a related problem about the distribution of *RSA pairs* (x, x^e) in the residue ring modulo m, see [93].

8.2. *Lattices and exponential sums*

8.2.1. *Introduction and notation*

In this section we describe how a rather unusual combination of two celebrated number-theoretic techniques, namely, bounds of *exponential sums* and *lattice reduction* algorithms, provides a powerful cryptographic tool. It can be applied to both proving several security results and designing new attacks.

For example, it has been used to prove certain bit security results for the Diffie-Hellman key exchange system, for the Shamir message passing scheme and for the XTR cryptosystem. It has also been used to design provably successful attacks on the Digital Signature Scheme and its modifications, including the Nyberg–Rueppel scheme, which are provably insecure under certain conditions.

Here we explain how these two techniques get together, outline several important applications and discuss some open problems on exponential sums which arise in this context and which need to be solved before any further progress in this area can be achieved.

Let p denote a prime number and let $\mathbb{F}_p$ denote the finite field of p elements. For integers s and $m \geq 1$ we denote by $\lfloor s \rfloor_m$ the remainder of s on division by m. For a prime p and $\ell > 0$ we denote by $\mathrm{MSB}_{\ell,p}(x)$ any integer u such that

$$\left| \lfloor x \rfloor_p - u \right| \leq p/2^{\ell+1}. \tag{7}$$

Roughly speaking, $\mathrm{MSB}_{\ell,p}(x)$ gives ℓ most significant bits of x, however, this definition is more flexible and suits better our purposes. In particular, we remark that ℓ in the inequality (7) need not be an integer.

Throughout this section, $\log z$ denotes the binary logarithm of $z > 0$.

The implied constants in symbols 'O' may occasionally, where obvious, depend on the small positive parameters ε and are absolute otherwise.

8.2.2. *Hidden number problem and lattices*

We start with a certain algorithmic problem, introduced in 1996 by Boneh and Venkatesan [8, 9], which seemingly has nothing in common with exponential sums. Namely we consider the following

HIDDEN NUMBER PROBLEM, **HNP**: *Recover a number* $\alpha \in \mathbb{F}_p$ *such that for many known random* $t \in \mathbb{F}_p^*$ *we are given* $\mathrm{MSB}_{\ell,p}(\alpha t)$ *for some* $\ell > 0$.

The paper [8] also contains a polynomial-time algorithm to solve this problem (with ℓ of order $\log^{1/2} p$). The most important ingredient of this algorithm is lattice reduction.

We briefly review a few results and definitions. For general references on lattice theory and its important cryptographic applications, we refer to the recent surveys [73, 74].

Let $\{\mathbf{b}_1, \ldots, \mathbf{b}_s\}$ be a set of linearly independent vectors in $\mathbb{R}^s$. The set of vectors

$$L = \left\{ \sum_{i=1}^{s} n_i \mathbf{b}_i \mid n_i \in \mathbb{Z} \right\}$$

is called an s-dimensional full rank lattice. The set $\{\mathbf{b}_1, \ldots, \mathbf{b}_s\}$ is called a *basis* of L, and L is said to be spanned by $\{\mathbf{b}_1, \ldots, \mathbf{b}_s\}$.

One of the most fundamental problems in this area is the *closest vector problem*, **CVP**: given a basis of a lattice L in $\mathbb{R}^s$ and a target vector $\mathbf{u} \in \mathbb{R}^s$, find a lattice vector $\mathbf{v} \in L$ which minimizes the Euclidean norm $\|\mathbf{u} - \mathbf{v}\|$ among all lattice vectors. It is well know that **CVP** is **NP**-hard (see [73, 74] for references). However, its approximate version [2] admits a polynomial-time algorithm which goes back to the lattice basis reduction algorithm of Lenstra, Lenstra and Lovász [55].

It has been remarked in Section 2.1 of [67] and then in Section 2.4 of [73] and Section 2.4 of [74] that the following statement holds, which is somewhat stronger than that usually used in the literature.

Theorem 8.1: *There exists a polynomial-time algorithm which, given an s-dimensional full rank lattice L and a vector $\mathbf{r} \in \mathbb{R}^s$, finds a lattice vector $\mathbf{v}$ satisfying the inequality*

$$\|\mathbf{v} - \mathbf{r}\| \leq 2^{O\left(s \log^2 \log s / \log s\right)} \min\left\{\|\mathbf{z} - \mathbf{r}\|, \quad \mathbf{z} \in L\right\}.$$

Proof: The statement is a combination of Schnorr's modification [82] of the lattice basis reduction algorithm of Lenstra, Lenstra and Lovász [55] with a result of Kannan [45] about reduction of the **CVP** to the approximate shortest vector problem.   □

One can also use a probabilistic analogue [1] of Theorem 8.1 which gives a slightly better constant.

We are now prepared to sketch the main ideas of [8] to solve the **HNP**. Let $d \geq 1$ be an integer. Given t_i, $u_i = \mathrm{MSB}_{\ell,p}(\alpha t_i)$, $i = 1, \ldots, d$, we build the lattice $\mathcal{L}(p, \ell, t_1, \ldots, t_d)$ spanned by the rows of the matrix:

$$\begin{pmatrix} p & 0 & \cdots & 0 & 0 \\ 0 & p & \ddots & \vdots & \vdots \\ \vdots & \ddots & \ddots & 0 & \vdots \\ 0 & 0 & \cdots & p & 0 \\ t_1 & t_2 & \cdots & t_d & 1/2^{\ell+1} \end{pmatrix}$$

and notice

$$\mathbf{w} = \left(\lfloor \alpha t_1 \rfloor_p, \ldots, \lfloor \alpha t_d \rfloor_p, \alpha/2^{\ell+1}\right) \in \mathcal{L}(p, \ell, t_1, \ldots, t_d).$$

This vector is very close to the *known* vector $\mathbf{u} = (u_1, \ldots, u_d, 0)$ (at the distance of order $p2^{-\ell}$). Thus, applying one of the lattice reduction algorithms one can *hope* to recover $\mathbf{v}$ and thus α. In order to make this algorithm *rigorous* one needs to show that (for almost all choices of $t_1, \ldots, t_d \in \mathbb{F}_p$) there is no other lattice vector which is close to $\mathbf{u}$. Namely, taking into account the "stretching" factor in the algorithm of Theorem 8.1, we have to show that there are very few d-tuples $(t_1, \ldots, t_d) \in \mathbb{F}_p^d$ for which the lattice $\mathcal{L}(p, \ell, t_1, \ldots, t_d)$ has a vector $\mathbf{v} \neq \mathbf{w}$ and such that

$$\|\mathbf{v} - \mathbf{u}\| \leq p2^{-\ell} \exp\left(O\left(\frac{d\log^2\log d}{\log d}\right)\right).$$

The last inequality implies that

$$\|\mathbf{v} - \mathbf{w}\| \leq p2^{-\ell} \exp\left(O\left(\frac{d\log^2 \log d}{\log d}\right)\right) \tag{8}$$

which is our main tool.

Any vector $\mathbf{v} \in \mathcal{L}(p, \ell, t_1, \ldots, t_d)$ is of the form

$$\mathbf{v} = \left(\beta t_1 - \lambda_1 p, \ldots, \beta t_d - \lambda_d p, \beta/2^{\ell+1}\right),$$

with some integers β and $\lambda_1, \ldots, \lambda_d$. Thus, (8) implies that for all $i = 1, \ldots, d$ we have

$$(\alpha - \beta)t_i \equiv y_i \pmod{p} \tag{9}$$

for some $y_i \in [-h, h]$, where

$$h = p2^{-\ell} \exp\left(O\left(\frac{d\log^2 \log d}{\log d}\right)\right).$$

The probability

$$\Pr_{y \in \mathbb{F}_p}\left[\gamma t = y \pmod{p} \mid y \in [-h, h]\right] \leq \frac{2h+1}{p} \tag{10}$$

for any $\gamma \neq 0$.

Therefore the probability P that the condition (9) holds for all $i = 1, \ldots, d$ and at least one $\beta \neq \alpha$, is at most

$$P \leq (p-1)\left(\frac{2h+1}{p}\right)^d \leq p(3h/p)^d = p2^{-\ell d} \exp\left(O\left(\frac{d^2 \log^2 \log d}{\log d}\right)\right).$$

Thus, if

$$\ell = \left\lceil C\frac{\log^{1/2} p \log\log\log p}{\log\log p}\right\rceil \quad \text{and} \quad d = 2\left\lceil\frac{\log p}{\ell}\right\rceil$$

with some absolute constant $C > 0$, then the lattice reduction algorithm returns $\mathbf{v}$ with probability exponentially close to 1.

8.2.3. *Extended hidden number problem, lattices and exponential sums*

It has turned out that for many applications, including some results about the bit security of Diffie-Hellman, Shamir and some other cryptosystems [31, 32, 61, 91, 92, 94] and rigorous results on attacks (following the heuristic arguments of [42, 69]) on the DSA and DSA-like signature schemes [23, 70, 71], the condition that t is selected uniformly at random from $\mathbb{F}_p$ is too restrictive.

It has been systematically exploited in [23, 31, 32, 61, 70, 71, 91, 92, 94] that the method of [8] can be extended to the case where t is selected from a sequence $\mathcal{T}$ having some uniformity of distribution property.

Accordingly, we consider the following:

EXTENDED HIDDEN NUMBER PROBLEM, **EHNP:** *Recover a number $\alpha \in \mathbb{F}_p$ such that for many known random $t \in \mathcal{T}$ we are given* $\mathrm{MSB}_{\ell,p}(\alpha t)$ *for some $\ell > 0$.*

If $\mathcal{T} = \mathbb{F}_p$, then rather simple counting arguments of Section 8.2.2 show that the number of d-tuples $(t_1, \dots, t_d) \in \mathbb{F}_p^d$ for which the algorithm of Theorem 8.1 returns a false vector is exponentially small. However, for other sequences $\mathcal{T}$ one needs a result about the uniformity of distribution of $\mathcal{T}$.

In the quantitative form which is based on best known lattice reduction algorithms [1, 2, 45, 46, 55, 73, 74, 82], this has been obtained in [70].

Recall that the *discrepancy* of an N-element sequence $\Gamma = \{\gamma_1, \dots, \gamma_N\}$ of elements of the interval $[0, 1]$ is defined as

$$\mathcal{D}(\Gamma) = \sup_{J \subseteq [0,1]} \left| \frac{A(J, N)}{N} - |J| \right|,$$

where the supremum is extended over all subintervals J of $[0, 1]$, $|J|$ is the length of J, and $A(J, N)$ denotes the number of points γ_n in J for $1 \leq n \leq N$.

We say that a finite sequence $\mathcal{T}$ of integers is Δ-*homogeneously distributed modulo p* if for any integer a with $\gcd(a, p) = 1$ the discrepancy of the sequence $\{\lfloor at \rfloor_p / p\}_{t \in \mathcal{T}}$ is at most Δ.

In this case the arguments of Section 8.2.2 go through with only one change, namely (10) becomes

$$\Pr_{y \in \mathcal{T}} \left[\gamma t = y \pmod{p} \mid y \in [-h, h] \right] \leq \frac{2h + 1}{p} + \Delta.$$

This leads to the following result from [70] which extends the algorithm of [8] to the **EHNP** with a general sequence $\mathcal{T}$.

Theorem 8.2: *For a prime p, define $\ell = \lceil \log^{1/2} p \rceil + \lceil \log \log p \rceil$ and $d = 2 \lceil \log^{1/2} p \rceil$. Let $\mathcal{T}$ be a $2^{-\log^{1/2} p}$-homogeneously distributed modulo p sequence of integer numbers. There exists a deterministic polynomial-time* algorithm $\mathcal{A}$ such that for any fixed integer α in the interval $[0, p-1]$, given a prime p and $2d$ integers

$$t_i \quad \text{and} \quad u_i = \text{MSB}_{\ell,p}(\alpha t_i), \qquad i = 1, \ldots, d,$$

its output satisfies for sufficiently large p,

$$\Pr_{t_1,\ldots,t_d \in \mathcal{T}} [\mathcal{A}(p, t_1, \ldots, t_d; u_1, \ldots, u_d) = \alpha] \geq 1 - 2^{-(\log p)^{1/2} \log \log p}$$

if $t_1, \ldots, t_d$ are chosen uniformly and independently at random from the elements of $\mathcal{T}$.

It follows from Corollary 3.11 of [76] that $\mathcal{T}$ is Δ-homogeneously distributed modulo p with

$$\Delta = O\left(\frac{\log p}{\#\mathcal{T}} \max_{c=1,\ldots,p-1} \left| \sum_{t \in \mathcal{T}} \exp\left(2\pi i c t / p\right) \right| \right). \tag{11}$$

The proof follows the same lines as the proof of Theorem 3.5 and is rather standard if one does not care about the hidden constant in the 'O'-symbol. However, in order to get a small constant there, one has to go through some technical complications, we refer to [76] for more details.

Therefore, in order to apply this result, one can establish the uniformity of distribution of various sequences $\mathcal{T}$ arising in cryptographic applications and thus one needs to estimate *exponential sums* with elements of $\mathcal{T}$. Thus, bounds of exponential sums enter the problem. It has turned out that in some cases relevant exponential sums are well studied in number theory, and thus the corresponding cryptographic result follows immediately, for example, see Section 8.2.4. On the other hand, in some cases the exponential sums are of very unusual structure which has no meaningful number-theoretic interpretations and thus they have required special treatment, for example, see Section 8.2.5.

8.2.4. *Bit security of the Diffie–Hellman secret key*

We recall the problem which underlies the Diffie–Hellman key exchange system: given an element g of order τ modulo p, find an efficient algorithm to recover the Diffie–Hellman secret key $K = \lfloor g^{xy} \rfloor_p$ from $\lfloor g^x \rfloor_p$ and $\lfloor g^y \rfloor_p$.

Typically, either $\tau = p - 1$ (thus g is a primitive root) or $\tau = q$, a large prime divisor of $p - 1$. The size of p and τ is determined by the present state of art in the *discrete logarithm problem*. Typically, p is at least about 500 bits, τ is at least about 160 bits.

However, after the common DH key $K = \lfloor g^{xy} \rfloor_p$ is established, only a small portion of bits of K will be used as a common key for some pre-agreed *private* key cryptosystem.

Thus, a natural question arises: *Assume that finding K is infeasible, is it still infeasible to find certain bits of K?*

In 1996, Boneh and Venkatesan [8] found very elegant links between the **EHNP** and the above problem.

Indeed, assume there is an efficient algorithm to find ℓ most significant bits of $\lfloor g^{xy} \rfloor_p$ from $X = \lfloor g^x \rfloor_p$ and $Y = \lfloor g^y \rfloor_p$. Then, given $A = \lfloor g^a \rfloor_p$ and $B = \lfloor g^b \rfloor_p$, one can select a random $u \in [0, \tau - 1]$ and one can apply the above algorithm to A and $U = \lfloor Bg^u \rfloor_p$, getting

$$\mathrm{MSB}_{\ell,p}\left(g^{a(b+u)}\right) = \mathrm{MSB}_{\ell,p}\left(\alpha g_a^u\right)$$

where $\alpha = \lfloor g^{ab} \rfloor_p$ and $g_a = g^a$. Thus, we have a special case of the **EHNP**. Unfortunately, the paper [8] has a minor gap in the proof of Theorem 2 of that paper. It is claimed that if g is a primitive root (that is, if $\tau = p - 1$), then the obtained problem is exactly the **HNP**. However, this is true only if g_a is a primitive root as well, thus if $\gcd(a, p - 1) = 1$.

To fix this gap and to extend the result to the case of $\tau < p - 1$, M. I. González Vasco and I. E. Shparlinski [31] have used the bounds of exponential sums from [52] which we have presented in Theorem 5.2 and Theorem 5.3.

Using (11) we see that under the conditions of Theorem 5.2 and Theorem 5.3 the sequence g^x, $x = 0, \ldots, T-1$, is $p^{-\delta}$-homogeneously distributed modulo p.

Combining this result with the above arguments and Theorem 8.2, one can obtain the following statement about the bit security of the Diffie–Hellman secret key.

For each integer $\ell \geq 1$ define the oracle $\mathcal{DH}_\ell$ as a 'black box' which, given the values of $X = \lfloor g^x \rfloor_p$ and $Y = \lfloor g^y \rfloor_p$, outputs the value of $\mathrm{MSB}_{\ell,p}(g^{xy})$.

Theorem 8.3: *Let Q be a sufficiently large integer. The following statement holds with $\vartheta = 1/3$ for all primes $p \in [Q, 2Q]$, and with $\vartheta = 0$ for all primes $p \in [Q, 2Q]$ except at most $Q^{5/6+\varepsilon}$ of them. Let $k = \left\lceil \log^{1/2} p \right\rceil + \lceil \log \log p \rceil$. For any $\varepsilon > 0$, sufficiently large p and any element $g \in \mathbb{F}_p^*$ of multiplicative order $T \geq p^{\vartheta+\varepsilon}$, there exists a probabilistic polynomial-time algorithm which for any pair $(a, b) \in [0, T-1]^2$, given the* values of $A = \lfloor g^a \rfloor_p$ and $B = \lfloor g^b \rfloor_p$, makes $O\left(\log^{1/2} p\right)$ calls of the oracle $\mathcal{DH}_k$ and computes $\lfloor g^{ab} \rfloor_p$ correctly with probability $1 + O\left(2^{-\log^{1/2} p}\right)$.

8.2.5. *Attack on the digital signature algorithm*

On the other hand, in some cases the corresponding exponential sums are new and require a separate study. For example, in [70] the sequence arising in the attack on the Digital Signature Algorithm (DSA) has been studied. We recall the DSA settings. Assume that q and p are primes with $q|p-1$ and that $g \in \mathbb{F}_p$ is a fixed element of multiplicative order q. Let $\mathcal{M}$ be the set of messages to be signed and let $h : \mathcal{M} \to \mathbb{F}_q$ be an arbitrary hash function. They all (that is, p, q, g, $\mathcal{M}$, h) are *publicly* known.

The *secret key* is an element $\alpha \in \mathbb{F}_q^*$ which is known only to the signer.

To sign a message $\mu \in \mathcal{M}$, the signer chooses a random integer $k \in \mathbb{F}_q^*$ usually called the *nonce*, and which must be kept secret. We define the following two elements of $\mathbb{F}_q$:

$$r(k) = \left\lfloor \lfloor g^k \rfloor_p \right\rfloor_q, \qquad s(k, \mu) = \lfloor k^{-1}\left(h(\mu) + \alpha r(k)\right) \rfloor_q.$$

The pair $(r(k), s(k, \mu))$ is the *DSA signature* of the message μ with a nonce k.

The attack on the DSA which has been developed in [69] (and which simplifies and improves the attack from [42]) is based on solving the **HNP** with the sequence

$$t(k, \mu) = \lfloor 2^{-\ell} r(k) s(k, \mu)^{-1} \rfloor_q, \qquad (k, \mu) \in \mathcal{S}, \tag{12}$$

where $\mathcal{S}$ is the set of pairs $(k, \mu) \in [1, q-1] \times \mathcal{M}$ with $s(k, \mu) \neq 0$.

Denote by W the number of solutions of the equation $h(\mu_1) = h(\mu_2)$, $\mu_1, \mu_2 \in \mathcal{M}$. Thus, $W/|\mathcal{M}|^2$ is the probability of collision and expected to be of order q^{-1} for any practically usable hash function.

In [71] the heuristic results of [69] have been made rigorous.

The central problem is bounding the exponential sums

$$T(c) = \sum_{(k,\mu)\in\mathcal{S}} \mathbf{e}_q\left(ct(k, \mu)\right)$$

where $\mathcal{S}$ the set of pairs $(k, \mu) \in [1, q-1] \times \mathcal{M}$ with $s(k, \mu) \neq 0$ (that is, the set of pairs (k, μ) for which $t(k, \mu)$ is defined).

The following bound of these sums uses

- bounds of exponential sums with exponential functions of S. V. Konyagin and I. E. Shparlinski [52] given by Theorems 5.2 and 5.3;
- **Weil's** bound given by Theorem 5.1;
- **Vinogradov's** method of estimates of double sums [102, 103].

The **main difficulty** is that the double reduction erases any number-theoretic structure among the values of $r(k)$.

Theorem 8.4: *Let Q be a sufficiently large integer. The following statement holds with $\vartheta = 1/3$ for all primes $p \in [Q, 2Q]$, and with $\vartheta = 0$ for all primes $p \in [Q, 2Q]$ except at most $Q^{5/6+\varepsilon}$ of them. For any $\varepsilon > 0$ there exists $\delta > 0$ such that for any $g \in \mathbb{F}_p$ of multiplicative order $q \geq p^{\vartheta+\varepsilon}$ and with the sequence (12) the bound*

$$\max_{\gcd(c,q)=1} |T(c)| = O\left(W^{1/2}q^{3/2-\delta}\right)$$

holds.

Proof: For $\lambda \in \mathbb{F}_q$ we denote by $H(\lambda)$ the number of $\mu \in \mathcal{M}$ with $h(\mu) = \lambda$.

We also define the integer $a \in [1, q-1]$ by the congruence $a \equiv 2^{-\ell}c_0$ (mod q).

We have

$$|T(c)| \leq \sum_{\lambda \in \mathbb{F}_q} H(\lambda) \left| \sum_{\substack{k=1 \\ ar(k) \not\equiv -\lambda \pmod q}}^{q-1} \mathbf{e}_q\left(a\frac{kr(k)}{\lambda + ar(k)}\right) \right|.$$

Applying the Cauchy inequality we obtain

$$|T(c)|^2 \leq \sum_{\lambda \in \mathbb{F}_q} H(\lambda)^2$$

$$\times \sum_{\lambda \in \mathbb{F}_q} \left| \sum_{\substack{k=1 \\ ar(k) \not\equiv -\lambda \pmod q}}^{q-1} \mathbf{e}_q\left(a\frac{kr(k)}{\lambda + ar(k)}\right) \right|^2.$$

The second sum does not depend on h anymore! (Vinogradov's trick) First of all we remark that

$$\sum_{\lambda \in \mathbb{F}_q} H(\lambda)^2 = W. \tag{13}$$

Furthermore,

$$\sum_{\lambda \in \mathbb{F}_q} \left| \sum_{\substack{k=1 \\ \alpha r(k) \not\equiv -\lambda \pmod{q}}}^{q-1} e_q\left(a\frac{kr(k)}{\lambda + \alpha r(k)}\right) \right|^2$$

$$= \sum_{\lambda \in \mathbb{F}_q} \sum_{\substack{k=1 \\ \alpha r(k) \not\equiv -\lambda \pmod{q}}}^{q-1} \sum_{\substack{m=1 \\ \alpha r(m) \not\equiv -\lambda \pmod{q}}}^{q-1} e_q\left(aF_{k,m}(\lambda)\right)$$

$$= \sum_{k,m=1}^{q-1} \sum_{\lambda \in \mathbb{F}_q} {}^* e_q\left(aF_{k,m}(\lambda)\right),$$

where

$$F_{k,m}(X) = \frac{kr(k)}{X + \alpha r(k)} - \frac{mr(m)}{X + \alpha r(m)}$$

and the symbol $\sum^*$ means that the summation in the inner sum is taken over all $\lambda \in \mathbb{F}_q$ with

$$\lambda \not\equiv -\alpha r(k) \pmod{q}, \qquad \lambda \not\equiv -\alpha r(m) \pmod{q}.$$

Thus, Theorem 5.1 applies (Weil's bound) **unless** $F_{k,m}(X)$ is constant in $\mathbb{F}_q$.

The function $F_{k,m}(X)$ is constant only if $k = m$ or $r(k) = r(m) = 0$ (in which case we use the trivial bound).

The condition $r(k) = 0$ is equivalent to

$$g^k \equiv qx \pmod{p}, \qquad k \in [1, q-1], \ x \in [0, L],$$

where

$$L = \left\lfloor \frac{p-1}{q} \right\rfloor.$$

Using Theorems 5.2 and 5.3 and the method of proof of Theorem 3.5, one can now prove that under the conditions of Theorem 8.4 the last congruence has $O(q^{1-\delta})$ solutions for some $\delta > 0$.

There are also q pairs $k = m$.

Putting everything together gives

$$|T(c)|^2 = O\left(W\left(q^2 \cdot q^{1/2} + (q + q^{2-2\delta})q\right)\right)$$

and the desired result follows. $\qquad\qquad\square$

Using (11) we see that under the conditions of Theorem 8.4 the sequence (12) is $q^{-\delta/3}$-homogeneously distributed modulo q, provided that

$$W \le \frac{(\#\mathcal{M})^2}{q^{1-\delta}}. \tag{14}$$

This result is based on a combination of the bounds of exponential sums with exponential functions from [52] given in Theorem 5.2 and Theorem 5.3, with the *Weil* bound (see [62]) and the Vinogradov method of estimates of double sums. As we have mentioned, the inequality (14) usually holds in the stronger form $W = O\left(|\mathcal{M}|^2/q\right)$.

Then the above arguments together with Theorem 8.2 imply the following statement.

For an integer ℓ we define the oracle $\mathcal{DSA}_\ell$ which, for any given DSA signature $(r(k), s(k, \mu))$, $k \in [0, q-1]$, $\mu \in \mathcal{M}$, returns the ℓ least significant bits of k.

Theorem 8.5: *Let Q be a sufficiently large integer. The following statement holds with $\vartheta = 1/3$ for all primes $p \in [Q, 2Q]$, and with $\vartheta = 0$ for all primes $p \in [Q, 2Q]$ except at most $Q^{5/6+\varepsilon}$ of them. For any $\varepsilon > 0$ there exists $\delta > 0$ such that for any element $g \in \mathbb{F}_p$ of multiplicative order q, where $q \ge p^{\vartheta+\varepsilon}$ is prime, and any hash function h satisfying (14), given an oracle $\mathcal{DSA}_\ell$ with $\ell = \left\lceil \log^{1/2} q \right\rceil + \lceil \log \log q \rceil$, there exists a probabilistic polynomial-time algorithm to recover the DSA secret key α, from $O\left(\log^{1/2} q\right)$ signatures $(r(k), s(k, \mu))$ with $k \in [0, q-1]$ and $\mu \in \mathcal{M}$ selected independently and uniformly at random. The probability of success is at least $1 - 2^{-(\log \log q) \log^{1/2} q}$.*

The same result holds for most significant bits and (in a marginally weaker form) for bit strings in the middle.

Practically: Numerical experiments (see [69, 70]) show that

- 4 bits of k are always enough,
- 3 bits are often enough,
- 2 bits are possibly enough as well.

Moral:

(1) Do not use **small** k (to cut the cost of exponentiation in $r(k)$). This can be very tempting because there is no $K^{1/2}$-attack (Shanks' *Baby–step–Giant–step* and Pollard's *Rho* do not apply) on $r(k)$ with $k \in [0, K]$.

(2) Protect your software/hardware against **timing/power attacks** when the attacker measures the time/power consumption and selects the signatures for which this value is smaller than "on average" – these signatures are likely to correspond to small values of k (because they correspond to faster exponentiation in $r(k)$, timing for other parts of the algorithm is about the same for all k and μ).

(3) Use quality **pseudorandom number generators** to generate k, biased generators are dangerous.

(4) Do not use **Arazi's cryptosystem** which combines DSA and Diffie-Hellman key exchange protocol – it leaks some bits of k (has just been noticed by *Don Brown & Alfred Menezes, 2001*).

(5) Do not buy CRYPTOLIB from **AT&T**: it always uses odd values of k, thus one bit is leaked immediately, one more and .... This was observed by Daniel Bleichenbacher and actually was the main motivation for studying this problem.

8.2.6. *Other applications and open questions*

The method of the proof of Theorem 8.3 can be used to establish the bit security of several other exponentiation based cryptographic algorithms. Several such schemes, including the *ElGamal cryptosystem* (see Section 8.4 in [66]) and the *Shamir message passing scheme* (see Protocol 12.22 of [66]), have been outlined in [8, 9]. As yet another example we also mention the *Matsumoto–Takachima–Imai key-agreement protocol*, see Section 12.6 of [66]. In fact, the treatment of the Shamir message passing scheme in [8] has the same gap as the treatment of the Diffie-Hellman scheme. Accordingly, using exponential sums this gap has been fixed in [31].

In [92] several results on the recently introduced *XTR cryptosystem* [56, 57] were shown. However, these results are substantially weaker than those known for the aforementioned. The main reason for this is that in studying the XTR the corresponding character sums are over small subgroups of *extension* fields and for such sums there is no analogue of Theorem 5.2 and Theorem 5.3. Accordingly, the paper [92] uses a different way of estimating the distribution of multipliers t of the corresponding FHNP. Unfortunately, this leads to a substantially weaker result. To be more precise, to apply an analogue of the approach of Section 8.2.4 to the XTR, one needs to improve the bounds of the exponential sums

$$\max_{\gamma \in \mathbb{F}_{p^6}^*} \left| \sum_{t \in \mathcal{G}} \exp\left(2\pi i \mathrm{Tr}\left(\gamma t\right)/p\right) \right| \leq p^3,$$

where $\mathrm{Tr}(z) = z + z^p + \ldots + z^{p^5}$ is the trace of $z \in \mathbb{F}_{p^6}$ in $\mathbb{F}_p$ and $\mathcal{G}$ is a subgroup of $\mathbb{F}_{p^6}^*$, see Theorem 8.78 in [62] (combined with Theorem 8.24 of the same work) or the bound (3.15) in [52]. This bound is trivial for $\#\mathcal{G} \leq p^3$ while the subgroups relevant to XTR are of size of order p^2. Thus, in [92] an alternative approach has been used which is based on the fact, even if the sequence $\mathcal{T}$ is not known to be homogeneously distributed but at least admits a non-trivial upper bound for the number of its elements in an interval, one can still obtain some analogues of (10). Then the upper bound from [13] on the number of zeros of sparse polynomials can used to extract such information. However, the ball is now back to the exponential sum technique. Using some new bounds of short exponential sums in finite fields, W. W.-C. Li, M. Näslund and I. E. Shparlinski [61] proved for XTR a result of about the same strength as that known for the Diffie-Hellman scheme.

The result of Theorem 8.5 has been extended to other DSA-like signature schemes, including the *elliptic curve* version of DSA in [23,71]. In particular, the bound of [51] provides an analogue of Theorem 5.2 for exponential sums over an orbit generated by a point on an elliptic curve, see [71]. However, some interesting questions still remain open. For example, for the *Nyberg–Rueppel* signature scheme the range of p and q in which the results of [23] are nontrivial are narrower than in practical applications. It is shown in [23] that the attack designed in that paper on the Nyberg–Rueppel signature scheme can be reduced to **EHNP** with the sequence of multipliers

$$r(k,\mu) = \left\lfloor \left\lfloor h(\mu)g^k \right\rfloor_p \right\rfloor_q, \qquad (k,\mu) \in [1, q-1] \times \mathcal{M}.$$

Unfortunately, it is not clear how to estimate the exponential sums

$$\sum_{\mu \in \mathcal{M}} \sum_{k \in \mathbb{F}_q^*} \exp\left(2\pi i c r(k,\mu)\right), \qquad c \in [1, q-1],$$

and obtaining such a bound is an interesting open question. Using a rather indirect method, it has been shown in [23] that the sequence $r(k,\mu)$ is $2^{-\log^{1/2} q}$ homogeneously distributed modulo q, provided that

$$W \leq \frac{(\#\mathcal{M})^2 q^{3-\delta}}{p^3}$$

for some $\delta > 0$. We remark that in the settings of the Nyberg–Rueppel signature scheme it is natural to assume that h is bijective, that is, $W =$

$\#\mathcal{M}$. Also, if the message set $\mathcal{M}$ is "dense" (that is, $\#\mathcal{M}$ is of order p), then the above result holds for $q \geq p^{2/3+\delta}$. It would be very interesting to lower this bound.

Yet another modification of the **HNP** has recently been introduced in [41]. Namely, that paper introduces the following

HIDDEN NUMBER PROBLEM WITH HIDDEN MULTIPLIER, **HNP-HM**: *Recover a number $\alpha \in \mathbb{F}_p$ such that for many unknown random $t \in \mathcal{T}$ we are given* $\mathrm{MSB}_{\ell,p}(\alpha t)$, $\mathrm{MSB}_{\ell,p}(t)$ *and* $\mathrm{MSB}_{\ell,p}(\alpha)$ *for some $\ell > 0$.*

In the case $\mathcal{T} = \mathbb{F}_p^*$ and $\ell \geq (4/5 + \varepsilon) \log p$ the paper [41] provides a polynomial-time algorithm for the **HNP-HM**. In fact, it also works in more general residue rings (which is important for applications to [80]). As one can see, this result is substantially weaker than those known for **HNP** and **EHNP** where one can take ℓ of order $\log^{1/2} p$. However, using exponential sums, it has been shown in [41] that indeed for **HNP-HM** to have a unique solution the value of ℓ must be very large. Namely for $\ell \leq (1/2 + \varepsilon) \log p$ there can be exponentially many possibilities for α.

The aforementioned algorithm has been used in [41] to establish a certain bit security result for the "timed-release crypto" introduced by Rivest, Shamir and Wagner [80] and also to design a "correcting" algorithm for noisy exponentiation black-boxes.

It is an interesting and challenging problem to study **HNP-HM** for more general sequences $\mathcal{T}$, in particular for subgroups of $\mathbb{F}_p^*$.

In the case $\mathcal{T} = \mathbb{F}_p^*$ the paper [9] provides a *non-uniform* polynomial-time algorithm for the **HNP** which works with $\ell = O(\log \log p)$. We recall that non-uniformity means that the algorithm exists, but to actually design this algorithm one may need exponential time (thus such algorithms are of rather limited value). Nevertheless, it would be of interest to extend this result to subgroups of $\mathbb{F}_p^*$. In order to get such a generalisation, one needs an analogue of Lemma 2.4 of [9] for subgroups and this seems to be a rather feasible task, taking into account the bounds of exponential sums of Theorem 5.2 and Theorem 5.3.

Finally, several more modifications of the **HNP** have been considered in the papers [7, 30, 50, 61, 95, 96, 101]. However, they are of more algebraic than geometric nature and lattices have not been involved in their study.

9. Applications to Algorithms

9.1. *Primitive roots*

The **main problem** in this area can be described as follows: Given a finite field $\mathbb{F}_q$, find a primitive root of $\mathbb{F}_q$.

Unfortunately, obtaining a deterministic polynomial-time algorithm for this problem seems to be out of reach nowadays. In particular, just primitivity testing already seems infeasible without the knowledge of the prime factorization of $q - 1$.

Thus, one can try to compromise and consider a presumably simpler problem: Given a field $\mathbb{F}_q$, find a small set $M \subset \mathbb{F}_q$ containing at least one primitive root of $\mathbb{F}_q$.

In fact, for many applications one can just use all elements from M without testing which one is primitive.

Fortunately, for this problem some efficient algorithms have been designed by Shoup [84] and Shparlinski [85] who proved that for any p and n, in time $pn^{O(1)}$ one can find a set $M \subseteq \mathbb{F}_{p^n}$ of size

$$|M| = O(pn^{6+\varepsilon})$$

containing at least one primitive root of $\mathbb{F}_{p^n}$.

This result has been slightly improved in [47] where it has been shown that for any p and n, in time $p^{1/2}n^{O(1)}$ one can find a set $M \subseteq \mathbb{F}_{p^n}$ of size

$$|M| = O(p^{1/2}n^{O(1)})$$

containing at least one primitive root of $\mathbb{F}_{p^n}$.

Several more related results can also be found in [87].

In particular, if p is fixed (for example, $p = 2$), then the set M in the above constructions is of polynomial size.

Certainly there is no need to stress that exponential and character sums play a central role in the aforementioned constructions.

More precisely, they rely on the following bound obtained by Carlitz [15] and then rediscovered by Katz [48].

Let r be a prime power and let α be a root of an irreducible polynomial of degree k over $\mathbb{F}_r$ and let χ be a multiplicative character of $\mathbb{F}_{r^k}$. Then

$$\left| \sum_{t \in \mathbb{F}_r} \chi(\alpha + t) \right| \le kr^{1/2}. \tag{15}$$

The bound is nontrivial for $k \le r^{1/2-\varepsilon}$. For k of this order the sum is very **short** compared to the field size. Therefore, we have a "small" set with a non-trivial bound of character sums; thus we can study the distribution of primitive roots in such sets. In [79] this bound has been extended to sums over sequences of consecutive integers of length $h < r$ (where r is a prime number).

It is very tempting to try to fix a small subfield $\mathbb{F}_r \subset \mathbb{F}_q$ (with, say, $r \sim \log^6 q$), find an irreducible polynomial $f \in \mathbb{F}_r[X]$ of degree $k = \log q / \log r$ and put $M = \mathbb{F}_r + \alpha$, $f(\alpha) = 0$.

Certainly this *naive way* has an obvious flaw — the required subfield may not exist.

However, there is a way to get around this problem.

Let $q = p^n$. Select

$$k = \left\lfloor \frac{\log q}{6 \log \log q} \right\rfloor,$$

find an irreducible polynomial $f \in \mathbb{F}_q[X]$ of degree k and construct $\mathbb{F}_{q^k}$. Then we have $\mathbb{F}_{p^k} \subset \mathbb{F}_{q^k}$ and the field $\mathbb{F}_{p^k}$ is of the required size, so our naive approach applies to the field $\mathbb{F}_{q^k}$, producing a small set R containing a primitive root of $\mathbb{F}_{q^k}$. And now we "return" to $\mathbb{F}_q$ by putting

$$M = \{\rho^{(q^k-1)/(q-1)} \; : \; \rho \in R\}.$$

Obviously, if ρ is a primitive root of $\mathbb{F}_{q^k}$, then $\rho^{(q^k-1)/(q-1)}$ is a primitive root of $\mathbb{F}_q$. Hence M contains a primitive root.

Although we still cannot identify this primitive root among the elements of M, the above approach can be useful for several problems in coding theory, cryptography, graph theory, combinatorial designs, pseudorandom number generators, sparse polynomial interpolation and some other areas.

9.2. *Pseudorandom regular graphs*

One of the most challenging problems in this area is finding explicit constructions of "sparse" regular graphs of small diameter. This problem is closely related to the problem of constructing "sparse" regular graphs with small second largest eigenvalue.

Such graphs have numerous applications in combinatorics, networking, coding theory, complexity theory ... and they are just nice.

Let us fix a set $S = \{s_1, \ldots, s_r\} \subseteq \mathbb{Z}/m\mathbb{Z}$.

The *difference graph* $G(S, m)$ is an m-vertex directed graph such that vertices i and j are connected if and only if the residue of $i - j$ modulo m is in S.

Similarly one can define undirected *sum graphs*.

Here we consider only difference graphs.

It is easy to show by using the properties of circulant matrices that the *eigenvalues* of $G(S, m)$ are given by

$$\lambda_{k+1} = \sum_{\nu=1}^{r} \exp(2\pi i k s_\nu / m), \qquad k = 0, \ldots, m - 1.$$

The following construction has been proposed by F. R. K. Chung [16], see also [17].

Let $f \in \mathbb{F}_q[x]$ be an irreducible polynomial of degree $\deg f = n$. Fix a root $\alpha \in \mathbb{F}_{q^n}$ of f, thus $\mathbb{F}_q(\alpha) = \mathbb{F}_{q^n}$.

Then the graph $G(f, n, q)$ is defined as follows: we identify the vertices of $G(f, n, q)$ with elements of $\mathbb{F}_{q^n}^*$ and we connect the vertices $\tau, \mu \in \mathbb{F}_{q^n}^*$ if and only if $\tau = \mu(\alpha + t)$ for some $t \in \mathbb{F}_q$.

It has been shown in [16] that the bound (15) implies the following result:

Theorem 9.1: *If $q^{1/2} > n - 1$, then $G(f, n, q)$ is a connected q-regular graph with $|G(f, n, q)| = q^n - 1$ vertices and the diameter*

$$D(G(f, n, q)) \leq 2n + 1 + \frac{4n \log n}{\log q - 2 \log(n - 1)}.$$

Moreover, for the second largest eigenvalue the bound

$$\lambda(G(f, n, q)) \leq (n - 1)q^{1/2}$$

holds.

The above construction has been generalised in [86]. For a prime number p and an integer h with $1 \leq h < p$ the graph $G(f, n, p, h)$ is defined as follows: we identify the vertices of $G(f, n, p, h)$ with elements of $\mathbb{F}_{p^n}^*$ and we connect the vertices $\tau, \mu \in \mathbb{F}_{q^n}^*$ if and only if $\tau = \mu(\alpha + t)$ for some $t \in \{0, \ldots, h - 1\}$.

It has been shown in [86] that the bound of exponential sums of [79], generalising (15), allows to obtain non-trivial results for such graphs, provided that $p^{1/2+\varepsilon} \leq h \leq p$. In particular, for the second largest eigenvalue of $G(f, n, p, h)$ the bound

$$\lambda(G(f, n, p, h)) = O(np^{1/2} \log p)$$

holds.

Despite these and many other important applications of exponential sums to graph theory, sometimes other number-theoretic methods give more exact results. For example, for very large q a better bound on the diameter (about n rather than $2n$) has been obtained by S. D. Cohen [18, 19]. The method is based on more sophisticated tools, namely on the Lang–Weil bound for algebraic varieties rather than on the Weil bound for curves, see also [49].

Several more exciting links between exponential sums and graph theory can be found in [59, 60].

9.3. *Polynomial factorisation*

A nice application of bounds of character sums to polynomial factorisation over finite fields has been found by V. Shoup [83].

It is well known that the polynomial factorisation problem can be easily reduced to factorisation of squarefree polynomials over prime fields.

The algorithm is very simple, to factor a squarefree polynomial $f \in \mathbb{F}_p[X]$ we compute

$$L_t(X) = \gcd\left((X+t)^{(p-1)/2} - 1, f(X)\right), \quad t = 0, 1, \ldots, Q,$$

where Q is the main parameter of the algorithm, hoping that at least one polynomial L_t is *nontrivial*, that is, is equal to neither 1 nor f.

For each t the polynomial L_t can be computed in a very efficient way, if one uses repeated squaring to compute

$$g_t(X) \equiv (X+t)^{(p-1)/2} \pmod{f(X)}, \qquad \deg g_t < \deg f,$$

and then computes

$$L_t(X) = \gcd\left(g_t(X) - 1, f(X)\right)$$

via the Euclidean algorithm.

We recall that for $x \in \mathbb{F}_p$, the equation $x^{(p-1)/2} = 1$ holds if and only if x is a quadratic residue modulo p.

Hence, if L_t is trivial, then for any two distinct roots a, b of f we have

$$\chi(a+t) = \chi(b+t), \quad t = 0, 1, \ldots, Q,$$

where χ is the quadratic character. Because $a \neq b$, the case $\chi(a+t) = \chi(b+t) = 0$ is not possible. Therefore, if all our attempts fail, then

$$\sum_{t=0}^{Q} \chi\left((a+t)(b+t)\right) = Q+1.$$

On the other hand, V. Shoup [83] has noticed that the Weil bound implies that sums of this type are of order $p^{1/2} \log p$.

Therefore, for some $Q = O(p^{1/2} \log p)$ one of the L_t is nontrivial!

It has been shown in [88] that in fact the same statement holds for some $Q = O(p^{1/2})$. This leads to the best known **deterministic** polynomial factorisation algorithm.

Moreover, J. von zur Gathen and I. E. Shparlinski [27] have shown that the same technique leads to a **deterministic** algorithm for finding all rational points of a plane curve in polynomial time "on average" per point. This may have applications to algebraic-geometry codes and maybe to some other areas.

9.4. *Complexity lower bounds*

Exponential sums can be an efficient tool not only in algorithm design and analysis, but in establishing lower complexity bounds of some problems as well.

For example, it has been shown by J. von zur Gathen and I. E. Shparlinski [28] that, for some absolute constant $c > 0$, if the modulus m is not highly composite (for example, if m is prime), then computing the inversion x^{-1} (mod m) takes at least $c \log \log m$ for the parallel time on an exclusive-write parallel random access machine (CREW PRAM). It is remarkable that if m has many small prime divisors (that is, it is highly composite), then one can compute x^{-1} (mod m) in $O(\log \log m)$ on a CREW PRAM, see [26]. Although generally speaking these lower bounds and algorithms require somewhat opposite properties of the moduli, there is a wide class of moduli where they both apply and match each other, thus giving a very rare example of a nontrivial complexity-theory problem where the lower and upper bounds coincide. For example, this holds for moduli $m = p_1 \cdots p_k$, where $p_1, \ldots, p_k$ are any $k = \lceil s/\log s \rceil$ prime numbers between s^3 and $2s^3$.

Applications of exponential sums to estimating Fourier coefficients of various Boolean functions related to several cryptographic and number-theoretic problems can be found in [20, 89, 90].

10. Tutorial Problems

Problem 10.1: Let

$$S(a) = \sum_{x=1}^{p-1} e_p(ax^n).$$

From the bound

$$\max_{1 \le a \le p-1} |S(a)| \le np^{1/2}$$

derive that the number of the n-th degree residues (that is, integers $a \not\equiv 0 \pmod{p}$ for which the congruence $a \equiv z^n \pmod{p}$ is solvable) in any interval $[k+1, k+h]$ of length $1 \le h \le p$ is $h/n + O(np^{1/2} \log p)$.

Problem 10.2: Show that for a fixed n and sufficiently large p, any c can be represented as

$$c \equiv x^n + y^n + z^n \pmod{p}, \quad 0 \le x, y, z \le p - 1.$$

Hint: For $c \equiv 0 \pmod{p}$ this is obvious. For $c \not\equiv 0 \pmod{p}$ the last congruence is solvable if and only if $cw^n \equiv x^n + y^n + z^n \pmod{p}$, with some $0 \le x, y, z \le p - 1$, $1 \le w \le p - 1$.

Problem 10.3: Let

$$S(a, b) = \sum_{x=1}^{p-1} e_p(ax^n + bx).$$

Prove that

$$\sum_{u,v=0}^{p-1} |S(u, v)|^4 \le 2np^4.$$

Problem 10.4: Show that for $b \not\equiv 0 \pmod{p}$

$$|S(a, b)| \le 2n^{1/4}p^{3/4}.$$

Hint: For any $y \not\equiv 0 \pmod{p}$, $S(a, b) = S(ay^n, by)$, therefore

$$(p - 1)|S(a, b)|^4 \le \sum_{u,v=0}^{p-1} |S(u, v)|^4.$$

Problem 10.5: Let $n | p - 1$. Prove that for $b \not\equiv 0 \pmod{p}$

$$|S(a, b)| \le p/n^{1/2}.$$

Hint: Let $k = (p - 1)/n$. For $y \not\equiv 0 \pmod{p}$,

$$S(a, b) = \sum_{x=1}^{p-1} e_p\left(a(xy^k)^n + bxy^k\right)$$

$$= \sum_{x=1}^{p-1} e_p\left(ax^n + bxy^k\right).$$

Thus

$$(p-1)|S(a,b)| = \left| \sum_{x=1}^{p-1} \mathbf{e}_p\left(ax^n\right) \sum_{y=1}^{p-1} \mathbf{e}_p\left(bxy^k\right) \right|$$

$$\leq \sum_{x=1}^{p-1} \left| \sum_{y=1}^{p-1} \mathbf{e}_p\left(bxy^k\right) \right|$$

$$\leq \left(p \sum_{x=1}^{p-1} \left| \sum_{y=1}^{p-1} \mathbf{e}_p\left(bxy^k\right) \right|^2 \right)^{1/2}.$$

Problem 10.6: Combine the previous bound with the Weil bound

$$|S(a,b)| \leq np^{1/2}$$

and show that for *any* $n|p-1$

$$|S(a,b)| \leq p^{5/6}.$$

Problem 10.7: Show that for any quadratic character χ and $a \not\equiv b$ (mod p)

$$\sum_{x=0}^{p} \chi(x+a)\chi(x+b) = -1.$$

Problem 10.8: Show that for any nontrivial multiplicative character χ and $a \not\equiv b$ (mod p)

$$\sum_{x=0}^{p} \chi(x+a)\overline{\chi(x+b)} = -1,$$

where $\overline{z}$ denotes the complex conjugation.

Problem 10.9: Show that for any arbitrary subsets $\mathcal{X}, \mathcal{Y}$ of $\mathbb{F}_p$ and any nontrivial multiplicative character χ,

$$\left| \sum_{x \in \mathcal{X}} \sum_{y \in \mathcal{Y}} \chi(x+y) \right| \leq (p \# \mathcal{X} \# \mathcal{Y})^{1/2}.$$

Problem 10.10: Show that for any nontrivial multiplicative character χ and $a \not\equiv 0$ (mod p)

$$\left| \sum_{x=0}^{p} \chi(x)\mathbf{e}_p(ax) \right| = p^{1/2}.$$

Hint: For any $y \not\equiv 0 \pmod{p}$,

$$\sum_{x=0}^{p} \chi(x)e_p(ax) = \sum_{x=0}^{p} \chi(xy)e_p(ayx).$$

Therefore

$$(p-1)\left|\sum_{x=0}^{p} \chi(x)e_p(ax)\right|^2 = \sum_{b=1}^{p-1}\left|\sum_{x=0}^{p} \chi(x)e_p(bx)\right|^2.$$

Problem 10.11: Let $n|p-1$ and Ω_n be the set of all multiplicative characters χ for which χ^n is the trivial character, $\chi^n = \chi_0$. Prove that $|\Omega_n| = n$ and that

$$\sum_{\chi \in \Omega_n} \chi(u) = \begin{cases} n, & \text{if } u \equiv x^n \pmod{p} \text{ is solvable,} \\ 0, & \text{otherwise.} \end{cases}$$

Problem 10.12: Let $n|p-1$. Prove that

$$\max_{1 \le a \le p-1}\left|\sum_{x=1}^{p-1} e_p(ax^n)\right| \le np^{1/2}.$$

Hint: Show that

$$\sum_{x=1}^{p-1} e_p(ax^n) = \sum_{x=1}^{p-1} e_p(ax) \sum_{\chi \in \Omega_n} \chi(x).$$

Problem 10.13: The following sums are known as *Kloosterman sums*

$$K(a,b) = \sum_{x=1}^{p} e_p(ax + bx^{-1}),$$

where x^{-1} is the multiplicative inverse modulo p of x. Using the Weil bound

$$\max_{\gcd(a,b,p)=1} |K(a,b)| \le 2p^{1/2},$$

derive an upper bound on incomplete sums

$$K_{M,N}(b) = \sum_{x=M+1}^{M+N} e_p(bx^{-1})$$

and then the asymptotic formula for the number of $x \in [M+1, M+N]$ for which

$$x^{-1} \pmod{p} \in [k+1, k+h],$$

for integers M, N, k, h, where $1 \le h, N \le p$.

References

1. M. Ajtai, R. Kumar and D. Sivakumar, 'A sieve algorithm for the shortest lattice vector problem', *Proc. 33rd ACM Symp. on Theory of Comput.*, Crete, Greece, July 6-8, 2001, 601–610.
2. L. Babai, 'On Lovász' lattice reduction and the nearest lattice point problem', *Combinatorica*, **6** (1986), 1–13.
3. A. M. Barg, 'Incomplete sums, *DC*-constrained codes, and codes that maintain synchronization', *Designs, Codes and Cryptography*, **3** (1993), 105–116.
4. A. M. Barg, 'A large family of sequences with low periodic correlation', *Discr. Math.*, **176** (1997), 21–27.
5. A. M. Barg and S. N. Litsyn, 'On small families of sequences with low periodic correlation', *Lect. Notes in Comp. Sci.*, Springer-Verlag, Berlin, **781** (1994), 154–158.
6. L. A. Bassalygo and V. A. Zinoviev, 'Polynomials of special form over a finite field with maximum modulus of the trigonometric sum', *Uspechi Matem. Nauk*, **52** (1997) 2, 31–44 (in Russian).
7. D. Boneh and I. E. Shparlinski, 'On the unpredictability of bits of the elliptic curve Diffie–Hellman scheme', *Lect. Notes in Comp. Sci.*, Springer-Verlag, Berlin, **2139** (2001), 201–212.
8. D. Boneh and R. Venkatesan, 'Hardness of computing the most significant bits of secret keys in Diffie–Hellman and related schemes', *Lect. Notes in Comp. Sci.*, Springer-Verlag, Berlin, **1109** (1996), 129–142.
9. D. Boneh and R. Venkatesan, 'Rounding in lattices and its cryptographic applications', *Proc. 8th Annual ACM-SIAM Symp. on Discr. Algorithms*, ACM, New York, 1997, 675–681.
10. V. Boyko, M. Peinado and R. Venkatesan, 'Speeding up discrete log and factoring based schemes via precomputations', *Lect. Notes in Comp. Sci.*, Springer-Verlag, Berlin, **1403** (1998), 221–234.
11. E. Brickell, D.M. Gordon, K.S. McCurley and D. Wilson, 'Fast exponentiation with precomputation', *Lect. Notes in Comp. Sci.*, Springer-Verlag, Berlin, **658** (1993), 200–207.
12. D. A. Burgess, 'The distribution of quadratic residues and non-residues', *Mathematika*, **4** (1957), 106–112.
13. R. Canetti, J. B. Friedlander, S. V. Konyagin, M. Larsen, D. Lieman and I. E. Shparlinski, 'On the statistical properties of Diffie–Hellman distributions', *Israel J. Math.*, **120** (2000), 23–46.
14. R. Canetti, J. B. Friedlander and I. E. Shparlinski, 'On certain exponential sums and the distribution of Diffie–Hellman triples', *J. London Math. Soc.*, **59** (1999), 799–812.
15. L. Carlitz, 'Distribution of primitive roots in a finite field', *Quart. J. Math. Oxford*, **4** (1953), 4–10.
16. F. R. K. Chung, 'Diameters and eigenvalues', *J. Amer. Math. Soc.*, **2** (1989), 187–196.
17. F. R. K. Chung, *Spectral graph theory*, Regional Conf. Series in Math., Vol. 92, Amer. Math. Soc., Providence, RI, 1997.

18. S. D. Cohen, 'Polynomial factorization, graphs, designs and codes', *Contemp. Math.*, Vol. 168, Amer. Math. Soc., Providence, RI, 1994, 23–32.

19. S. D. Cohen, 'Polynomial factorization and an application to regular directed graphs', *Finite Fields and Their Appl.*, 4 (1998), 316–346.

20. D. Coppersmith and I. E. Shparlinski, 'On polynomial approximation of the discrete logarithm and the Diffie–Hellman mapping', *J. Cryptology*, **13** (2000), 339–360.

21. T. W. Cusick and H. Dobbertin, 'Some new three-valued correlation functions for binary sequences', *IEEE Trans. Inform. Theory*, **42** (1996), 1238–1240.

22. P. Deligne, 'La conjecture de Weil, I', *Inst. Hautes Etudes Sci. Publ. Math.*, **43** (1974), 273–307.

23. E. El Mahassni, P. Q. Nguyen and I. E. Shparlinski, 'The insecurity of Nyberg–Rueppel and other DSA-like signature schemes with partially known nonces', *Lect. Notes in Comp. Sci.*, Springer-Verlag, Berlin, **2146** (2001), 97–109.

24. J. B. Friedlander, M. Larsen, D. Lieman and I. E. Shparlinski, 'On correlation of binary M-sequences', *Designs, Codes and Cryptography*, **16** (1999), 249–256.

25. J. B. Friedlander and I. E. Shparlinski, 'On the distribution of Diffie–Hellman triples with sparse exponents', *SIAM J. Discr. Math.*, **14** (2001), 162–169.

26. J. von zur Gathen, 'Computing powers in parallel', *SIAM J. Comp.*, **16** (1987), 930–945.

27. J. von zur Gathen and I. E. Shparlinski, 'Finding points on curves over finite fields', *Proc. 36th IEEE Symposium on Foundations of Computer Science*, Milwaukee, 1995, IEEE Press, 1995, 284–292.

28. J. von zur Gathen and I. E. Shparlinski, 'The CREW PRAM complexity of modular inversion', *SIAM J. Comp.*, **29** (1999), 1839–1857.

29. C. F. Gauss, *Disquisitiones arithmeticae*, Fleischer, Leipzig, 1801.

30. M. I. González Vasco, M. Näslund and I. E. Shparlinski, 'The hidden number problem in extension fields and its applications', *Lect. Notes in Comp. Sci.*, Springer-Verlag, Berlin, **2286** (2002), 105–117.

31. M. I. González Vasco and I. E. Shparlinski, 'On the security of Diffie–Hellman bits', *Proc. Workshop on Cryptography and Computational Number Theory*, Singapore 1999, Birkhäuser, 2001, 257–268.

32. M. I. González Vasco and I. E. Shparlinski, 'Security of the most significant bits of the Shamir message passing scheme', *Math. Comp.*, **71** (2002), 333–342.

33. G. H. Hardy and J. E. Littlewood, 'Some problems of "Partitio Numerorum". I A new solution of Waring's problem', *Göttingen Nachrichten*, 1920, 231–267.

34. D. R. Heath-Brown and S. Konyagin, 'New bounds for Gauss sums derived from kth powers, and for Heilbronn's exponential sum', *Quart. J. Math.*, **51** (2000), 221–235.

35. T. Helleseth, 'Some results about the cross-correlation function between two maximal linear sequences', *Discr. Math.*, **16** (1976), 209–232.

36. T. Helleseth, 'A note on the cross-correlation function between two binary maximal length linear sequences', *Discr. Math.*, **23** (1978), 301–307.

37. T. Helleseth, 'Correlation of m-sequences and related topics', *Proc. Intern. Conf. on Sequences and their Applications (SETA'98)*, Singapore, 1998, Springer-Verlag, London, 1999, 49–66.

38. T. Helleseth, 'On the crosscorrelation of m-sequences and related sequences with ideal autocorrelation', *Proc. Intern. Conf. on Sequences and their Applications (SETA'01)*, Bergen, 2001, Springer-Verlag, London, 2002, 34–45.

39. T. Helleseth and P. V. Kumar, 'Sequences with low correlation', *Handbook of Coding Theory*, Elsevier, Amsterdam, 1998, 1765–1853.

40. T. Helleseth and K. Yang, 'On binary sequences of period $n = p^m - 1$ with optimal autocorrelation', *Proc. Intern. Conf. on Sequences and their Applications (SETA'01)*, Bergen, 2001, Springer-Verlag, London, 2002, 209–217.

41. N. A. Howgrave-Graham, P. Q. Nguyen and I. E. Shparlinski, 'Hidden number problem with hidden multipliers, timed-release crypto and noisy exponentiation', *Math. Comp.*, (to appear).

42. N. A. Howgrave-Graham and N. P. Smart, 'Lattice attacks on digital signature schemes', *Designs, Codes and Cryptography*, **23** (2001), 283–290.

43. L.-K. Hua, 'On an exponential sum', *J. Chinese Math. Soc.*, **2** (1940), 301–312.

44. L.-K. Hua, *Abschätzungen von Exponentialsummen und ihre Anwendung in der Zahlentheorie*, Teubner-Verlag, Leipzig, 1959.

45. R. Kannan, 'Algorithmic geometry of numbers', *Annual Review of Comp. Sci.*, **2** (1987), 231–267.

46. R. Kannan, 'Minkowski's convex body theorem and integer programming', *Math. of Oper. Research*, **12** (1987), 231–267.

47. M. Karpinski and I. E. Shparlinski, 'On some approximation problems concerning sparse polynomials over finite fields', *Theor. Comp. Sci.*, **157** (1996), 259–266.

48. N. M. Katz, 'An estimate for character sums', *J. Amer. Math. Soc.*, **2** (1989), 197–200.

49. N. M. Katz, 'Factoring polynomials in finite fields: An application of Lang-Weil to a problem in graph theory', *Math. Ann.*, **286** (1990), 625–637.

50. E. Kiltz, 'A primitive for proving the security of every bit and about universal hash functions & hard core bits', *Preprint*, 2001, 1–19.

51. D. R. Kohel and I. E. Shparlinski, 'Exponential sums and group generators for elliptic curves over finite fields', *Lect. Notes in Comp. Sci.*, Springer-Verlag, Berlin, **1838** (2000), 395–404.

52. S. V. Konyagin and I. Shparlinski, *Character sums with exponential functions and their applications*, Cambridge Univ. Press, Cambridge, 1999.

53. N. M. Korobov, 'On the distribution of digits in periodic fractions', *Matem. Sbornik*, **89** (1972), 654–670 (in Russian).

54. N. M. Korobov, *Exponential sums and their applications*, Kluwer Acad. Publ., Dordrecht, 1992.
55. A. K. Lenstra, H. W. Lenstra and L. Lovász, 'Factoring polynomials with rational coefficients', *Mathematische Annalen*, **261** (1982), 515–534.
56. A. K. Lenstra and E. R. Verheul, 'The XTR public key system', *Lect. Notes in Comp. Sci.*, Springer-Verlag, Berlin, **1880** (2000), 1–19.
57. A. K. Lenstra and E. R. Verheul, 'Key improvements to XTR', *Lect. Notes in Comp. Sci.*, Springer-Verlag, Berlin, **1976** (2000), 220–233.
58. V. I. Levenshtein, 'Bounds for packing in metric spaces and certain applications', *Problemy Kibernetiki*, **40** (1983), 44–110 (in Russian).
59. W.-C. W. Li, *Character sums and abelian Ramanujan graphs*, J. Number Theory, **41** (1992), 199–217.
60. W.-C. W. Li, *Number theory with applications*, World Scientific, Singapore, 1996.
61. W.-C. W. Li, M. Näslund and I. E. Shparlinski, 'The hidden number problem with the trace and bit security of XTR and LUC', *Lect. Notes in Comp. Sci.*, Springer-Verlag, Berlin, **2442** (2002).
62. R. Lidl and H. Niederreiter, *Finite fields*, Cambridge University Press, Cambridge, 1997.
63. J. E. Littlewood, 'Research in the theory of Riemann ζ-function', *Proc. Lond. Math. Soc.*, **20** (1922) (2), XXII–XXVIII.
64. F. J. MacWilliams and N. J. A. Sloane, *The theory of error-correcting codes*, North-Holland, Amsterdam, 1977.
65. L. Mazur, 'On some codes correcting asymmetrical errors', *Problemy Peredachi Inform.*, **10** (1974), 40–46 (in Russian).
66. A. J. Menezes, P. C. van Oorschot and S. A. Vanstone, *Handbook of Applied Cryptography*, CRC Press, Boca Raton, FL, 1996.
67. D. Micciancio, 'On the hardness of the shortest vector problem', *PhD Thesis*, MIT, 1998.
68. L. J. Mordell, 'On a sum analogous to a Gauss sum', *Quart. J. Math. Oxford*, **3** (1932), 161–167.
69. P. Q. Nguyen, 'The dark side of the hidden number problem: Lattice attacks on DSA', *Proc. Workshop on Cryptography and Computational Number Theory*, Singapore 1999, Birkhäuser, 2001, 321–330.
70. P. Q. Nguyen and I. E. Shparlinski, 'The insecurity of the Digital Signature Algorithm with partially known nonces', *J. Cryptology*, **15** (2002), 151–176.
71. P. Q. Nguyen and I. E. Shparlinski, 'The insecurity of the elliptic curve Digital Signature Algorithm with partially known nonces', *Designs, Codes and Cryptography*, (to appear).
72. P. Q. Nguyen, I. E. Shparlinski and J. Stern, 'Distribution of modular sums and the security of the server aided exponentiation', *Proc. Workshop on Cryptography and Computational Number Theory*, Singapore 1999, Birkhäuser, 2001, 331–342.
73. P. Q. Nguyen and J. Stern, 'Lattice reduction in cryptology: An update', *Lect. Notes in Comp. Sci.*, Springer-Verlag, Berlin, **1838** (2000), 85–112.

74. P. Q. Nguyen and J. Stern, 'The two faces of lattices in cryptology', *Lect. Notes in Comp. Sci.*, Springer-Verlag, Berlin, **2146** (2001), 146–180.
75. H. Niederreiter, 'Quasi-Monte Carlo methods and pseudo-random numbers', *Bull. Amer. Math. Soc.*, **84** (1978), 957–1041.
76. H. Niederreiter, *Random number generation and quasi–Monte Carlo methods*, SIAM, Philadelphia, 1992.
77. H. Niederreiter and I. E. Shparlinski, 'Recent advances in the theory of nonlinear pseudorandom number generators', *Proc. Conf. on Monte Carlo and Quasi-Monte Carlo Methods 2000*, Springer, Berlin, 2002, 86–102.
78. F. Özbudak, 'On lower bounds on incomplete character sums over finite fields', *Finite Fields and Their Appl.*, **2** (1996) 173–191.
79. G. I. Perel'muter and I. E. Shparlinski, 'On the distribution of primitive roots in finite fields', *Uspechi Matem. Nauk*, **45** (1990)1, 185–186 (in Russian).
80. R. L. Rivest, A. Shamir and D. A. Wagner, 'Time-lock puzzles and timed-release crypto', *Preprint*, 1996, 1–9.
81. F. Rodier, 'Minoration de certaines sommes exponentielles, 2', *Arithmetic, Geometry and Coding Theory*, Walter de Gruyter, Berlin, 1996, 185–198.
82. C. P. Schnorr, 'A hierarchy of polynomial time basis reduction algorithms', *Theor. Comp. Sci.*, **53** (1987), 201–224.
83. V. Shoup, 'On the deterministic complexity of factoring polynomials over finite fields', *Inform. Proc. Letters*, **33** (1990), 261–267.
84. V. Shoup, 'Searching for primitive roots in finite fields', *Math. Comp.*, **58** (1992), 369–380.
85. I. E. Shparlinski, 'On primitive elements in finite fields and on elliptic curves', *Matem. Sbornik*, **181** (1990), 1196–1206 (in Russian).
86. I. E. Shparlinski, 'On parameters of some graphs from finite fields', *European J. Combinatorics*, **14** (1993), 589–591.
87. I. E. Shparlinski, 'On finding primitive roots in finite fields', *Theor. Comp. Sci.*, **157** (1996), 273–275.
88. I. E. Shparlinski, *Finite fields: Theory and computation*, Kluwer Acad. Publ., Dordrecht, 1999.
89. I. E. Shparlinski, *Number theoretic methods in cryptography: Complexity lower bounds*, Birkhäuser, Basel, 1999.
90. I. E. Shparlinski, 'Communication complexity and Fourier coefficients of the Diffie–Hellman key', *Lect. Notes in Comp. Sci.*, Springer-Verlag, Berlin, **1776** (2000), 259–268.
91. I. E. Shparlinski, 'Sparse polynomial approximation in finite fields', *Proc. 33rd ACM Symp. on Theory of Comput.*, Crete, Greece, July 6-8, 2001, 209–215.
92. I. E. Shparlinski, 'On the generalised hidden number problem and bit security of XTR', *Lect. Notes in Comp. Sci.*, Springer-Verlag, Berlin, **2227** (2001), 268–277.
93. I. E. Shparlinski, 'On the uniformity of distribution of the RSA pairs', *Math. Comp.*, **70** (2001), 801–808.

94. I. E. Shparlinski, 'Security of most significant bits of g^{x^2}', *Inform. Proc. Letters*, **83** (2002), 109–113.

95. I. E. Shparlinski, 'Playing "Hide-and-Seek" in finite fields: Hidden number problem and its applications', *Proc. 7th Spanish Meeting on Cryptology and Information Security*, Univ. of Oviedo, 2002, (to appear).

96. I. E. Shparlinski, 'Exponential sums and lattice reduction: Applications to cryptography', *Proc. 6th Conference of Finite Fields and their Applications*, Oxaca, 2001, (to appear).

97. S. B. Stečkin, 'An estimate of a complete rational exponential sum', *Proc. Math. Inst. Acad. Sci. USSR*, Moscow, **143** (1977), 188–207 (in Russian).

98. S. A. Stepanov, 'Character sums and coding theory', *Finite Fields and Applications*, London Math. Soc. Lect. Notes Ser., Vol. 233, Cambridge Univ. Press, Cambridge, 1996, 355–378.

99. S. A. Stepanov, 'Character sums, algebraic curves and coding theory', *Lect. Notes in Pure and Appl. Math.*, Marcel Dekker, NY, **193** (1997), 313–345.

100. R. C. Vaughan, *The Hardy–Littlewood method*, Cambridge Univ. Press, Cambridge, 1981.

101. E. R. Verheul, 'Certificates of recoverability with scalable recovery agent security', *Lect. Notes in Comp. Sci.*, Springer-Verlag, Berlin, **1751** (2000), 258–275.

102. I. M. Vinogradov, 'On Weyl's sums', *Matem. Sbornik*, **42** (1935), 258–275 (in Russian).

103. I. M. Vinogradov, 'Representation of an odd number as a sum of three primes', *Doklady Russian Acad. Sci.*, **15** (1937), 291–294 (in Russian).

104. A. Weil, 'On some exponential sums', *Proc. Nat. Sci. Acad. Sci U.S.A.*, **34** (1948), 204-207.

105. H. Weyl, 'Über die Gleichverteilung von Zahlen mod Eins', *Math. Ann.*, **77** (1916), 313–352.

DISTRIBUTED AUTHORIZATION: PRINCIPLES AND PRACTICE

Vijay Varadharajan

Microsoft Chair in Computing, Macquarie University
Sydney, NSW 2109, Australia
E-mail: vijay@ics.mq.edu.au

Design and management of authorization policies and service in distributed systems pose several challenges in practice. In this paper, first we discuss some of the key architectural principles involved in the design of a distributed authorization service. We consider the different types of authorization information involved, the ways of propagating them and their verification and management. Then we consider a language based approach to policy specification and outline the constructs of a simple language that can be used to specify a range of commonly used access policies. Then we propose a distributed authorization architecture and outline the various components involved in the provision of the authorization service. We discuss the issues involved in formulating a policy management framework and highlight the various design options and alternatives in a mechanism independent manner. Finally we apply this framework to a specific case of role based access control to illustrate its applicability.

1. Introduction

Security plays a vital role in the design, development and practical use of the distributed and networked computing environment, for greater availability and access to information in turn imply that systems are more prone to attacks. Recently, there have been several well-publicized attacks on and successful penetrations of networks and distributed systems. Increasingly, we are living in a world in which users and processes interact with each other by communicating over both wireless and wired networks, using both fixed and mobile applications through both fixed and mobile devices. The main objective is to deliver useful services to users anywhere at anytime, in a transparent manner. Achieving this objective in a pervasive mobile networked environment in a reliable and secure manner poses several challenges. First, security itself is pervasive in that it is needed in a variety of technologies such as operating systems, networks, databases, applications as well as system hardware. Also there are multiple platforms in each of these technologies, each of which may have its own security mechanisms. Then there are different security policies arising from different business segments, different parts of which may be managed by different distributed entities. Finally, there are numerous security standards that may not always inter-operate and there are new emerging technologies and hence new standards. All of this makes the design of security services and their management in pervasive mobile networked computing environment not an easy task.

In a distributed system, when one principal requests a service from an-other, the receiving principal needs to address at least two questions. Is the requesting principal the one it claims to be and does the requesting principal have appropriate privileges for the requested service? These two questions relate to the issues of authentication and authorization. There are also other security concerns such as auditing, secure communication, avail-ability and accountability. The authorization requirements in distributed applications [15] are much richer than authentication both in terms of the types of privileges required and the nature and degree of interactions be-tween participating entities. In this paper, we consider some key aspects involved in the design of a scalable distributed authorization service and its practical implementation. The approach to authorization presented in this paper is aimed towards achieving consistent policies to be implemented across applications throughout an organization. Policy managers may be able to use the authorization service for managing multiple applications and the users that have access to all of those applications. We will also discuss the characteristics of a policy language that can be used to specify a variety of commonly used access policies, ranging from simple identity based policies to role based policies to delegation to dynamic separation of duty and joint action policies. We will describe a distributed authoriza-tion architecture and discuss a security policy management framework for a large distributed system.

The paper is organized as follows. Section 2 addresses the architec-tural principles involved in the design of a distributed authorization service. Section 3 considers a language-based approach to authorization policy spec-ification and gives examples of authorization policies specified using this language. Section 4 considers the operation of the authorization system. In

Section 5, we consider the various design options for formulating a policy management framework. Finally, Section 6 concludes the article with some further remarks and open issues.

2. Authorization Architecture Principles

A security architecture provides a way of mapping security requirements to security services and how the security services can be provided to meet customers' security needs. It also gives a means of describing components in terms of the services they support, which in turn helps to identify how components fit together. Hence, in general, in outlining a security architecture, we need to start from the security threats faced in a distributed environment, which then gives rise to the set of security requirements that the architecture should address. Then we need to consider the types of security services and mechanisms that the architecture can support and their interactions. We can then identify generic security components of the architecture and describe how the security services are provided by these components and how they can be used to achieve the required security solutions.

Currently most applications make authorization decisions using application specific access control functions in conjunction with local files, which are often inflexible and inadequate. In this paper, we consider a three tier architecture which will address the different types of authorization information, the different forms of access control mechanisms required to support the authorization service and the design of management authorities and design options for formulating policy management framework. It provides application developers with facilities to perform simple access control checks to complex business logic authorization. Administration is provided with an easy to use graphical user interface that enables administrators to manage relatively large numbers of users with consistent policies across applications. Before we look at the design of the distributed authorization service, let us start with the requirements from different perspectives.

2.1. *Authorization perspectives*

In general, the authorization architectural framework should be aimed at addressing the needs of several classes of users. The users within an enterprise that have to deal with security include the developer, the security administrator, the policy setter and the end user.

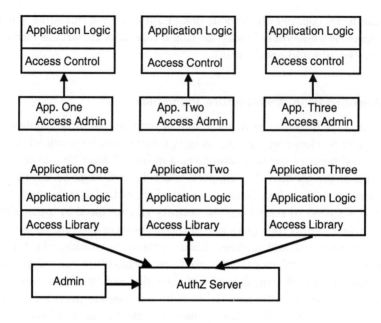

Fig. 1.   Authorization Approach.

2.1.1. *The developer's view*

Much of application security today is implemented as part of the programming effort. The security logic is written for each application. There are unique policies set per application based on the developer's knowledge and skill level. The aim of our approach is to remove security processing from the application logic. This is accomplished by defining appropriate authorization attributes and formulating rules that specify how the authorization decision is computed and processed outside the application logic. There are several benefits that arise from this strict separation. An important practical consequence is that it results in a consistent application programming interface across all applications and platforms, thereby allowing common policies across them. Furthermore, checking of access rights is abstracted away from a particular application. This enables the developers to concentrate on the business logic instead of reinventing the security portion of the application. Figure 1 shows the common API and how the security portion of each application can be condensed using such a common security solution.

2.1.2. *The administrator's view*

Often each application is administered uniquely. A custom security process for administering the required security information, such as user names, passwords and privilege capabilities, is required. Enterprise wide security policies are difficult to implement consistently across all applications and these policies are even more difficult to verify. Certain security tasks require more than one administrator to check the correctness of the task. Our architecture approach will help to consolidate security administration. Rather than dealing with each application subsystem and its security component separately through a different tool, it offers a consistent view of principals, profiles and their privileges for a given application. We will define these terms precisely later in the paper.

2.1.3. *The end user view*

End users today suffer most from the fact that each system and each application implements its own security mechanisms. Authentication mechanisms may not be consistent, which causes the user to obtain and remember many different user names and passwords. When each application provides its own authorization facilities, the user can be confused by the different policies in the application. For example, why is a bank manager's limit in one banking application $1500 and in a related application it is $2000? It may be necessary for the security related attributes of a user to be consistent throughout the range of similar applications. We consider facilities to modify the set of attributes effectively with low administrative overhead.

2.1.4. *The policy setter view*

Policy setters require management tools to ensure that the policies are correctly implemented. Our authorization approach proposes the notion of profiles, which provide a way of grouping privileges to establish common access rights for sets of users. The policy setters are given tools to set and test authorization policies.

2.2. *Authorization service design principles*

The design of any security service involves at least the following aspects: security information used in the provision of the security service, the security mechanisms that are required to support the service and the authorities involved in the management of the service. In the case of authorization service, the security information used ranges from user identities to group identities to role information to location information to actions and parameters associated with the actions. In fact, from an architecture point of view, it is important to understand the characteristics of the different types of the authorization security information. At a conceptual level, we can classify the various types of information as follows: Some security information are generic and static in nature. For instance, typically identity based information falls into this category. Then there is security information that are specific but still somewhat static in nature. For instance, role based access information falls into this category. Roles are specific to organizations and they are reasonably static in the sense that they are unlikely to change on a day to day or even on a monthly (or even yearly) basis. In fact, one of the main benefits of access control system based on role based information is to reduce the effect of the changes in the user population on the management of access privileges. Then we have security information that is specific and dynamic in nature. These are specific in the sense they may relate to specific applications or parts of applications or files. They are dynamic in the sense they are prone to state changes.

Such a classification of different types of authorization information in turn helps to understand better the requirements on their management. First, the authorities involved in the management of these different types of authorization information are likely to be different. Not only the strategies with respect to when the changes and updates to these information take place are likely to be different, but also who are allowed to make these changes are likely to vary. For instance, the specification and changes to the role information in an organization will be the responsibility of a certain group of people who can be different to those responsible for setting the privileges for a specific file or application in a server.

From an architectural point of view, such a characterization also leads to some basic design principles. First, it is appropriate to store the static and generic authorization information in some form of a central server within a domain, which is responsible for a collection of clients and server principals. Second, the dynamic and specific authorization information can be stored near or in the target, enabling the target system authorities to be involved in their management. From a distribution point of view, static and generic authorization information (for instance, stored in a central server within a domain) can be "pushed" by the client principal to the target principal. In fact, static and specific information can also be pushed in a similar fashion. Finally, specific and dynamic authorization information needs to be "pulled" at the time of the decision process.

Based on these principles, one can argue that conceptually there can be multiple types of trusted management authorities involved. For instance, one dealing with generic and static security information and another dealing with specific and static information, both of which can be architected as centralized servers (one per domain) servicing a number of clients and targets within the domain. Furthermore, specific and dynamic security information needs to be managed at the target (or at some entity attached to or close to the target), as this type of information is often dependent on the state of the application or resource under consideration. Such information may include attributes associated with specific rights in the application. We use these architectural design principles in the design and implementation of a distributed authorization service presented below.

2.3. *Authorization mechanisms*

Let us now briefly review some of the basic access control mechanisms that are commonly used to support the authorization service.

Access Control Lists (ACLs) has been one of the traditional mechanisms that have and are being used in many existing systems and applications. They associate lists of users to every type of access for a resource. The access types are typically actions such as read, write, execute, create, append, modify and delete. Any user on the list for a particular action is allowed to perform that action. ACLs are very common in file systems. The benefits and problems of ACLs are very well known and hence we will not describe them here. We will just illustrate one or two issues that are relevant for this paper. For instance, ACLs have limitations in expressing the types of controls necessary in many applications. For example, it is difficult to create numerical limits for transactions with ACLs. Moreover, ACLs are inherently resource based. Most ACL systems today store the lists at the resource (within the application). A common nightmare with ACL based systems is the difficulty of their maintenance especially if the administrative tools are not common across the ACLs in all applications. If the authorization information is distributed which is the case in distributed applications, it makes answering questions about a user's access rights very difficult. For example, to remove an employee who has left the company from all ACLs usually requires the administrator to edit the ACLs in each application to determine if the user has access rights, and if so, remove those rights. This can be a very time consuming process. Our authorization system focuses on the users and groups of users making it a more natural and intuitive approach for many business requirements.

The other well known mechanism that has been used in the provision of the authorization service is the capabilities. A capability is essentially a ticket that provides the holder with some type of access right. A capability system requires the user to acquire this ticket before performing the access; the ticket is presented at the time of access. If all the information necessary for computing the access rights is available outside of the application, then a capability system can work well. However in applications where some of the information necessary to make the access determination is inside the application, the capability system does not have all the information it needs. For instance, in an Automated Teller Machine (ATM) system, a capability system might be able to determine that the ATM card looks valid, the check digit on the account is correct, the bank information looks valid and so on, but only the bank applications know if the request to withdraw money is valid since only the bank applications know whether the account has enough money in the account and if the user is over his daily withdrawal limit. The decision on whether or not to allow the request to take place must be made from within the application and is dependent on its state.

Based on the architectural design principles outlined earlier, let us now consider a variation on the access mechanism. This mechanism uses a separate object to define the access rights for a particular action. The access object, referred to as an entitlement, contains a rule that defines the access and operands (attributes) used in the rule evaluation. Once the entitlement has been defined, it can be assigned to individuals or groups of individuals. This assignment also includes providing values for some of the attributes in the rules. Separating the authorization control into a separate object provides flexibility in the way authorization is administered and the manner in which it is evaluated. The administration is controlled by the owner/administrator of the entitlement. Access to the application can be provided to any individual to whom the administrator assigns the entitlement. In addition, logically the evaluation can be outsourced to a separate module that is optimized to perform evaluations on entitlements. This frees the applications from performing the authorization evaluation and the developer from having to create this evaluation code. This access mechanism forms the basis of the authorization service proposed in this paper and the design of the Authorization Server.

2.4. *Types of authorization checks*

Another fundamental aspect that needs to be considered in the design of the distributed authorization service is the type of authorization checks and their location. In practice, there are different places at which authorization checks can be made. The closer the check is to the application, the more integrated the authorization system is into the business logic. In general, there are three basic approaches.

- Application start-up/connection time : In this approach, the check may be performed by the application or by some component that is a front end to the application. This check determines whether or not a user has access rights to the applications. Once the application is started, this authorization system performs no further checks.
- Function Access Check: This check is made purely on the type of function/transaction being requested. This may or may not include parameters from the request in the authorization decision. As with "application start-up/connection time" above, this check can be embedded in the application or performed by some front-end component. However, in the case of a front-end component, more sophistication is required for the front-end to recognize one function/transaction request from another.
- Business Logic: This evaluation is performed not only on the function/transaction being requested, but it also includes the parameters sent by the user, information from the authorization database, and parameters from calculations from within the application program. This is the most complete check that can be performed and it must be embedded in the business logic of the application.

The Function Access Checks provide the ability to verify that a principal is allowed to execute a specific function or transaction. The Business Logic Checks can be as simple or as complex as needed for the specific application. All of the information input by the user or available to the program in any way can be built into a rule and used to verify access. Both the Function Access and the Business Logic checks can be used to protect any code segment, transaction or action.

2.5. *Authorization service stages*

Let us now briefly outline the two distinct stages involved in the design of the distributed authorization service. Conceptually, there are two stages, namely the administration phase and the runtime or the evaluation phase.

The administration phase involves facilities and services for the specification of authorization policies, updating and deleting of policies and their administration. The runtime phase is concerned with the use of these authorization policies in the evaluation of the access requests. The representation of the authorization policy information can be different in these phases. In fact, there are at least two arguments for maintaining distinct representations of authorization information in these two phases. First, there is the information captured by the language that can be compiled out before access decision time. For example, the meaning of inheritance and override depends only on the user and not on the server and transaction attributes. This enables the authorization system to make the access decision faster, by avoiding searching the inheritance hierarchy. Secondly, it is possible to envisage different strategies for replication of the information needed for administration versus the information needed for access evaluation decisions at the application servers. This in turn has implications in terms of interfaces and components required in the authorization service design. For instance, in terms of interfaces, there will be at least two interfaces, namely the administration interface used by the administrators and the policy setters and the runtime evaluation interface used by the clients and application servers during access decision time. There may also be another interface used for auditing purposes. For instance, the configuration and the management of the authorization server may require an independent auditor to monitor whether the access policy meets the required controls. In terms of components, the authorization service requires an administration component where the authorization policies are entered and stored in one representation, and a runtime evaluation component, which stores the authorization rules at a different representation for runtime access.

3. Authorization Policy Language

Having considered some of the principles involved in the design of a distributed authorization architecture, let us now consider the design and specification of authorization policies to be supported by the authorization service.

Languages have long been recognized in computing as ideal vehicles for dealing with expression and structuring of complex and dynamic relationships. While in theory a general-purpose language could be used to specify authorization policies, often a special-purpose language providing for optimizations and domain specific structures considerably improves the efficiency. We strongly believe that a language-based approach to authorization is required not only for supporting a range of access control policies, but also in separating out the policy representation from policy enforcement. The policy language enables administrators and managers to specify and update authorization polices required in their systems. In Section 5, we will address the issue of multiple managers when we consider the policy management framework. In this section, let us consider the constructs required in a policy language and how they can be used to specify a range of authorization policies.

Over the years several languages have been proposed, some mathematical logic and graph based and some programming language based [8, 1, 5, 4, 16]. We have developed dedicated languages of both types for specifying authorization policies. In this paper, we will describe one such language [15] and give access policy specifications using this language. The two fundamental requirements of any such language are as follows: on the one hand, it should be simple, thereby enabling the administrators and policy setters to use the language in specifying their policies, while on the other hand, it should have sufficient expressive and analytical power to represent and evaluate a range of policies used in practical systems. We will show that the language considered in this section is powerful enough to specify the commonly used policies such as discretionary access control, role-based access control [11], separation of duty [3], delegation [14] and joint action policies [13]. We will then consider some simple policy examples using the language. In terms of the ability to specify policies in practice, our authorization approach uses a graphical user interface allowing the security administrators and the policy setters to easily define their policies. We will not be describing the user interface aspects in this paper. Recently, we have also defined a language based on XML (eXtensible Markup Language) that provides similar constructs.

3.1. *Language constructs*

The basic constructs of the language include the following: Principal Records, Entitlements, Privileges, Attributes, Rules, Profiles, Time-Related Parameters and Privilege Operators. Let us now consider each of these elements in turn.

3.1.1. *Principal record*

A principal in the system is any entity that can be authenticated and that wishes to access certain resources. It could be a user, a client or a server application. A principal record has several elements associated with it, namely the principal name, the principal type, the principal identity value, profile membership list, privileges list, and enabled flag and information. The principal name identifies the name of the principal and the principal type is used to qualify both the name and the identity value. The profile membership list contains a list of the names of the profiles of which this principal is a member. We give the definition of a profile below. A principal inherits all the privileges found in all the profiles in this list. The privileges list contains the privileges the principal has for entitlements. The enabled flag is a Boolean flag that indicates if the principal record is to be activated. The date and time specification information associated with this flag indicate the range of dates and times during which the principal is enabled. The earliest is called the enable time and the latest is called the disable time.

3.1.2. *Entitlements*

An entitlement defines the authority granted to a principal to perform a certain task or access some information. For example, in a financial application an administrator can create an entitlement called Withdrawal to allow authorized principals to withdraw cash from a system. A Withdrawal entitlement could specify for example that "a user Fred can withdraw $5000 between Monday to Friday from a Cash Management Account". Alternatively, a medical application might have entitlements such as Read-Patient-Record and Discharge-Patient. In general, an entitlement has the following components: entitlement name, transaction attributes, privilege attributes, the access rule, default privileges, enabled flag and information, and entitlement privileges. Let us look at each of these components in turn.

All entitlement names are unique in the authorization system. There are two types of attributes: transaction attributes and privilege attributes. Each of these attributes has type information associated with them. Transaction attributes of the entitlement are passed from the application server to the authorization system, which are used in the decision making. These correspond to parameters in the access decision rule. Such parameters include values such as the day of the week on which the transaction is being attempted. The other type of attributes is the privilege attributes. The privilege attributes are values defined by the security administrator when specifying a privilege for an entitlement. For example, consider again the entitlement Withdrawal. This may have privilege attributes such as Valid-Currency (which has values such as dollar) and Amount-Limit (which can have values such as 5000). These are parameters of an access rule and these have been determined at the administration time when the access rule was specified.

There are two ways of adding a privilege. One way is to add it to a profile and principal records. We will consider these later when we define the profile and principal record. The other way is to add them as default values in the entitlement. This is useful when a specific privilege is required by many principals.

A subset of the privilege attributes are designated as key attributes. Key attributes are used when multiple sets of privileges must be assigned to the principals for a particular entitlement. For instance, if a set of users are performing transactions on a variety of different currencies, then they can be assigned different limits for each type of currency. Here the currency type can be designated as a key attribute. Note that values of the key attribute must be unique, though the values for other privilege attributes may not be.

In general, the transaction attributes are compared with the privilege attributes in determining whether an access request is to be granted or not. This is achieved by the rule in the entitlement. The rule is used by the authorization system to determine if a particular transaction should be authorized for a given principal. It is a logical expression that specifies how transaction attributes are compared with the privilege attributes. Rule elements are composed of the following : a transaction attribute, a comparator (such as $=, >, <, \geq, \leq$ and $<>$) and a privilege attribute. Rule elements in an entitlement rule can be combined using AND and OR logical operators and nested parenthesized expressions. The AND operator has higher precedence than the OR operator when no parentheses are used. This leads to the specification of general logical expressions in the entitlement rule. Let us consider a simple example of entitlement.

3.1.3. *Example entitlement*

Entitlement Name : Withdrawal

> Transaction Attributes :
> > Acc-Type : string
> > Check-Amount : bignum (type large integer)
> > New-Balance : bignum
>
> Key Privilege Attributes :
> > Valid-Acc : string

Non-Key Privilege Attributes :
    Check-Limit : bignum
    Credit-Limit : bignum
    Max-Amount : bignum

Rule : Valid-Acc = Acc-Type & (Check-Amount $\leq$ Check-Limit
              | New-Balance $\leq$ Credit-Limit)

The objective of this rule is to grant authority to withdraw money from a given account type according to one of the limiting factors: either a specific withdraw amount limit (Credit-Limit) or the sum of the withdrawal amount plus the customer's outstanding balance is less than or equal to the Credit-Limit.

The entitlement specification also has a Boolean enable flag, which indicates whether the entitlement is activated or not. If the flag is true, then the date and time specifications in the rule can be used in the decision making. If an entitlement is not enabled when a request for authorization is made (that is, the flag is false or the date and time of the transaction are outside the range of the enable time and disable time), then the request is denied, regardless of any privileges held by the principal attempting the transaction.

Finally, an administrator can also define a principal-independent entitlement. This entitlement is not associated with any particular principal (that is, it does not have a principal identity value associated with it) and is automatically applied to all principals. This is useful in many commercial situations in policy definition. An example of such a situation is given later when we consider some example policy specifications.

3.1.4. *Profiles*

The profile construct allows to group entitlements for a class of principals. It helps to simplify the process of assigning privileges to principals. It is clear that roles can be easily defined using the profile construct. First, a set of privileges is grouped together in terms of profiles and these are then associated with one or more principals. For instance, profiles can be used to provide the same privileges to a class of principals say all senior bank managers. If groups of principals require the same privileges, it is easier to create a profile and add it to their profile membership list rather than specify the individual privileges for each principal. Profiles can also be used to group privileges for related sets of responsibilities; for example, the privileges required to perform transactions in Japanese Yen.

A profile has the following components: Profile Name, Profile Membership List and Privilege List. The name of the profile is unique within the authorization system. The profile membership list contains the names of the profiles that this profile inherits. A principal that is a member of a given profile (that is, a principal that lists this profile in its list of profile memberships) is granted all the privileges in all the profiles inherited by that profile, and all the privileges listed in the profile. The Privilege List contains the list of privileges that this profile has for entitlements. In this language, the order the profiles are listed is significant as well as the order the privileges are listed in the privileges list is significant. We will address this aspect when we consider conflict resolution of privileges and profiles.

A key concept in the formulation of profiles and assignment of privileges is the notion of inheritance. Inheritance makes it easier to build and manage sets of privileges so that the same privileges need not be defined repeatedly. For example, consider the following: an administrator creates a profile called Bank-Manager that defines all the privileges granted to bank managers. A principal that becomes a member of the Bank-Manager profile automatically has all the privileges defined in it. Let us now create another profile called Senior-Bank-Manager that defines all the privileges granted to say senior bank managers. Let the Senior-Bank-Manager profile include the Bank-Manager profile within its specification. Then the users who are members of the Senior-Bank-Manager and Bank-Manager profile gain all the privileges defined in Bank-Manager and Senior-Bank-Manager profiles. Inheritance can be supported to varying levels of depth. In general, such an inheritance process can create conflicts in privileges. We need to resolve these conflicts using overrides, which are described below.

3.1.5. *Time-related parameters*

The authorization language allows administrators to specify the periods during which the entitlements, privileges and principals are active. Each privilege has two associated date and time pairs, "effective from" and "effective to" times. Each pair consists of a starting date and time and ending date and time. Each principal and entitlement has two associated date and time pairs, "enabled from" and "enabled to". Each pair consists of a starting date and an ending date and time. Each principal and each entitlement has an associated enabled/disabled flag. These time-related parameters provide the administrator the flexibility to specify a range of policies. For instance, the administrator can specify a valid time range when a basic element is active. So a branch manager's privileges can be granted to an assistant manager only for the time period the branch manager is away on vacation. This is an example of delegation with a time-out. Administrators can also specify the time a group of related changes becomes active. For example, consider the situation when bringing a new application on-line, an administrator wishes to create the entitlement(s) for the application, modify the existing profile(s) to grant privileges for the application, and modify the principal records to add user-specific privileges. All these changes to the authorization system are to become effective at the same specific time in the future.

An administrator can enable or disable entitlements, privileges and principals' records. Enabling an entitlement allows privilege authorization, while enabling a principal's record authorizes that individual's privileges. When an entitlement or a principal is enabled, either start and stop times are specified or the default values (which are "as soon as possible" and "no stop time") are accepted. Overriding the defaults provides definitive control and is useful for temporary needs; for example, the administrator can set the stop time of a principal record for a temporary employee. Disabling an entitlement inactivates any of the privileges that were granted to profiles or principals. Disabling essentially allows the administrator to turn off an entitlement or a principal record without having to delete it. When a principal record is disabled, all authorization requests from the principal are denied.

3.1.6. *Privilege operators*

In the language, we define a set of privilege operators to specify, modify and override privileges assigned to principals in the authorization system. These operators include addition or replacement (AR), replacement (RE) and deletion (DE). AR is the most commonly used operator. If no matching is found, then the specified privilege is added to the privilege list. If a matching is found, this privilege overrides the matching privilege. A privilege marked with an RE operator can only override another privilege. If no matching privilege is found, in this case, it is not added to the privilege list. If the matching privilege is deleted at a later time, the corresponding privilege marked with RE is also deleted. The DE operator is used to remove a specific privilege. A specific form of the DE operator DE:all is used to delete all privileges that a principal has for a particular entitlement. Let us consider an example to illustrate the use of these privilege operators in policy specifications. Consider again the entitlement Withdrawal.

Entitlement Name : Withdrawal

Key Privilege Attributes :

Valid-Currency : fstr (type - full string of up to 256 characters)

Valid-Start-Day : wkday (type - a string with a single digit 1 (Sunday)
and 7 (Saturday))

Valid-End-Day : wkday

Privilege Attributes : Amount-Limit : sint (type - a base 10
representation of a 32-bit signed integer)

The AR privilege operator is used to grant a particular privilege to a principal or profile. For example, consider

AR : Withdrawal : Start-Time : End-Time : Valid-Currency="USD",
Valid-Start-Day="2", Valid-End-Day="6", Amount-Limit="10000"

This is a privilege for the Withdrawal entitlement that grants the principal the ability to withdraw up to $10000 from Monday through Friday. The AR operator means that this privilege overrides any matching privilege encountered in the authorization administration and is added to the list if no matching privilege is found.

The RE privilege operator is similar to the AR operator except that the privilege can only override another privilege. This feature of the RE operator makes it most useful when the override is to limit the authority of the principal rather than extend it. For instance, let the profile Teller contain the above privilege and a principal Jeff was granted the Teller profile. Jeff was then given the following privilege in his principal record privilege list:

RE : Withdrawal : Start-Time : End-Time : Valid-Currency="USD",
Valid-Start-Day="2", Valid-End-Day="6", Amount-Limit="5000"

In this case, the RE privilege will override the AR privilege; hence Jeff will only be able to withdraw $5000 and any attempt to withdraw a higher amount will be denied. If the AR privilege in the Teller profile is later deleted, since the RE privilege does not add any privilege, the effect is that Jeff will have no withdrawal privilege. If the AR operator had been used, the privilege to withdraw $5000 would be added to the list and Jeff would still have that privilege even if the privilege to withdraw up to $10000 was deleted from the Teller profile.

The DE privilege operator is used to remove a specific privilege. Any privilege that matches the effective times and key privilege attribute values

of this privilege will be deleted from the list of privileges. For example, let
the Teller profile contain the following privilege:

> AR : Withdrawal : Start-Time : End-Time : Valid-Currency="USD",
> Valid-Start-Day="2", Valid-End-Day="6", Amount-Limit="10000"

Assume that the principal Jeff was granted the Bank-Manager profile
and then given the following privilege in his principal record privilege list:

> DE : Withdrawal : Start-Time : End-Time : Valid-Currency="USD",
> Valid-Start-Day="2", Valid-End-Day="6"

Note that only the values for key privilege attributes are necessary for
privileges using the DE operator. Hence it is not necessary to specify a
value for the Amount-Limit. The DE operator causes the AR privilege to
be deleted. Thus, Jeff will no longer have a privilege for Withdrawal and
any attempt to withdraw funds will be denied.

3.2. *Some simple examples*

The basic constructs of the authorization policy language given above can
be used to specify a range of access control policies. This section gives a
number of examples of common access control situations. In the interest
of space, the examples do not include all the necessary preliminary decla-
rations and initializations. Nonetheless, we hope they convey the required
information.

3.2.1. *Privileges using logical rules*

Consider the entitlement Withdrawal below.

> Entitlement Name : Withdrawal
> Transaction Attributes :
> Acc-Type : string
> Amount : bignum
> Key Privilege Attributes :
> Valid-Acc : string
> Non-Key Privilege Attributes
> Max-Amount : bignum
>
> Rule :
> Account-Type := Valid-Acc
> Amount ≤ Max-Amount

Assume that the policy requires the administrator to grant a principal John multiple privileges for this entitlement in order to authorize him to withdraw up to $5000 from his savings account and up to $1000 from his cheque account. This can be achieved by granting to John the following:

Withdrawal :
From-Time : To-Time : Valid-Acc = Savings, Max-Amount = "5000"

Withdrawal:
From-Time : To-Time : Valid-Acc = Cheque, Max-Amount = "1000"

3.2.2. *Delegation using time-related parameters*

The dates and times of privileges and dates, times and flags of principal and entitlement interact to determine whether or not a given privilege is in effect for a given entitlement and a given principal. This interaction is a three-step process:

(1) The effective times of the privilege define a range specific to that privilege.
(2) The enabled to and from times of the principal are reconciled with the effective times of the privilege. The resulting times are: (a) the later of the principal's enabled from time and the privilege's effective from time; (b) the earlier of the principal's enabled to time and the privilege's effective to time. The goal is to grant the principal the least amount of authorization time.
(3) Finally, regardless of the effective range for this privilege, the overriding factors are the enable flag and the time range of the entitlement. If the entitlement is disabled (either because the enabled flag is turned off or because the enabled flag is turned on and the enabled to time has passed), the privileges are not evaluated. If the entitlement is enabled, the effective times in the privileges are used.

Temporary Time-Limited Delegation

These effective times can be used to temporarily grant authority to a principal for whom no privileges exist. When the effective times are enabled for the principal and entitlement, the principal will be authorized from the beginning effective time to the ending effective time. For example, Wayne needs authority to perform withdrawals for a week while his superior is on vacation. We can give Wayne a temporary privilege, with an effective start time of May 18, 1997, at 8:00 AM and an effective ending time of May 22,

1997, at 5:00 PM. As of the effective start time, Wayne will be authorized to perform withdrawals. Of course, all the other parameters of the transaction must be valid for the authorization request to be granted. As of the effective ending time, Wayne will not be authorized to make withdrawals, and any requests he makes for withdrawals will be denied.

One can provide a variety of other features such as using times for overriding privileges, using overlapping privileges to grant temporary authority and using multiple privileges to temporarily remove authority. Let us consider another example.

Temporary Removal or Reduction of Privileges

To temporarily remove or reduce an authority of a principal, one may need to create multiple privileges for the same authority, but with coinciding effective times and the correct limits. Consider a user Tom having a privilege to withdraw up to $5000. To reduce his ability to withdraw to $100 for one week from May 18 to May 22, the following can be done:

- Change the existing privilege so that its ending effective time is May 18 at 8:00 AM.
- Create a new privilege for withdrawal with a limit of $100 and effective times of May 18 at 8:01 AM through May 22 at 5:00 PM.
- Create another privilege for withdrawal with the original limit of $5000 and effective times of May 22 at 5:00 PM through "No end date".

Using the same example, one can completely remove Tom's authority to withdraw dollars for the same period, instead of simply reducing his limit. This can be accomplished by not creating the second privilege shown above. This will result in Tom not having any effective privilege for withdrawal during the same period specified.

3.2.3. *Role hierarchy*

One of the important advantages claimed for the role-based approach is that it can model the structures of real world organizations. A typical part of such structures is a hierarchical ordering of responsibilities, with more senior positions encompassing all the privileges of the more junior positions plus some extra privileges. For example, consider a hypothetical structure for a bank branch where there are Branch Managers, Accounts Managers, Loans Managers and Tellers. Let the profile Teller have all the entitlements that are needed for a teller principal to carry out his/her functions. Let

E1 and E2 be the extra entitlements required for the Accounts Manager profile. Let E3 and E4 be the extra entitlements required for the Loans Manager profile. Let E5 and E6 be the extra entitlements required for the Branch Manager. The profiles for Accounts Manager, Loans Manager and Branch Manager can be created as follows:

Profile Name : Accounts Manager

Profile Membership : Teller

Privileges :
  1. E1;
  2. E2;

Profile Name : Loans Manager

Profile Membership : Teller

Privileges :
  1. E3;
  2. E4;

Profile Name : Branch Manager

Profile Membership : Accounts Manager, Loans Manager

Privileges :
  1. E5;
  2. E6;

The issue of conflicts between privileges in profiles and their resolution is addressed later.

3.2.4. *Separation of duty*

Finally, let us consider a simple example of separation of duty. This is where one or a group of principals is not allowed to perform two classes of actions [10]. For instance, a principal which has the privilege to create a document should not have the privilege to certify the document say for external release. This can be specified as follows:

Let Create-Doc and Certify-Doc be the respective entitlements and let Creator and Certifier be the respective profiles.

Profile Name : Creator
Profile Membership : None
Privileges :
AR : Create-Doc : Start-Time : End-Time : Valid-Doc = "Memo"

Profile Name : Certifier
Profile Membership : None
Privileges :
AR : Certify-Doc : Start-Time : End-Time : Valid-Doc = "Memo"

One way of achieving the separation of duty is to specify the restriction in the principal record privilege list. So for instance, if a principal Jane was granted the Creator profile, then she can be restricted from doing the certification by adding the following in her principal record privilege list:

DE : Certify-Doc: Start-Time : End-Time : Valid-Doc = "Memo"

Similarly, in the case of a principal having the Certifier profile can be restricted from creating the document using

DE : Create-Doc : Start-Time : End-Time : Valid-Doc = "Memo".

Alternatively, it is also possible to define this separation at the profile level instead of at the principal record level.

4. Authorization System Design

Let us now address the practical design of the authorization system. Figure 2 shows the components of the Authorization Server and how they fit together to provide the authorization service. At this stage, we will assume that a logical Authorization Server system component exists in each domain. In Section 5, we will consider multiple such management authorities and the design options for structuring them. Additional subsystems in the overall security architecture include Authentication Server and their associated libraries. The Authorization Server depends on the authentication service for reliable identification of principals. This can be provided by either symmetric key based systems such as Kerberos [7] and DCE (Distributed Computing Environment) [10] or using some form of public key based systems such as those based on X.509 protocols.

Based on the types of authorization checks given earlier, let us consider an authorization system that supports three levels of checks. Level I and Level II authorization checks are concerned with the permission for a client to have interface and operation access to a server. Level I check is made at the client side. Level II check is made at the server side. Interface corresponds to an application server. For instance, for DCE applications, this is equivalent to the name of the interface specified in the IDL file. For GSS-API [6] applications, this is application-defined, however typically it

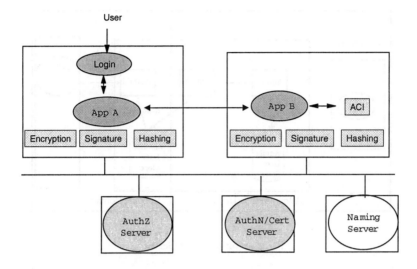

Fig. 2. Distributed System Security.

is the process name. Operation corresponds to an application subroutine or procedural call. For DCE applications, this is equivalent to the name of the procedure specified in the IDL file. For GSS-API applications, this is application-defined, however typically it corresponds to a single named client request.

Level III checks are business authorization rule checks. Applications ask the Authorization Server System for a Yes or No decision on a specific authorization rule. The application passes the parameters (transaction attributes and values) that are compared with the authorization information stored in the Authorization Server (privilege attributes and values) for a given authorization rule (entitlement).

4.1. *Administration and runtime components*

The Authorization Server is composed of three processes which are grouped into the administrative and runtime domains (see Figure 3). There are two data store components, namely administration database and runtime database that store the authorization information.

The administrative domain is concerned with the setting up and management of privileges and profiles granted to principals. The AuthU server controls the administrative database, which stores information describing principals, entitlements and profiles. The interface to this store is imple-

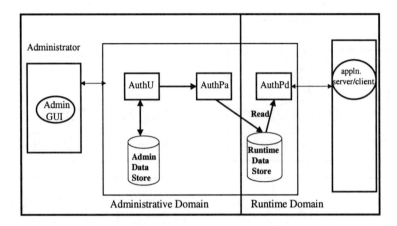

Fig. 3.   Authorization Server.

mented by the administration store access library and is used by the administration client GUI or through a batch utility. AuthU information is the master set of information and can be used to generate the runtime database if necessary (see below Runtime Domain). AuthU causes items to be added to or deleted from the runtime database. For example, when an entitlement is created, AuthU calls AuthPa to add an entry to the rule set and to create a table to contain the privileges for the new entitlement. When an entitlement is deleted, its entry in the rule set is removed and the table for the entitlement is deleted. When a principal is created and granted privileges, AuthU tells AuthPa to add those privileges to the appropriate entitlement tables. Each time a change is made to the list of privileges, enable times or profiles of a principal or another profile, AuthU goes through a process of evaluating and resolving the privileges and profiles in order to keep the runtime database up to date. Changes to Authorization Server databases may require two administrators referred to as the Maker/Checker, one to make the request (Maker) and the other to check it (Checker). A check table, created when the Authorization Server is configured is used to determine whether a request to modify the information requires approval by a second administrator.

The runtime domain consists of an evaluation engine server called AuthPd and the runtime database. The runtime database is written to by AuthPa and read by AuthPd. It is optimized for efficient processing of the authorization request. AuthPd uses the information stored in the runtime database to decide on authorization requests.

4.2. *Authorization server interfaces*

Consider an application server which is a client of the Authorization Server. There is an application library component that resides in the application server and acts as an interface to the Authorization Server. It uses an authorization evaluation interface to pass the authorization request to the Authorization Server.

The administration client uses the administration store interface implemented by administration store access library to enter authorization information into the Administration Store. This is characterized by AuthU on the server side.

At administration time, rules and privilege attribute values need to be statically checked and converted to the runtime (evaluation) time representation. Furthermore, as mentioned above, the semantics of overrides and inheritance can be compiled out. Logically there can be three libraries involved, namely a compiler library, a dispatcher library and an administration type checker library. Then this information is entered into the runtime database via an insertion interface implemented by runtime store insertion library. All of these libraries and interfaces are characterized by AuthPa.

At runtime, the evaluation engine library evaluates all the information to give the access decision. There is a query interface which allows the engine to fetch information from the runtime database. The types of the attributes are checked using a runtime type checker library. This differs from the administration time type in that it checks runtime representation of values and rules instead of administration time representation. In general, there may be a need for translation of the attributes presented by the authorization evaluation interface (from the application server) into a suitable representation for the authorization engine. These libraries and interfaces are characterized by AuthPd.

The basic authorization system operation is as follows: In the process of designing the application, the application designer and the security administrator define the entitlements and the set of transaction attributes and privilege attributes and the rules that express the security constraints. Authorization decisions are made by evaluating the list of entitlements. At evaluation time, the application passes the transaction attribute values to the authorization system where the evaluation engine compares these values with the privilege attributes in the rules and grants or denies the access request (see Figure 4).

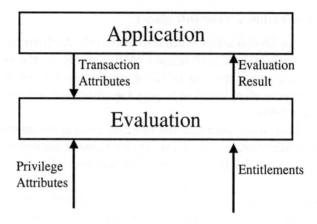

Fig. 4.   Authorization Evaluation.

4.3. Privilege evaluation and resolution

When entitlements, privileges and profiles are changed, the privileges and
profiles must be evaluated and resolution is required for the changes that
may affect the privileges in other places. The administration component
of the authorization system determines how each new privilege or profile
affects the privileges that are already in other lists and determines the
changes that must be made to the runtime database. This is done by following the inheritance tree of each profile and determining which privileges
to add or delete. The resulting privilege lists are then used to update the
authorization runtime information. Principal privilege lists must be evaluated whenever any one of a principal's privileges is changed and whenever
there is a change to a profile where the principal is a member. Privilege resolution is the process of evaluating and determining which privileges apply
to a principal or profile and therefore should be applied at authorization
runtime.

   In general, there are several approaches to privilege and conflict resolution. Common ones include the following: (a) Specific overrides generic
privileges: For instance, if a privilege given to an individual user conflicts
with a privilege derived from being a member of a group, then the individual
privilege overrides the group one. (b) Recent privileges override older ones:
For instance, a privilege given today overrides a privilege given yesterday.
(c) Negative privileges override positive ones: For instance, a negative privilege stating that "John cannot withdraw" overrides the positive that John

can. (d) Higher priority privileges override lower priority ones: For instance, one can associate priority labels with privilege rules and resolution can be achieved based on the level of these priority labels. (e) Ordering of privilege rules governs the resolution process. For instance, privileges specified later in the list override the ones specified earlier in the list.

We have implemented a combination of the above schemes in our designed authorization system. In this paper, we will restrict our discussion to the simple one based on ordering of privileges. The authorization administration component comprises principal, profile and entitlement tables. During resolution, these are changed into the runtime schema that consists of a rule table and a number (n) of different entitlement tables. Each row of an entitlement table corresponds to a privilege for a particular principal (indexed by principal identity value). Each privilege consists of zero or more privilege attributes. During the resolution process, privilege overrides may occur if two privileges match. Privileges match if they have the same entitlement name, values for all the key privilege attributes, and beginning and ending effective date and times.

The order of evaluation is specified as follows:

- Privileges found later in a privilege list override matching privileges found earlier in the Privilege list
- Privileges from profiles that are evaluated later in a profile list override matching privileges from profiles found earlier in a profile list.
- Profile lists are evaluated before a principal's privilege list. Therefore, privileges for principals take precedence over privileges for profiles.

The steps used in the evaluation are as follows:

- First the principal record to be resolved is extracted.
- The hierarchy of any profiles a principal has in its profile list and principal list is then traversed and resolved each in order, from the first in the list to the last. The result is a list of privileges for all the profiles.
- Then the privilege operators are applied. For each privilege in the list, the operators are applied for each privilege. The result is a cumulative list of privileges.

  — The enable flags on principals and entitlements, and effective times of privileges are then applied to see which privileges are not relevant because of time. This results in an update list.
  — The changes are then consolidated. The list of current privileges is compared with the list of privileges on the update list. Any duplicates

and privileges that are not operative because of current privileges are deleted. This results in the list of privileges to enter into the runtime database.

— Finally, the runtime information is updated.

Let us illustrate the above profile and privilege resolution by considering a simple example. Consider an entitlement Deposit with one key attribute Acct with three different values (Cheque, Savings, Credit). The principal called Very-Senior-Teller is a member of four profiles and has two additional privileges. The four profiles grant privileges to Very-Senior-Teller for the Deposit entitlement and a key attribute of Acct.

Profile Name : Very-Senior-Teller
Profile Membership : Junior-Teller, Teller, Senior-Teller, Chief-Teller
Privileges :
1. AR : Deposit : Acct = Cheque, Amount = 10000;
2. AR : Deposit : Acct = Savings, Amount = 10000;

The four profiles that Very-Senior-Teller belongs to have the following profile membership and privileges.

Profile Name : Junior-Teller
Profile Membership : None
Privileges:
1. AR : Deposit : Acct = Cheque, Amount = 500;
2. AR : Deposit : Acct = Savings, Amount = 500

Profile Name : Teller
Profile Membership : None
Privileges :
1. AR : Deposit : Acct = Cheque, Amount = 1000;
2. AR : Deposit : Acct = Savings, Amount = 1000

Profile Name : Senior-Teller
Profile Membership : None
Privileges :
1. AR : Deposit : Acct = Cheque, Amount = 2000;
2. AR : Deposit : Acct = Savings, Amount = 2000

Profile Name : Chief-Teller
Profile Membership : Senior-Teller
Privileges :
1. AR : Deposit : Acct = Credit, Amount = 2000;

The steps involved in resolving the privileges for this principal are as follows:

- First profile Junior-Teller profile has no inherited profiles. Since the privileges for Junior-Teller use the AR privilege operator and the key privilege attribute (Acct) is not the same, these two privileges now become those of Very-Senior-Teller:

  AR : Deposit : Acct = Cheque, Amount = 500;
  AR : Deposit : Acct = Savings, Amount = 500

- The next profile, Teller, has no inherited profiles. Since the privilege operator is AR and the key attribute value is the same, both privileges for Teller override the privileges granted by Junior-Teller. Therefore the privileges are as follows:

  AR : Deposit : Acct = Cheque, Amount = 1000;
  AR : Deposit : Acct = Savings, Amount = 1000

- The next profile, Senior-Teller, has no inherited profiles. Since the privilege operator is AR and the key attribute value is the same, both privileges for Senior-Teller override the privileges granted by Teller. Therefore the privileges for Very-Senior-Teller are as follows:

  AR : Deposit : Acct = Cheque, Amount = 2000;
  AR : Deposit : Acct = Savings, Amount = 2000

- The fourth profile, Chief-Teller, inherits a profile Senior-Teller which needs to be evaluated. The Senior-Teller has no inherited profiles. The privileges for Senior-Teller match those already given to the Very-Senior-Teller.

  AR : Deposit : Acct = Cheque, Amount = 2000;
  AR : Deposit : Acct = Savings, Amount = 2000

- The evaluation process continues by returning to the profile Chief-Teller. The specific privileges granted to Chief-Teller is now added to the privileges for Very-Senior-Teller which are now as follows:

  AR : Deposit : Acct = Cheque, Amount = 2000;
  AR : Deposit : Acct = Savings, Amount = 2000;
  AR : Deposit : Acct = Credit, Amount = 20000

- Finally, the evaluation process returns to the principal record (Very-Senior-Teller) and the privileges that have been granted to Very-Senior-Teller. The final resolved privileges are :

  AR : Deposit : Acct = Cheque, Amount = 10000;
  AR : Deposit : Acct = Savings, Amount = 10000;
  AR : Deposit : Acct = Credit, Amount = 20000

4.4. *Using the authorization system to specify policies*

Let us now consider the steps involved in using the authorization system described above. These include the following:

- Definition and Creation of Entitlements
- Creation of Profiles
- Creation of Principals and Assignment of Privileges
- Enabling of Entitlements

Definition and Creation of Entitlements

First entitlements are defined; the names must be unique across a given Authorization Server and across a cell (a collection of clients and servers). Then it is necessary to determine which information is required to control the authorization for each entitlement. That is, it is necessary to determine the transaction and privilege attribute names, types and the entitlement rule. For instance, if an entitlement such as Withdrawal is defined, then transaction attributes could be the amount, the currency, the day of the week, the location, the time of day and the type of account from which the funds can be withdrawn. Then the privilege attributes need to be defined for each transaction. For instance, Weekday can be a privilege attribute for the Withdrawal transaction with the range Monday to Friday. There can also be other privilege attributes such as Amount Limit, Valid Currency and so on. Then it is necessary to consider whether there are any key privilege attributes. Next, the rule for the entitlement needs to be defined. This may be for instance, the transaction is only allowed on weekdays from certain starting time to some finishing time, for a certain type of currency, with a certain limit on the value, from a certain type of account. Then the default privileges are to be specified. If a principal is assigned a privilege for entitlement and no default privilege attribute values are specified, the principal is given the set of default privileges. Now the entitlement can be created using the administrative client interface and the entitlement stored in the administration database.

Creation of Profiles

The profile creation groups privileges for an entitlement. Again suppose we want to assign principals the privileges to withdraw dollars from cheque accounts with a base limit of $500. From the administration client GUI the Create Profile item is selected and the following information is entered in the dialog boxes. This example creates a profile named Withdraw-Cheque containing two privileges which can be to any principal who is a member of this profile. The first privilege allows the principal to access the Funds-Withdrawal operation. The second privilege authorizes the principal to access the server to withdraw up to $500 from check account between 8 AM and 5 PM, Monday through Friday. The effective times have been set to default values. Unless there is some need to specify some particular effective times, it is better to use default values. This makes it easy to override privileges because it is easier to match effective times.

Profile Name : Withdraw-Cheque
Profile Membership : None
Privileges :
Privilege Operator = AR ("add or override")
Entitlement Name = "Funds Withdrawal"
Beginning Effective Date/Time = "As soon as possible"
Ending Effective Date/Time = "No end date"
Valid-Currency = "USD"
Valid-Acct-Type = "Check"
Amount-Limit = "500"
Valid-Start-Day = "Monday"
Valid-End-Day = "Friday"
Valid-Start-Time = "8:00 AM"
Valid-End-Time = "5:00 PM"

Other profiles can be created in a similar manner, each of which represents groups of privileges to be granted in blocks. One can also create some combination profiles to make it easy to grant privileges for multiple types of similar access to the application server. For instance, we can create a profile containing privileges for all types of withdrawals from cheque, savings and money market accounts.

Profile Name = "Withdraw"
Profile Memberships :
"Withdraw-Cheque"
"Withdraw-Savings"
"Withdraw-Moneymkt"
Privileges : None

One can also add to profiles that are intended to represent classes of users. For example, one can add Withdraw profile to the list of profile memberships in the Teller profile. This would grant the necessary privileges to all members of Teller.

Creation of Principals and Assignment of Privileges

Having created the profile, principals are created in the Authorization Server database. Then the privileges need to be assigned to the principals authorized to access the server. One can assign privileges to a new principal when creating it or modify the privileges and profiles of existing principals. In practice, it turns out to be more effective to assign privileges using profiles created earlier. To grant privileges to a specific principal, one can simply add one of the profiles containing the necessary privileges to the principal's list of profile memberships. To grant privileges for the new entitlement to an existing class of principals, one can add one of the profiles to the list of profile memberships of a profile already available to the principals in that class. Every principal that is a member of that profile (it is listed in the list of profile memberships) is modified to include those privileges. The final way to grant privileges is to add one of the profiles to the list of profile memberships of a profile that is inherited by other existing profiles.

For example, to arrange that all principals who have any teller responsibilities have the basic authority to perform withdrawals, one can set up a hierarchy to represent differing levels of responsibility. A Senior-Teller profile may inherit the Teller profile but adds some extra privileges and overrides some others. Other profiles may also inherit the Teller profile. Adding one of the profiles containing Funds-Withdrawal privileges to the Teller profile grants those privileges to all the principals who are members of Senior-Teller.

Enabling of Entitlements

The final step is to enable the created entitlements via the administration GUI. Until an entitlement is enabled, requests for authorization on that entitlement are denied. For instance, when adding a new application server, using this flag mechanism one can ensure that the requests for authority for the new entitlement are not valid until the server is online.

5. Security Policy Management Framework

Let us now consider the more general situation where we have a number of such authorization management authorities (say, one per each domain) and look at some of the design issues involved in structuring these authorities in a large distributed environment. This policy management framework should capture the relationship between managers and the policy objects they manage and the arrangement of those objects into separate policy administration domains. The framework should be flexible and should not prescribe fixed answers to fundamental questions of policy administration, such as who is able to change the domain structure and which managers can change the placement of an object within the structure. The identification and explicit recognition of these meta-policy design choices provide a greater understanding of policy management.

As with the policies themselves, their management can vary in complexity. Comparatively little attention has been given to the topic of policy management, especially in terms of the management structure. Some work, such as [12], considers the management of policy solely as an extension of a given mechanism or propose a new mechanism [2], but do not consider the possible forms the management structures may take. Many access control proposals (e.g. [9]) consider policy management to be centralized, with all decisions being made by a security administrator or equivalent. This allows good control and regulation of what polices exist in the system, but does constrict users. Even Unix-like systems, which do not have a centralized policy manager, have a very simple model of policy management, where each user administers the polices for objects they own, with any super user being able to override these settings. This presents a simple two-level hierarchy of management. With the increasing complexity of systems it appears worthwhile to examine a more extensive range of options, beyond the limits of commonly used models. In the first instance this should be done in a mechanism-independent manner, as this will simplify the analysis, while recognizing that in practice the policy mechanism chosen will affect the policy management. The results of such analysis should be a better understanding of the options available for policy management.

Let us now examine possible structures for the management of authorization policies. We start by identifying some central issues that distinguish policy management models. The majority of these issues do not submit to single answers. From these issues a number of options are identified which could realistically be employed. This should allow comparisons to be made between various systems and proposals. The framework developed should further allow other practical arrangements for the management of access control policy to be identified. First we present our basic definition of a policy and policy management domains. Then we formulate a framework and discuss the issues of policy management in a mechanism-independent manner. We use this to handle some differentiation in the power of managers, in order to demonstrate its flexibility. Then we apply the framework to a specific instance, namely role-based access control policies.

5.1. *Policy management domains*

Regardless of the means of expression, for a policy to actually allow access it must be related to one or more objects. This defines the scope of the policy. The linkage between policy and object may be explicit, to one or more objects, or implicit. Unix, for example, explicitly links a policy to a single object, by having a separate access control list for each file and directory. Other systems allow a single policy to apply to a number of explicitly named objects. Still further proposals allow the use of policy domains, where policies within the domain apply to all objects also within the domain, making the linkage more implicit than explicit. For the purposes of this paper we will confine ourselves to systems where the linkage between policy and object is explicit.

In a given system there will be a set of policies that a user has expressed and objects to which those policies can be made to apply. We call such a set a policy management domain as it consists of the entities under the policy control of that user. Obviously, in certain systems and for some users, this set will be empty. In many systems we can identify a single user (or group of users) who can, potentially at least, write policies that apply to all objects within the system. While it is unlikely, except in MAC (Mandatory Access Control) systems, that any given policy will be so broad in actual scope, the potential exists and defines a notional limit. The objects within such a limit (and the policies applying to them) form a single management domain. This domain can be considered the root management domain for the system. In some systems it is the only management domain and such tasks are limited to a single policy administrator. While this may be tempting to those that favour centralized control, this approach is not suitable for a large distributed system and may simply be deemed undesirable from an organizational point of view.

As an alternative to centralized management, some policy management functions may be devolved to other users. If the management of certain policies and objects is entirely removed from the ambit of the chief administrator, then, for our purposes, those entities so separated can be considered to constitute a separate system and we need not consider this for the moment. Instead, we need to consider the case where policy management duties are given to other users, but without lessening the abilities of the grantor. As an example, Unix may be considered to act in this way, with policy management domains forming a two-level hierarchy, the super user domain at the top level, general users at the bottom.

More generally, the structure of policy management domains can be viewed as a hierarchical tree (or possibly a directed graph). The depth, conceptually at least, is unlimited. Once such a structure is in place, we need to consider how it is to be managed. More specifically, how is the domain structure managed and altered, and how is the placement of policies and objects handled. There will be a need for defined answers to these questions. For example, in Unix, the placement of objects is altered by use of the *chown* operation and the target user of a *chown* operation can be any user in the system. This is a very unrestrictive rule. The two cases of a single entity controlling all policy management decisions (centralized) or all users having an independent area of authority (fully distributed) should not be viewed in terms of an either/or choice, but rather as two ends of a spectrum. Some limits may be desired on the management choices available to users. For example, more restrictive rules might be as follows:

- The target user of a *chown* operation issued by a member of group1 must be another member of group1
- The target user of a *chown* operation issued by a member of group2 must be a member of group2 or group3 or group4

It may appear that the above two rules should not appear in the same hierarchy. If the arrangement of policy management domains into a hierarchy is to have any meaning, then it appears sensible to have uniform rules governing its management (Unix for one certainly follows this guideline). However, it should be clear that the division of policy management into a number of domains lends itself to uniform policies. For example, the two rules above need not imply a one-to-one mapping between groups and policy management domains. If the second rule also applies to members of groups 3 and 4, then we could have two domains, one consisting of the objects of group1, the other consisting of the objects of groups 2, 3 and 4. Alternatively, if there is a one-to-one mapping between groups and domains in our hypothetical organization, then the domain of group2 may be the parent of those of groups 3 and 4.

Either way, we can postulate the single rule as follows:

- The target user of a *chown* operation issued by a member of a domain must be another member of that domain or one of its descendant domains.

Implicit in the above rule is the idea of multiple managers for a policy management domain. For simplicity, from this point, we consider each policy management domain to have a single, unique manager. It should be noted that a policy management domain includes the objects and policies of its descendant domains. That is, the manager of a policy management domain can exercise policy management authority over the objects and policies in their domain and in the descendant domains of that domain.

Some authority needs to decide upon the rules to be applied throughout the hierarchy. The obvious candidate is the manager of the root domain. We note that this model does not reflect all possible situations, specifically those where no central management entity is desired, or where it is possible to limit the rights of a central management entity. However, we believe that the situation we have described is general enough to merit discussion. A policy management domain hierarchy may or may not correspond to a single network or intranet. However, as it embodies a single management approach, a single intranet (or other organizational network) would consist of one or more, identifiable, policy management domain hierarchies.

Note that the rules that govern the policy management domains can be regarded as policies themselves (or, more properly, meta-policies). It may even be possible, in some systems, to implement them using the mechanism for expressing policy on general objects. While this may be so, for purposes of analysis it is worthwhile considering them separately. What we have described so far is independent of a particular mechanism for expressing the access control policy. This allows us, potentially, to compare the management structures of hierarchies regardless of the actual policy expression mechanism.

5.2. Root policy manager

Within a single policy management hierarchy, ultimate authority resides with a single management entity, the manager of the root domain. It is not important whether this management entity corresponds to one or a number of real world system users. However, the number of managers is considered to be much smaller than the total number of users of the system. We shall refer to a single root domain manager, while realizing that this could correspond to a number of actual users. The root domain manager has total control over the rules that govern the hierarchy. This also gives the root domain manager control over all policy decisions, as control over the hierarchy allows them to take control of any or all policy decisions.

While it may appear from our definition of a root domain manager that we are advocating a centralized approach, two points must be remembered. First is that the root domain manager may not be a single real world entity. Second, a 'system' (such as an intranet or a LAN) may consist of a number of policy management hierarchies. Each hierarchy could have its own root domain manager and rules, allowing for any desired level of decentralization. This may be desirable for some organizations, which do not wish a single entity to have the authority over all decisions. It could also be provided for by removing automatic authority over descendant domains. However, in this paper, we will not discuss these options for two reasons. Firstly, it complicates the discussion beyond what can comfortably be discussed in the space allowed here. Secondly, the notion of a root policy management manager that cannot administer all policies seems rather contradictory. Especially when one considers that such a manager would still be able to assign which users can manage these policies (and so include themselves at any time). It would be better to consider these as separate policy management hierarchies, and provide mechanisms for agreeing on meta-policies across multiple domains. Some of the issues involved in such a structure are discussed below.

5.3. *Delegation*

The hierarchical arrangement of policy management described above can be looked at as a question of delegation. Consider an initial state of a system, where all policy decision authority ultimately rests with a single root management entity. Over time that entity decides to delegate some of this authority to particular users. That is, the root authority creates new domains, moves objects to those domains and appoints managers for the domains. These entities can, if they wish, delegate the authority further, in the same way. Viewed in this way, the relation to delegation is clear. Even Unix can be viewed in this manner, as the system manager (super user) can change the policy settings for any object, but usually delegates most of this to the users, by way of the ownership property.

Let us consider a single method to handle delegation, including both static (long term) and dynamic (short term) delegation. This is the ability to move objects between domains and will be further discussed below. As a domain manager can add objects to a descendant domain (i.e. delegate) and remove them from such domains (i.e. revoke), the ability to delegate is straightforward. We acknowledge that this is a fairly simple approach to delegation, but it is sufficient for our purposes. A more thorough discussion would require consideration of possible limitations being placed on the authority delegated (for example, restrictions being placed on the recipient of the delegated object moving it to another domain), but we will not consider these issues here in this paper.

5.4. *Policy domain representation*

Let us now consider a single policy management hierarchy. That is, we are considering a system or organization that can be regarded as having a single management structure for the management of access control policies. The policies for this organization are organized in a hierarchy of policy management domains. After this we will discuss some issues related to multiple hierarchies.

Each domain is under the control of a management entity. It would be possible to call each node of the tree a sub-domain and the entire hierarchy a policy management domain, but the meaning of the term domain in the following should be clear from the context. The identity and nature of the management entities is not further specified at the moment, but left for later discussion. All that we will assume for now is that each management entity manages no more than one domain (node) within the hierarchy. Note that one domain being the child (or descendant) of another in the hierarchy means, at least in some sense, that the child domain is a part of the parent domain. An alternative view of the situation is domains residing inside their parent. The hierarchy is still apparent. Every object in the system resides in a domain. We consider policies to be a subset of the objects that exist within a system.

The manager of the root domain is called the root domain manager. We can consider a newly initialized system to simply consist of the root domain, some objects within that domain, and the domain's manager. What meaning follows from the location of an object within a domain? That is, what can the manager of that domain do with (and to) the object? These questions are simplistic statements of the questions of formulating policy and meta-policy for an object. In terms of policy, the obvious answer to the question is that if an object resides within a domain, then the manager of that domain may express policy for that object. It is worth noting that this means the entities that can express policy for an object are then the manager of the domain in which it directly resides and the managers of all the ancestors of that domain. There are two basic issues to consider in the realm of meta-policy:

(1) By whom is it decided in which domain an object resides (and how is the location changed)?
(2) By whom is the shape of the tree itself decided?

5.5. *Domain location*

Consider the example in Figure 5. Object O1 resides in the domain 3 managed by Manager 3. Who has the authority to move an object to another domain? For example, which manager or managers can move object O1 to the domain managed by manager 5? We can immediately rule out managers 4 and 5. Manager 4 has no authority over O1 at all. Manager 5 has no current authority over O1 (even if the point of the exercise is to give it the authority) and managers should not be able to arbitrarily assume authority over another domain's objects. Can Manager 3 make the transfer? This operation is obviously related to the Unix *chown* operation and other equivalent mechanisms. In many popular variants of Unix there are no restrictions on the target user of a *chown* operation. Indeed, ownership of an object in these systems can be transferred to a user without consulting that user, in fact without their knowledge. So, if we are considering a Unix-like system, Manager 3's ability to make the transfer is a viable option.

Managers 1 and 2 may also be able to make this change. Object O1 resides in their domains, so it could be argued that they should be able to carry out the operation. In fact, it could well be argued that only they could do it, as once a manager is given responsibility for an object they can

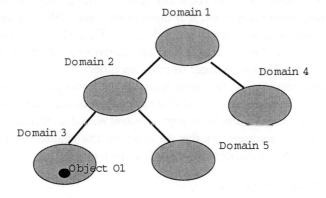

Fig. 5.   Policy Management Framework.

not divest themselves of it (ie, Manager 3 should not be able to make the transfer). Alternatively, it may be desired to restrict the authority to make all such changes to only a single entity (in this case, Manager 1).

So we have three possible options, each reflecting different organizational approaches:

(1) The managers of the ancestor domains of the domain in which the object resides.
(2) Both the manager of the immediate domain in which the object resides and the managers of the ancestor domains of the domain in which the object resides.
(3) The manager of the root domain.

It could be argued that a fourth option exists, that only the manager of the immediate domain can change the mapping, but as the meaning of the hierarchy is that an object resides in both its immediate domain and in the ancestor domains of that domain, this seems counter-intuitive. For our example, if only Manager 3 was to have authority, then it should not be placed subordinate to the other managers. Note that option 1 presents a problem in the case of objects in the domain of manager 1 (the root domain). This can be solved either by not allowing any objects to actually reside in that domain or by making an exception for that domain (effectively combining options 1 and 3). In some sense the manager of the root domain is a super user. Some organizations may not wish such a user to exist, but for our purposes this could be handled by multiple separate hierarchies, each representing the responsibilities given to each user.

The next question is, are there any limits on the destination of the object? Again, considering Unix, it is obvious that one viable answer is that an object can be moved to any other domain. While this offers great flexibility, it also offers little control. Other alternatives are available. The most restrictive is to allow a manager to move an object only to a domain that the manager controls (i.e. the immediate domain managed and any descendants of that domain). Less restrictive options are to allow an object to be passed to the parent domain or any ancestral domains. Organizationally, this relates to a manager passing an issue up the chain of control. The target domain when an object is moved in the tree may be:

(1) Any domain in the tree
(2) The domain of the manager making the move or any descendant domains

(3) As in 2 plus the immediate parent domain

(4) As in 2 plus any ancestor domains

Combining this with the example above, we would have the rule that a manager can relocate any object which resides in their domain or any descendant domain and that the target domain can be the domain managed, any descendant domain or any ancestor domain. Note that this also gives us a rule for the domain in which a new object could be placed when created by a user.

Let us again consider our example, from Figure 5, of moving O1 from domain 3 to domain 5. Under rule 3 above only managers 1 or 2 could perform the operation. Only under option 2 on which manager can make the move and option 1 on destination could Manager 3 move the object. Even with option 2 on who can make the move, any other choice but 1 for destination control prevents Manager 3 making the change. Other examples can easily be found which illustrate the differences between the various combinations. We leave construction of such examples to the reader. Another possible option might be that objects could be transferred to sibling domains. This would, in our example, allow Manager 3 to move object O1 to the domain of Manager 5. We find it likely that some organizations would not allow such transfers of responsibility without the involvement of more senior authority. However, it would allow structures to be created where objects could be transferred between the domains of a specified group of managers without allowing unrestricted transfers to any point in the tree. This could be combined with any of options 2, 3 and 4 above.

5.6. *Writing policy*

Membership of an object in a policy management domain means that the manager of that domain can write policy for that object. Note that a policy contains no indication of the objects it governs. This link must be made separately. We assume that a new policy (like a new object) can be placed in any management domain to which a user could move an object. As policies do not in themselves specify objects (and therefore independently grant no access), there appears to be no need to restrict the writing of policies. It is the association of policies with actual objects that is important. As discussed above, we only consider explicit linkage of policies and objects, not policy domains. We do allow the mapping between policies and objects to be many-to-many. Note that this may require some conflict resolution mechanism, but the issue of conflicting policies can be considered separately to who is actually allowed to set policies. The act of making a policy apply to an object can be regarded as the basic function of policy management. It seems clear that, for a user to do this, both the policy and the object should be under their control. That is, the policy and the object must reside in the domain managed by the user or one of that domain's descendants (but they do not have to reside in the same domain). A similar rule can be written for removing an object from the scope of a rule. It may seem counter-intuitive to consider policy objects that do not exist separately from objects. However, making this separation allows for a greater choice of policy management structures. Making the policy-to-object mapping a many-to-one (or even one-to-one) relation can simply represent a structure where policies are directly related to objects. Access control list systems, such as Unix, can easily be viewed in this way. Separating the policy expression from the assignment of that policy to objects allows the policy-to-object mapping to be many-to-many and enables our framework to encompass systems which abstract over objects and are hence easier to manage due to the smaller number of actual policy objects (or expressions of policy). This approach that the mapping can only be made by a manager in whose domain both the policy and the object exist may appear overly restrictive. Some organizations may find this overly burdensome on their higher-level policy administrators. For example, referring to Figure 5, the organization may wish there to be a mechanism for mapping an object in the domain of Manager 3 to a policy in the domain of Manager 5 without having to consult either Manager 1 or Manager 2.

A number of options exist. We could remove the restriction on the policy having to be within the domain. In this case, the manager has no obvious policy authority over the policy object (it is not in the manager's domain) and hence this option appears undesirable. It would be possible to allow two managers to jointly make the decision by some means of consensus. This would require a degree of elaboration beyond the space available in this paper and we leave topics of negotiation and consensus for future work. A more promising option is to alter our tree structure for domains to that of a directed graph. For instance, in our example, a new domain could be created, subordinate to the domains of Managers 3 and 5 and the policy and object placed therein. While this is feasible, it raises a number of issues, such as who is allowed to create such a domain (as it effects the domains of the ancestor managers) and who manages the domain. Note that a graph would solve other problems, such as allowing all managers to attach objects to a given policy by having a domain containing all such policies at the bottom of the graph.

5.7. *Domain management*

Now that we have established how to relate policies to objects, let us consider which users can access these policies. A policy, as considered above, is simply an expression of what form of access is allowed to an object, with no direct indication of who can access. We could represent users with objects, in the same way that policies are, apply the same rules that govern linkage of policies and objects to the linkage of users and policies. Unfortunately, there are a number of drawbacks to treating users in this fashion. Most significantly, if a user is represented by an object and placed in a domain in the hierarchy, then only the manager of that domain, and the managers of ancestor domains, may map policies to that user. This makes it impossible to represent even as simple a system as Unix, as all users can write policy for all other users. It would also make it impossible for a manager to give access to objects to their superiors. Instead, we can take advantage of the requirements we introduced earlier, that all users must manage at least one domain. We then pose the question, the managers of which domains (i.e., users) can a manager map to a policy under his/her control? The options are in fact the same as those for the target movement of an object.

(1) Any user
(2) The manager or any user who is a manager of a descendant domain of that in which the policy resides
(3) As in 2 plus the manager of the immediate parent domain
(4) As in 2 plus any the managers of any ancestor domains

Let us now briefly consider the creation and deletion of domains (and the appointment of managers for newly created domains). A manager should only be able to create and delete descendants of the domain that they manage. Manipulating domains for which the manager has no responsibility is obviously undesirable. A simple rule could be that the manipulation of the shape of the domain tree be restricted to the root manager. A similar rule can be written for deleting domains.

Once a domain has been created, a manager for it needs to be appointed. One simple answer is that a manager is simply appointed, with no restrictions (beyond our implicit assumption that each management entity manages only one domain). If we assume that managers are (some subset of) the users of the system and that an object, for management purposes, represents the users, then a more sophisticated approach is possible. It is that the creator of a new domain can only appoint a manager from amongst those users that they manage. As managing a user means setting policy for that user, this appears natural. As a real-world analogy, consider a team leader. The team leader is given management responsibility over the team members, which can include defining domains over which each member has policy management authority. Transfer of management of a domain from one entity to another can be considered as a process of creating a new domain, transferring all objects from the old domain to the new domain and the deletion of the old domain. Hence we will not discuss this further.

5.8. *Relationship to mandatory and discretionary access control*

At a superficial level it may appear that the model outlined above is directly related to DAC (Discretionary Access Control). The concept of authority to write policy is a central part of the DAC approach, often encapsulated by a notion of ownership (as in Unix). The framework abstracts the notion of ownership into the authority to write policy for an object, captured by the presence of an object in a domain.

The framework can also encompass MAC systems. A simple MAC system, where a uniform set of access control policies is enforced on all objects, can be modelled by a single domain. The model can also handle combined MAC/DAC systems. As a simple example, the MAC policies could be located in the root domain, and be mapped to all objects. The DAC policies could appear in the lower level domains. The framework thereby captures the essence of the policy management distinction (as opposed to policy enforcement) between MAC and DAC. In both system types policy must be set and administered. Unlike others, e.g. [14], we do not see the mechanism used to express policy as essential to distinguishing between MAC and DAC. The essential feature of MAC is that certain policy decisions cannot be amended or rescinded by users, regardless of mechanism. The domain hierarchy can easily represent this notion.

5.9. *Possible extensions to the basic framework*

Up to this point we have limited ourselves to a single hierarchy. This is representative of an organization (or sub-unit of an organization) that wishes to have uniform meta-policies on the structure of policy management. To extend our discussion to multiple hierarchies and relationships between them, we need to consider several other issues. If we have multiple hierarchies, then it would be reasonable to assume that the meta-policy settings may differ from hierarchy to hierarchy. For example, the rule governing the possible domains to which a manager may move an object to (as discussed before) may differ between the hierarchies. The basic issues discussed above (such as which manager can move an object and to which domain) also need to be revisited. However, the answers need not be the same as those adopted within any of the involved hierarchies. For example, inter-hierarchy moves may be limited to root managers only or some other limited subset of the domain managers.

The fundamental issue in considering multiple hierarchies is that there is no longer any single source of authority in an equivalent sense to that provided by a root manager within a single hierarchy. All provisions for interactions between the hierarchies would have to be negotiated between the relevant authorities. This may require some form of cross-certification between the hierarchies. This opens up another whole area which we will not address in this paper.

The concept of negotiation or joint responsibility for setting policies also applies if we wish to allow multiple managers of a single domain. We have limited our discussion above to domains being managed by single unique entities. If policy management authority is to be shared equally (as opposed to superior/inferior division of the hierarchy), then provision needs to be made for negotiation between the managers or for mandating what would result from one manager attempting to override the settings of another. A somewhat simpler extension than either of the above is allowing entities to manage more than one domain. Initially all that is required is to make the mapping manager to domain a one-to-many mapping rather than a one-to-one mapping. This will allow managers to be responsible for multiple domains, with these domains possibly occupying quite separate parts of the hierarchy. The framework would need to reflect the separate areas of responsibility of such managers. Intuitively each of the domains managed by one entity should be treated separately and no interaction allowed between them that is not authorized by the relevant ancestor managers. The necessary alterations to the rules presented above are, in outline at least, reasonably clear.

5.10. *Role based access control*

The description of design options in the preceding sections applies to policy management frameworks in general. Let us now apply this framework to Role Based Access Control (RBAC). The essential elements of an RBAC system are the mappings between users and roles, and permissions and roles. Earlier, we considered the policy setting to be a unitary action, requiring only authority over an object (that is, the object to reside within the domain of the manager who wishes to express the policy) and a policy. While in RBAC systems the former is equivalent to creating (or changing) a permission, this in itself is insufficient. For a policy (permission) in an RBAC system to be useful, the permission must be mapped to at least one role. There must also be a mapping between that role and at least one user. Let us now consider the management of these mappings. However, it will become evident that the fundamental property of these mappings is very similar to that between object and policy, indicating a useful generality of our approach.

Permissions and Roles

A flexible approach to the creation of roles and permissions is to allow any manager to create a new (empty) role or permission. A role expresses meaningful policy when it has users and permissions mapped to it and a permission is meaningful only when it actually applies to some objects. Hence this flexibility does not allow arbitrary policy decisions. The placement of a new role or permission should depend on the organizational decision as to where an object can be moved within the domain hierarchy, as with policies and objects above. The same argument can be made about restricting this ability to certain users. Domain membership for an object means that the manager of that domain can write policy about that object. In an RBAC system this implies having a permission which includes that object in its scope. Obviously, a manager should only be able to add objects to the scope of a permission it manages (i.e. is within its domain). Therefore, being able to write policy for an object in an RBAC system means that a manager can add it to (or subtract it from) the scope of a permission, as long as that permission is also within the manager's domain. This mirrors the discussion of policy and objects in the general framework above.

Let us now consider the mapping between roles and permissions. If the role and permission exist in the same domain, then the manager of that domain should be able to make the mapping. Given our definition of domain, it seems sensible that this should also be open to the ancestors of this domain. More generally, the mapping can be made if both role and permission exist somewhere within the domain of the manager. This means that domain membership for a permission means that a manager can add objects to its scope and map it to a role, as long as the object and role are also within the domain. Similarly for subtracting objects and removing the permissions from a role. The meaning of a role being in a domain is now clear.

Consider the situation in Figure 6. Manager 1 could map object O1 to permission P1. Manager 2 could, at a later point, map permission P1 to role R1. The result of this is that object O1 is now accessible to any member of role R1 and, most importantly, a manager whose domain did not include O1 made this decision. This appears to be a violation of our basic rule that policy decisions can only be made about an object if that object lies within the domain of the manager making the decision. We could prevent this from happening by having a variant rule which requires all objects indirectly linked to a role by a role-permission mapping to fall within the domain of the manager making the mapping. It could be argued that this is overly restrictive and that by mapping the object to the permission in this example, Manager 1 has delegated to lower-level managers the ability to map the object (indirectly) to roles. This is another case of two different, but both potentially valid, approaches. Adoption of one or the other would depend upon organizational requirements.

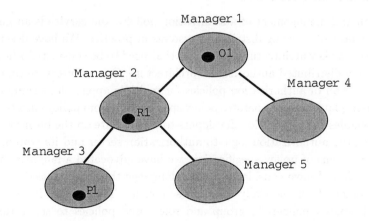

Fig. 6.   Policy Management Framework and RBAC.

Similar rules could be written for the role-to-user mapping if we treat users as objects (or at least represent the ability to set policy for a user by having an object which represents a user). However, it should be apparent that the fundamental principle is that where we have two policy-relevant objects (such as a permission and object or permission and role or role and user), then a direct mapping between the two can only be made by a manager whose domain includes both objects. This general principle may appear obvious, but does neatly encapsulate policy management.

Role Hierarchies

The effect of a role hierarchy is to make permissions available to the users of a role without the permissions being directly mapped to that role. Establishing an inheritance relationship between two roles obviously relates to the management of policy. In the example above, we will refer to the role that will be inheriting as R1 and the role that will be inherited from as R2. We need to consider which managers can make such a relationship. Obviously, if both roles and all permissions of the role to be inherited are in the domain of a manager, then that manager can assign one role to inherit from the other. As the manager controls the permissions, the manager could directly assign them to R1 and hence assigning them via inheritance seems sensible. If the permissions are not within the domain of the manager, then we have similar arguments to those discussed before.

6. Concluding Remarks

Design and management of authorization policies and service is an important aspect of securing distributed systems in practice. We have described some of the key architectural principles that need to be considered when designing a distributed authorization architecture. Our architecture supports specification of authorization policies for multiple applications running on different platforms. It separates policy specification from policy enforcement and enables the application developers to concentrate on the business logic leaving the authorization logic to authorization server and its administrators. In terms of policy specification, we have advocated a language-based approach and have considered a small language that can be used to specify a range of commercially used access policies. These range from simple identity-based policies to group and role based policies to separation of duty, delegation and joint action policies. This language has been used in practice to specify access control policies in large-scale financial and medical systems. The authorization service design and the logical components presented in this paper have been developed and they have been used in commercial environments in real systems.

Then we extended the architecture to consider multiple policy management domains in a large-scale distributed system. We outlined a policy management framework and highlighted the various design options and alternatives for the structuring of management authorities, in a mechanism-independent manner. It encapsulated an intuitive basic principle that explicit relationships between policy-significant objects can only be made by an entity with policy authority over both objects. It is sufficiently flexible to allow for indirect relationships to be handled this way. Then we illustrated the applicability of the policy framework by applying it to the specific case of role based access control. There are several other important aspects that are worth pursuing further, such as co-operative policy management, trust relationships between multiple hierarchies and a graph rather than tree based policy management structure.

References

1. Bai, Y. & Varadharajan, V., 'A Logic for State Transformations in Authorization Policies', Proceedings of the IEEE Computer Security Foundations Workshop, 1997.
2. Barkley, J. & Cincotta, A., 'Managing role/permission relationships using object access types', Proceedings of the 3rd ACM workshop on Role-based access control, Fairfax, USA, 1998, pp. 73–80.
3. Brewer, D. & Nash, M., 'The Chinese Wall Security Policy', Proceeedings of the IEEE Symposium on Security and Privacy, 1989, pp. 206–214.
4. Damianou, N., Dulay, N., Lupu, E. & Sloman, M., 'The Ponder Policy Specification Language', Proceedings of International Workshop on Policies for Distributed Systems and Networks, Bristol, UK, January, 2001, pp. 18–38.
5. Hitchens, M. & Varadharajan, V., 'Tower: A Language for Role Based Access Control', Proceedings of International Workshop on Policies for Distributed Systems and Networks, Bristol, UK, January, 2001, pp. 88–106.
6. IETF RFC 1508, RFC 1509: Generic Security Service Application Programming Interface, 1993.
7. IETF RFC 1510: The Kerberos Network Authentication Service, 1993.
8. Jajodia, S., Samarati, P. & Subrahmanian, V.S., 'A Logical Language for Expressing Authorizations', Proceedings of the IEEE Symposium on Security and Privacy, 1997.
9. Marshall, I. & McKee, P., 'A Policy Based Management Architecture for Large Scale Active Communication Systems', Proceedings of International Workshop on Policies for Distributed Systems and Networks, Bristol, UK, January, 2001, pp. 202–213.
10. Open Software Foundation (OSF), Distributed Computing Environment, Security Services, Practical Programming and User Documentation, 1995.
11. Sandhu, R., Coyne, E.J. & Feinstein, H.L., 'Role based access control models', IEEE Computer, 1996, 29, (2), pp. 38–47.
12. Sandhu, R. & Munawer, Q., 'The ARBAC99 Model for Administration of Roles', Proceedings of 15th Annual Computer Security Applications Conference, Phoenix, USA, 1999.

13. Varadharajan, V. & Allen, P., 'Joint Action based Authorization Schemes', ACM Operating Systems Review Journal, Vol. 30, No. 3, pp. 32–45.
14. Varadharajan, V., Allen, P. & Black, S., 'An Analysis of the Proxy Problem in Distributed Systems', Proceedings of the IEEE Symposium on Security and Privacy, 1990.
15. Varadharajan, V., Crall, C. & Pato, J., 'Authorization for Enterprise wide Distributed Systems', Proceedings of the IEEE Computer Security Applications Conference, ACSAC'98, 1998, USA.
16. Woo, T.Y.C. & Lam, S.S., 'Authorization in Distributed Systems: A Formal Approach', Proceedings of the IEEE Symposium on Security and Privacy, 1992, pp. 33–50.

INTRODUCTION TO ALGEBRAIC GEOMETRY CODES

Chaoping Xing

Department of Mathematics, National University of Singapore
2 Science Drive 2, Singapore 117543, Republic of Singapore
E-mail: matxcp@nus.edu.sg

Goppa's construction of algebraic geometry codes is introduced in this paper. Some basic results on Goppa geometry codes are presented. Following Goppa's construction, we introduce two more constructions of linear codes based on curves over finite fields. These two constructions allow us to use closed points of higher degree. One interesting consequence is that many new codes are found from these two constructions.

1. Introduction

Algebraic geometry codes were discovered by V. D. Goppa. This beautiful discovery came as a result of many years of thinking over possible generalizations of Reed-Solomon, BCH and classical Goppa codes. Nowadays, algebraic geometry codes constructed by Goppa are usually called Goppa geometry codes to distinguish them from classical Goppa codes and other codes constructed from algebraic curves or varieties.

Goppa geometry codes open rather exciting possibilities in coding theory and other topics such as sphere packing, cryptography and low-discrepancy sequences [2, 4]. These codes also lead to some deep and interesting problems in algebraic geometry and number theory such as curves over finite fields with many rational points [2].

In this paper, we introduce some basic results on Goppa's construction of algebraic geometry codes as well as other constructions of linear codes based on algebraic curves. For the details on this subject, the reader may refer to the books [2, 3, 4].

The present paper is divided into seven sections. Codes are introduced in Sections 2 and 3. The fourth section gives some necessary material on algebraic curves over finite fields to understand our constructions. Sections 5 and 6 contribute to Goppa's construction of algebraic geometry codes. In the last section, two new constructions are presented and some new codes are obtained.

2. Introduction to Codes

Let A be a finite set of q elements and $A^n = \{(a_1, \ldots, a_n) | a_i \in A\}$ the affine space of dimension n over A.

Definition 2.1: A subset C of A^n is called a *q-ary code* of length n over A. The set A is called the *code alphabet*. We say that C is a q-ary (n, M)-code if the size of C is M.

Example 2.2: The code

$$C = \{00000, 01110, 10111, 11001\}$$

is a binary $(5, 4)$-code.

The information rate of a q-ary (n, M)-code is defined by

$$R = \frac{\log_q M}{n}.$$

The information rate measures the transmission efficiency of a code.

For two words $\mathbf{x}$ and $\mathbf{y}$ of A^n, the *Hamming distance* between $\mathbf{x}$ and $\mathbf{y}$ is defined to be the number of positions where $\mathbf{x}$ and $\mathbf{y}$ differ.

Example 2.3: Consider the words

$$\mathbf{x} = (12201), \quad \mathbf{y} = (01211).$$

Then $d(\mathbf{x}, \mathbf{y}) = 3$.

Definition 2.4: The *minimum distance* or *distance* $d(C)$ of a q-ary (n, M)-code C $(M \geq 2)$ is defined to be

$$d(C) = \min\{d(\mathbf{x}, \mathbf{y}) | \mathbf{x} \neq \mathbf{y} \in C\}.$$

An (n, M)-code with distance d is denoted by (n, M, d).

Example 2.5: For the code in Example 2.2, the distance is 3. It is a binary $(5, 4, 3)$-code.

The following result tells us that the error correcting capability of a code is determined by its distance

Theorem 2.6: *A q-ary (n, M, d)-code can correct up to $\lfloor (d-1)/2 \rfloor$ errors and detect up to $d - 1$ errors.*

The reader may find the proof of the above theorem from any coding theory textbook.

For instance, the code in Example 2.2 has distance 3. It can correct 1 error.

For a q-ary (n, M, d)-code, the error-correcting rate is defined by

$$\delta = \frac{d-1}{n}.$$

Goal: For a q-ary (n, M, d)-code, we want both $R = (\log_q M)/n$ and $\delta = (d-1)/n$ as large as possible.

However, the above goal cannot be easily achieved, as we will see from the following result that R and δ are not compatible with each other.

Theorem 2.7: *(Singleton Bound) Let R and δ be the information rate and error-correcting rate, respectively. Then*

$$R + \delta \leq 1. \tag{1}$$

Proof: For a q-ary (n, M, d) code C, bound (1) is equivalent to

$$M \leq q^{n-d+1}. \tag{2}$$

By deleting the last $d - 1$ positions for each codeword of C, we get a new code C'. It is clear that C' still has M distinct codewords as the distance of C is d. As $C' \subseteq A^{n-d+1}$, we have

$$M = |C| = |C'| \leq |A^{n-d+1}| = q^{n-d+1}.$$

This completes the proof. □

Therefore, from the Singleton bound we have to find a balance between information rate and error-correcting rate.

3. Linear Codes

Now we equip C with some algebraic structure. Let $\mathbf{F}_q$ denote the finite field with q elements.

Definition 3.1: A subspace C of the vector space $\mathbf{F}_q^n$ is called a linear code. The dimension of C as a linear space over $\mathbf{F}_q$ is called the *dimension* of the code C. A linear code of length n, dimension k and distance d is denoted by $[n, k, d]$.

Remark 3.2: For a q-ary $[n, k, d]$-linear code C

(1) The information rate is k/n since $|C| = q^k$.
(2) The Singleton bound for C is

$$k + d \le n + 1.$$

The Hamming weight $\mathrm{wt}(\mathbf{u})$ of a vector $\mathbf{u}$ of $\mathbf{F}_q^n$ is defined to be the number of non-zero positions.

Example 3.3: Consider the vectors

$$\mathbf{x} = (12201), \quad \mathbf{y} = (01211), \quad \mathbf{u} = (00021) \in \mathbf{F}_3^5.$$

Then

$$\mathrm{wt}(\mathbf{x}) = \mathrm{wt}(\mathbf{y}) = 4 \quad \text{and} \quad \mathrm{wt}(\mathbf{u}) = 2.$$

For a linear code, the distance is actually determined by its smallest nonzero weight.

Lemma 3.4: *Let C be a linear code over $\mathbf{F}_q$ of length n, then*

$$d(C) = \min\{d(\mathbf{x}, \mathbf{y}) | \mathbf{x} \neq \mathbf{y} \in C\}$$
$$= \mathrm{wt}(C) := \min\{\mathrm{wt}(\mathbf{z}) | \mathbf{z} \in C \setminus \{\mathbf{0}\}\}.$$

Proof: For any words $\mathbf{x}, \mathbf{y}$, we have $d(\mathbf{x}, \mathbf{y}) = \mathrm{wt}(\mathbf{x} - \mathbf{y})$.
By definition, there exist $\mathbf{x}', \mathbf{y}' \in C$ such that $d(\mathbf{x}', \mathbf{y}') = d(C)$, so

$$d(C) = d(\mathbf{x}', \mathbf{y}') = \mathrm{wt}(\mathbf{x}' - \mathbf{y}') \ge \mathrm{wt}(C),$$

since $\mathbf{x}' - \mathbf{y}' \in C$.
Conversely, there is a $\mathbf{z} \in C \setminus \{\mathbf{0}\}$ such that $\mathrm{wt}(C) = \mathrm{wt}(\mathbf{z})$, so

$$\mathrm{wt}(C) = \mathrm{wt}(\mathbf{z}) = d(\mathbf{z}, \mathbf{0}) \ge d(C). \qquad \square$$

4. Background on Curves

For the finite field $\mathbf{F}_q$, let $\mathcal{X}$ be a smooth, absolutely irreducible, projective algebraic curve defined over $\mathbf{F}_q$. We express this fact by simply saying that $\mathcal{X}/\mathbf{F}_q$ is a curve. The genus of $\mathcal{X}/\mathbf{F}_q$ is denoted by g. A point on $\mathcal{X}$ is called *rational* if all of its coordinates belong to $\mathbf{F}_q$. Let $N(\mathcal{X}/\mathbf{F}_q)$ be the number of rational points on $\mathcal{X}$. For our purpose, we are interested in curves with many rational points.

A divisor G of $\mathcal{X}$ is called *rational* if

$$G^\sigma = G$$

for any automorphism $\sigma \in \mathrm{Gal}(\overline{\mathbf{F}_q}/\mathbf{F}_q)$. In this paper we always mean a rational divisor whenever a divisor is mentioned.

We denote by $\mathbf{F}_q(\mathcal{X})$ the function field of $\mathcal{X}$. An element of $\mathbf{F}_q(\mathcal{X})$ is called a function. We write ν_P for the normalized discrete valuation corresponding to the point P of $\mathcal{X}/\mathbf{F}_q$. Let $x \in \mathbf{F}_q(\mathcal{X})\backslash\{0\}$ and denote by $Z(x)$, respectively $N(x)$, the set of zeros, respectively poles, of x. We define the *zero divisor* of x by

$$(x)_0 = \sum_{P \in Z(x)} \nu_P(x)P \tag{3}$$

and the *pole divisor* of x by

$$(x)_\infty = \sum_{P \in N(x)} (-\nu_P(x))P. \tag{4}$$

Then $(x)_0$ and $(x)_\infty$ are both rational divisors. Furthermore, the *principal divisor* of x is given by

$$\mathrm{div}(x) = (x)_0 - (x)_\infty. \tag{5}$$

The degree of $\mathrm{div}(x)$ is equal to zero, i.e.,

$$\deg((x)_0) = \sum_{P \in Z(x)} \nu_P(x) = \sum_{P \in N(x)} (-\nu_P(x)) = \deg((x)_\infty). \tag{6}$$

For an arbitrary divisor $G = \sum m_P P$ of $\mathcal{X}$, we denote by $\nu_P(G)$ the coefficient m_P of P. Then

$$G = \sum \nu_P(G)P.$$

For such a divisor G we form the vector space

$$\mathcal{L}(G) = \{x \in \mathbf{F}_q(\mathcal{X})\backslash\{0\}|\mathrm{div}(x) + G \geq 0\} \cup \{0\}.$$

Then $\mathcal{L}(G)$ is a finite-dimensional vector space over $\mathbf{F}_q$, and we denote its dimension by $\ell(G)$. By the Riemann-Roch theorem we have

$$\ell(G) \geq \deg(G) + 1 - g, \tag{7}$$

and equality holds if $\deg(G) \geq 2g - 1$.

By a closed point, we mean a set of $\mathbf{F}_q$-conjugate points. The *degree* of a closed point is the number of points in the set.

5. Goppa Geometry Codes

Example 5.1: (*Reed-Solomon Codes*) For $1 \leq k \leq n \leq q$, consider a subset $\{\alpha_1, \alpha_2, \ldots, \alpha_n\}$ of $\mathbf{F}_q$ and construct a linear code in the following way:

$$C = \{(f(\alpha_1), \ldots, f(\alpha_n)) \mid f \in \mathbf{F}_q[x], \deg(f) \leq k - 1\}.$$

It is easy to show that the linear code C constructed as above has parameters $[n, k, d]$ with $d = n + 1 - k$, i.e., $k + d = n + 1$. Thus, it achieves the Singleton bound.

Geometric interpretation of the polynomial codes

Let ∞ be the point of a projective line $\mathcal{X}/\mathbf{F}_q$ corresponding to the degree valuation of the rational function field $\mathbf{F}_q(\mathcal{X}) = \mathbf{F}_q(x)$. Then a function $f(x)$ is a polynomial of degree $\leq k - 1$ if and only if

$$\nu_\infty(f) \geq -(k - 1), \text{ and } \nu_P(f) \geq 0$$

for all other places P, i.e.,

$$\operatorname{div}(f) + (k - 1)\infty \geq 0.$$

This is equivalent to $f(x)$ being in the Riemann-Roch space $\mathcal{L}((k - 1)\infty)$. Hence

$$\mathcal{L}((k - 1)\infty) = \{f \in \mathbf{F}_q[x] \mid \deg(f) \leq k - 1\},$$

and the Reed-Solomon codes can be expressed in the following way:

$$C = \{(f(\alpha_1), \ldots, f(\alpha_n)) \mid f \in \mathcal{L}((k - 1)\infty)\}.$$

Let $\mathcal{X}/\mathbf{F}_q$ be a curve over $\mathbf{F}_q$. Let $\infty, P_1, \ldots, P_n$ be $n+1$ rational points. For an integer $m < n$, we consider the map

$$\theta : \mathcal{L}(m\infty) \to \mathbf{F}_q^n, \quad f \mapsto (f(P_1), \ldots, f(P_n)).$$

The image of θ is a linear code of length n. It is called a *Goppa geometry code*, denoted by $C(P_1, \ldots, P_n; m\infty)$.

Theorem 5.2: *Let $\mathcal{X}/\mathbf{F}_q$ be a curve of genus g. Then the code $C(P_1, \ldots, P_n; m\infty)$ constructed above is a q-ary $[n, k, d]$-linear code with*

$$k \geq m - g + 1, \quad d \geq n - m.$$

Proof: Let f be a non-zero function in $\mathcal{L}(m\infty)$. Then f has at most m zeros as $(f)_\infty \leq m\infty$. Thus, $f(P_i) = 0$ for at most m points P_i's, i.e., $\theta(f)$ has weight at least $n - m$. This also implies that θ is injective. So the dimension of the code is equal to $\ell(m\infty)$ which is at least $m - g + 1$ by the Riemann-Roch theorem. □

Remark 5.3: From the above theorem, we find that for a Goppa geometry code based on a curve of genus g, we have

$$n + 1 - g \le k + d \le n + 1.$$

Example 5.4: Look at the curve

$$\mathcal{X}: \quad y^r + y = x^{r+1}$$

over $\mathbf{F}_q$ with $q = r^2$.

For each $x = \alpha \in \mathbf{F}_q$, we get r solutions for

$$y^r + y = \alpha^{r+1}.$$

Thus, the total number of rational points is $r^3 + 1$.

The genus of $\mathcal{X}$ is $r(r-1)/2$. Hence for any m satisfying

$$r(r-1) - 1 = 2g - 1 \le m < n = r^3,$$

the code $C(P_1, \ldots, P_n; m\infty)$ is a q-ary $[r^3, m - g + 1, r^3 - m]$-linear code.

For example, if $q = 4 = r^2$, we get $[8, m, 8 - m]$-linear codes over $\mathbf{F}_4$ for any $1 \le m < 8$.

6. The Fundamental Coding Problem

In coding theory, people are interested in the performance of long codes. In other words, we like to find the ratio

$$\limsup_{n \to \infty} \frac{k}{n} \quad \text{and} \quad \limsup_{n \to \infty} \frac{d-1}{n}$$

for a family of q-ary $[n, k, d]$-linear codes.

Definition 6.1: The q-ary linear code domain U_q^{lin} is the set of ordered pairs $(\delta, R) \in \mathbf{R}^2$ for which there exists an infinite sequence $C_1, C_2, \ldots$ of linear codes over $\mathbf{F}_q$ with $n(C_i) \to \infty$ and

$$R = \lim_{i \to \infty} \frac{k(C_i)}{n(C_i)}, \quad \delta = \lim_{i \to \infty} \frac{d(C_i) - 1}{n(C_i)}.$$

The following result can be found from the book [4].

Proposition 6.2: *There exists a continuous function* $\alpha_q^{\mathrm{lin}}(\delta)$, $\delta \in [0,1]$, *such that*

$$U_q^{\mathrm{lin}} = \{(\delta, R) \in \mathbf{R}^2 | 0 \le R \le \alpha_q^{\mathrm{lin}}(\delta),\ 0 \le \delta \le 1\}.$$

Moreover, $\alpha_q^{\mathrm{lin}}(0) = 1$, $\alpha_q^{\mathrm{lin}}(\delta) = 0$ *for* $\delta \in [(q-1)/q, 1]$, *and* $\alpha_q^{\mathrm{lin}}(\delta)$ *decreases on the interval* $[0, (q-1)/q]$.

For $0 < \delta < 1$ define the q-ary entropy function

$$H_q(\delta) = \delta \log_q(q-1) - \delta \log_q \delta - (1-\delta) \log_q(1-\delta),$$

where $\log_q$ is the logarithm to the base q, and put

$$R_{GV}(q, \delta) = 1 - H_q(\delta).$$

Then the Gilbert-Varshamov bound says that

$$\alpha_q^{\mathrm{lin}}(\delta) \ge R_{GV}(q, \delta) \quad \text{for all } \delta \in \left(0, \frac{q-1}{q}\right). \tag{8}$$

For any prime power q and any integer $g \ge 0$ put

$$N_q(g) = \max N(\mathcal{X}/\mathbf{F}_q),$$

where the maximum is extended over all curves $\mathcal{X}/\mathbf{F}_q$ with $g(\mathcal{X}) = g$.

We also define the following asymptotic quantity

$$A(q) := \limsup_{g \to \infty} \frac{N_q(g)}{g}.$$

Then we have the following result [2].

Theorem 6.3: *For any prime power* q, *one has*

$$\alpha_q^{\mathrm{lin}}(\delta) \ge 1 - \delta - \frac{1}{A(q)} \quad \text{for all } \delta \in [0, 1]. \tag{9}$$

Define the linear function

$$R_G(q, \delta) = 1 - \frac{1}{A(q)} - \delta$$

for any prime power q. Then Theorem 6.3 says that $\alpha_q^{\mathrm{lin}}(\delta) \ge R_G(q, \delta)$.

Proposition 6.4: (*see Chapter 5 of* [2])

(1) *If* q *is a square, then*

$$A(q) = \sqrt{q} - 1.$$

(2) *If q is a cube, then*

$$A(q) \geq \frac{\sqrt{q^{1/3}} + 1}{\lfloor 2\sqrt{2q^{1/3} + 2} \rfloor + 2}.$$

One consequence of the above results is the following.

Theorem 6.5:

(1) *If $q \geq 49$ is a square, then there exists an open interval $(\delta_1, \delta_2) \subseteq (0, 1)$ such that*

$$R_G(\delta) > R_{GV}(\delta) \quad \text{for any } \delta \in (\delta_1, \delta_2).$$

(2) *If $q > 15552$ is a cube, then there exists an open interval $(\delta_1, \delta_2) \subseteq (0, 1)$ such that*

$$R_G(\delta) > R_{GV}(\delta) \quad \text{for any } \delta \in (\delta_1, \delta_2).$$

7. New Codes from Algebraic Geometry Codes

We only outline some results in this section. For the detailed proofs, the reader may refer to [1, 5].

7.1. *Construction 1*

Rational points over $\mathbf{F}_q$ are not sufficient for curves of a given genus to obtain good q-ary codes. We will make use of points over $\mathbf{F}_{q^2}$.

First we fix some notations for this subsection.

$\mathcal{X}/\mathbf{F}_{q^2}$ – a curve over $\mathbf{F}_{q^2}$;
$g := g(\mathcal{X})$ – the genus of $\mathcal{X}/\mathbf{F}_{q^2}$;
$\mathbf{F}_{q^2}(\mathcal{X})$ – the function field of $\mathcal{X}/\mathbf{F}_{q^2}$;

We choose $n + 1$ points $\infty, P_1, \ldots, P_n$ over $\mathbf{F}_{q^2}$.
For $m \geq g$, we have

$$\dim_{\mathbf{F}_{q^2}}(\mathcal{L}(m\infty)) \geq m + 1 \quad g \geq 1$$

by the Riemann-Roch theorem. Therefore, we can find $m + 1 - g$ non-gaps $0 = v_1 < v_2 < \cdots < v_{m+1-g} \leq m$ of ∞, i.e, there exist $m + 1 - g$ functions $f_1, f_2, .., f_{m+1-g}$ in $\mathcal{L}(m\infty)$ such that

$$\nu_\infty(f_i) = -v_i \tag{10}$$

for all $1 \leq i \leq m+1-g$, where ν_∞ denotes the normalized discrete valuation of ∞. Since the v_i are strictly increasing, we obtain

$$v_i \leq v_{m+1-g} - ((m+1-g)-i) \leq m - (m+1-g-i) = g+i-1. \quad (11)$$

Put

$$e_{ij} = f_i^q f_j + f_i f_j^q$$

for all $1 \leq i < j \leq m+1-g$, and

$$e_{ii} = f_i^{q+1}$$

for all $1 \leq i \leq m+1-g$.

Choose an element γ of $\mathbf{F}_{q^2} \setminus \mathbf{F}_q$ and put

$$e'_{ij} = \gamma^q f_i^q f_j + \gamma f_i f_j^q$$

for all $1 \leq i < j \leq m+1-g$.

Lemma 7.1: *For any $\mathbf{F}_{q^2}$-rational point $P \neq \infty$ of $\mathcal{X}/\mathbf{F}_{q^2}$, both $e_{ij}(P)$ and $e'_{ij}(P)$ are elements of $\mathbf{F}_q$ for all i, j.*

Let U_m be the $\mathbf{F}_q$-linear span of the set

$$\{e_{ij}|1 \leq i \leq j \leq m+1-g\} \cup \{e'_{ij}|1 \leq i < j \leq m+1-g\}.$$

Lemma 7.2: *The dimension of the $\mathbf{F}_q$-linear space U_m is equal to $(m+1-g)^2$ if $m < q$.*

We construct a code as follows:

$$C_m = \{(f(P_1), f(P_2), ..., f(P_n))|f \in U_m\}.$$

This code is defined over $\mathbf{F}_q$ by Lemma 7.1 and it is clear that C_m is a linear code of length n.

In fact, U_m is a subspace of $\mathcal{L}(m(q+1)\infty)$. Hence,

$$C_m \subseteq C(P_1, P_2, \ldots, P_n; m(q+1)\infty).$$

Theorem 7.3: *Let $\mathcal{X}/\mathbf{F}_{q^2}$ be a curve of genus g with $n+1$ rational points $\infty, P_1, \ldots, P_n$. Suppose $g \leq m < \min\{q, n/(q+1)\}$. Then C_m is an $[n, (m+1-g)^2, d]$-linear code over $\mathbf{F}_q$ with*

$$d \geq n - m(q+1).$$

Example 7.4: Let $q = 8$. Consider the elliptic curve

$$y^2 + y = x^3$$

over $\mathbf{F}_{64}$. Then it has 81 points over $\mathbf{F}_{64}$ and the genus is 1. Thus, for any $1 \le m \le 7$ we have a q-ary $[80, m^2, d \ge 80 - 9m]$-linear code. In particular, we obtain the following new codes:

$$[80, 36, 26], \quad [80, 49, 17].$$

Corollary 7.5: *If* $1 \le m < q$, *then there exist*
(1) *a* q-ary $[q^2, (m + 1)^2, d \ge q^2 - m(q + 1)]$-linear code, and
(2) *a* q-ary $[q^2 + 2q, m^2, d \ge q^2 + 2q - m(q + 1)]$-linear code.

7.2. *Construction 2*

The ingredients for this construction are as follows.

$\mathcal{X}/\mathbf{F}_q$ — a curve over $\mathbf{F}_q$.
g — the genus of $\mathcal{X}/\mathbf{F}_q$.
$P_1, P_2, \ldots, P_s$ — s distinct closed points of F.
$k_i = \deg(P_i)$ — degree of P_i for $1 \le i \le s$.
G — divisor of F with $\text{Supp}(G) \cap \{P_1, P_2, \ldots, P_s\} = \emptyset$.
C_i — q-ary linear code of parameters $[n_i, k_i = \deg(P_i), d_i]$ with $k_i \ge d_i$ for $1 \le i \le s$.
π_i — fixed $\mathbf{F}_q$-linear isomorphism mapping the residue class field of P_i onto C_i for $1 \le i \le s$.

Then by concatenation we define the q-ary linear code

$$C := \{(\pi_1(f(P_1)), \pi_2(f(P_2)), \ldots \pi_s(f(P_s)))| f \in \mathcal{L}(G)\}.$$

Theorem 7.6: *With the notations above, suppose that* $g \le \deg(G) < \sum_{i=1}^{s} k_i$. *Then the code* C *defined above is a* q-ary $[n, k, d]$-linear code with

$$n = \sum_{i=1}^{s} n_i; \quad k \ge \deg(G) - g + 1; \quad d \ge \sum_{i=1}^{s} d_i - \deg(G).$$

Example 7.7: Consider the elliptic curve $\mathcal{X}/\mathbf{F}_4$ defined by

$$y^2 + y = x^3 + x.$$

Then $\mathcal{X}$ has 5 rational points, and hence its L-polynomial is equal to

$$4T^2 + 1.$$

It follows that there are 10 closed points of degree 2 and 20 closed points of degree 3. Since there exist linear codes over $\mathbf{F}_4$ with the following parameters

$$[1,1,1]; \quad [3,2,2]; \quad [5,3,3],$$

we can choose all of these closed points and get the codes with parameters

$$[135, k, d \geq 85 - k]$$

for any $1 \leq k \leq 84$ from Theorem 7.6. In particular, we obtain two new codes

$$[135, 29, d \geq 56], \quad [135, 30, d \geq 55].$$

References

1. C. S. Ding, H. Niederreiter and C. P. Xing, "Some new codes from algebraic curves", *IEEE Trans. Inform. Theory*, **46** (2000), 2638–2642.
2. H. Niederreiter and C. P. Xing, *Rational Points on Curves over Finite Fields: Theory and Applications*, Cambridge University Press, Cambridge, 2001.
3. H. Stichtenoth, *Algebraic Function Fields and Codes*, Springer, Berlin, 1993.
4. M. A. Tsfasman and S. G. Vlădut, *Algebraic-Geometric Codes*, Kluwer, Dordrecht, 1991.
5. C. P. Xing and S. Ling, "A class of linear codes with good parameters from algebraic curves", *IEEE Trans. Inform. Theory*, **46** (2000), 1527–1532.